NUMERICAL MATHEMATICS AND SCIENTIFIC COMPUTATION

Series Editors

NUMERICAL MATHEMATICS AND SCIENTIFIC COMPUTATION

---

*P. Dierckx: *Curve and surface fittings with splines*
*H. Wilkinson: *The algebraic eigenvalue problem*
*I. Duff, A. Erisman, and J. Reid: *Direct methods for sparse matrices*
*M. J. Baines: *Moving finite elements*
*J. D. Pryce: *Numerical solution of Sturm–Liouville problems*
K. Burrage: *Parallel and sequential methods for ordinary differential equations*
Y. Censor and S. A. Zenios: *Parallel optimization: theory, algorithms, and applications*
M. Ainsworth, J. Levesley, M. Marletta, and W. Light: *Wavelets, multilevel methods, and elliptic PDEs*
W. Freeden, T. Gervens, and M. Schreiner: *Constructive approximation on the sphere: theory and applications to geomathematics*
Ch. Schwab: *p- and hp- finite element methods: theory and applications to solid and fluid mechanics*
J. W. Jerome: *Modelling and computation for applications in mathematics, science, and engineering*
Alfio Quarteroni and Alberto Valli: *Domain decomposition methods for partial differential equations*
G. E. Karniadakis and S. J. Sherwin: *Spectral/hp element methods for CFD*
I. Babuška and T. Strouboulis: *The finite element method and its reliability*
B. Mohammadi and O. Pironneau: *Applied shape optimization for fluids*
S. Succi: *The lattice Boltzmann equation for fluid dynamics and beyond*

Monographs marked with an asterisk (*) appeared in the series 'Monographs in Numerical Analysis' which has been folded into, and is continued by, the current series

The Lattice Boltzmann Equation

for Fluid Dynamics and Beyond

Sauro Succi

Istituto Applicazioni Calcolo 'M. Picone'
Consiglio Nazionale delle Ricerche
Roma, Italy

Affiliated to The Physics Department, University of Rome

CLARENDON PRESS · OXFORD

2001

OXFORD
UNIVERSITY PRESS

Great Clarendon Street, Oxford OX2 6DP

Oxford University Press is a department of the University of Oxford.
It furthers the University's objective of excellence in research, scholarship,
and education by publishing worldwide in

Oxford New York

Auckland Bangkok Buenos Aires Cape Town Chennai
Dar es Salaam Delhi Hong Kong Istanbul Karachi Kolkata
Kuala Lumpur Madrid Melbourne Mexico City Mumbai Nairobi
São Paulo Shanghai Taipei Tokyo Toronto

Oxford is a registered trade mark of Oxford University Press
in the UK and in certain other countries

Published in the United States
by Oxford University Press Inc., New York

First published 2001
Reprinted 2002, 2004

A catalogue record for this book is available from the British Library

Library of Congress Cataloging in Publication Data

Succi, S.
The lattice Boltzmann equation for fluid dynamics and beyond / Sauro Succi.
(Numerical mathematics and scientific computation)
Includes bibliographical references and index.
1. Transport theory. 2. Kinetic theory of gases. 3. Evolution equations. I. Title. II. Series.
QC175.2.S78 2001 530.13′8—dc21 2001036220

ISBN 0 19 850398 9 (acid-free paper)

Printed in Great Britain by
Biddles Ltd, King's Lynn, Norfolk

To Claudia, Cate and to my parents

PREFACE

This book is about the *Lattice Boltzmann Equation* (LBE) a hyper-stylized version of the Boltzmann equation explicitly designed to solve fluid-dynamics problems, and beyond.

Initially developed as a spin-off of Lattice Gas Cellular Automata (LGCA), LBE rapidly evolved into a self-standing research subject within the theoretical framework of statistical mechanics, most notably (discrete) kinetic theory.

About ten years ago, mainly under the impact of lattice gas cellular automata, it was realized that phase-space lends itself to very drastic, and yet physically faithful, discretizations of momentum space such that a handful of appropriately chosen discrete momenta prove able to recover hydrodynamic equations in the continuum limit.

Besides its intrinsic intellectual charm, this discovery dissolved the computational barrier associated with the 'true' Boltzmann equation and paved the way to a series of fairly efficient numerical schemes for hydrodynamics (and beyond) based on discrete kinetic theory.

The mathematical 'stream-and-collide' structure of the Boltzmann equation proves instrumental to the success of LBE, both from analytical and computational points of view.

Computational aspects are by no means a minor point.

Most of the LBE computational work comes from the collision operator. Owing to the point-like nature of the interparticle collisions, this operator is completely local in configuration space, which means that different portions of physical space can be advanced concurrently: that spells ideal amenability to parallel computing, a key asset of LBE.

This computational connotation sets LBE somehow apart from its forerunners in discrete kinetic theory, namely discrete velocity models 'à-la' Broadwell.

The main focus of LBE is insight (mostly via numerical simulation) rather than exact analytic solutions of simplified model equations. Its targets, consequently, are the most pressing problems in fluid dynamics (and beyond), such as turbulence, multiphase flows in disordered media, and flows with suspensions.

This ties in nicely with a recent revival of kinetic theory, as beautifully expressed by M. Ernst, E. van Beijeren and F. Rochelle: (Dynamics: models and kinetic methods for non-equilibrium many-body systems, STATPHYS 20 Satellite Meeting, Leiden, July, 1998) *'Modern kinetic theory offers a unifying theoretical framework within which a great variety of seemingly unrelated physical systems that exhibit complex dynamical behavior can be explored in a coherent manner. For example, these methods are being applied in such diverse areas as the dynamics of colloidal suspensions, flows of granular material, transport of electrons in mesoscopic systems and the calculation of Lyapunov exponents and other characteristic properties of classical many-body systems that are categorized by chaotic behavior.'*

No words could better express the 'Beyond' part of this book. Given all this, it is natural to ask what real advantages is LBE supposed to bring about as compared with existing numerical techniques. This question spawns a tiny chapter on its own: 'Who needs LBE?'. There (very arbitrarily) I divide applications into four classes: *DU*, *CU*, *SU*, *MU*, standing for '*Don't Use*', '*Can Use*', '*Should Use*' and '*Must Use*', respectively. The DU class contains problems which (in this author's opinion) are better handled by other existing techniques (every respectable numerical scheme must have its own DU class!). The MU class, on the contrary, is the cradle of breakthroughs, problems which could not be solved by any other method. As we speak, I believe the SU class is decently populated and the much coveted MU class is very 'dilute' but perhaps not empty either. The subject is mature enough to deserve a book, but still in flux as far as advanced developments are concerned. It is therefore hoped that, upon reading the book, readers will feel bold enough to take the challenge of making their own entry in the MU class.

This book relies more on physical intuition and heuristic arguments than on detailed mathematical analysis, the idea being to provide an easy-going 'warm-up' to the subject for a broad audience rather than an exhaustive treatment for those aspiring to in-depth specialization in the field.[1] The latter category is consistently addressed in the current literature, for which this book is not meant to be a substitute.

The targeted audience of this book is graduated and advanced undergraduate students in physics, mathematics, engineering and computer science. More generally, it is hoped that researchers outside the lattice community will find enough material to judge for themselves whether or not the method is worth a try on their own research projects. To bias their mind towards the 'yes' side of the medal, an on-going elementary warm-up computer program is provided.

Some foreknowledge of basic statistical mechanics and kinetic theory makes Chapter 1 unnecessary. A similar statement applies to Chapter 2 for readers familiar with lattice gas cellular automata. These two chapters have been included for the sole purpose of self-consistency.

Acknowledgements

This book collects ten years of joyful research on the subject of lattice Boltzmann. Ten years that gave me the privilege of getting in personal touch with many distinguished colleagues, who taught me more than I can express. With apologies for inadvertent omissions: F. Abraham, B. Alder, G. Amati, M. Anile, G. Bella,

[1]Right in the final process of proof-reading this manuscript, I became aware of the very nice book by D. A. Wolf-Gladrow, *Lattice Gas Cellular Automata and Lattice Boltzmann Models*, Lecture Notes in Mathematics, Springer–Verlag, Berlin, 2000, which represents a useful complement to the present book for a more mathematically-inclined readership.

R. Benzi, M. Bernaschi, J. Bernsdorf, M. Bertsch, B. Boghosian, J. P. Boon, A. Brandt, G. Brenner, H. Cabannes, N. Cabibbo, P. Carnevali, F. Castiglione, C. Cercignani, H. Chen, S. Chen, Y. Chen, M. L. Chiofalo, B. Chopard, G. Ciccotti, S. Ciliberto, E. Cohen, P. Coveney, J. Dongarra, G. Doolen, M. Droz, D. d'Humières, T. Dupont, M. Ernst, B. Favini, O. Filippova, P. Franchi, D. Frenkel, U. Frisch, R. Gatignol, I. Ginzbourg, D. Gosman, P. Grosfils, J. Jimenez, F. Hayot, X. He, M. Henon, F. Higuera, A. Hoekstra, W. Hoover, L. Kadanoff, R. Kapral, I. Karlin, E. Kaxiras, W. Kohn, M. Krafczyk, A. Ladd, P. Lallemand, J. Lebowitz, L. S. Luo, M. Mareschal, N. Margolus, E. Marinari, F. Massaioli, W. Miller, P. Moin, K. Molvig, B. Nadiga, R. Natalini, H. C. Oettinger, I. Ohashi, S. Orszag, G. Parisi, L. Pietronero, R. Piva, Y. Pomeau, E. Presutti, M. Pulvirenti, Y. H. Qian, A. Quarteroni, C. Rebbi, S. Remondi, D. Rothman, D. Ruelle, P. Santangelo, G. Smith, J. Somers, D. Stauffer, C. Teixeira, T. Toffoli, F. Toschi, M. Tosi, R. Tripiccione, C. Tsallis, M. Vergassola, A. Vulpiani, V. Yakhot, J. Yeomans, J. Yepez, S. Zaleski, P. Zanella and G. Zanetti.

Special thanks to Dr Piero Sguazzero, for starting off my LBE adventure by suggesting I use the HPP automaton as a testbed for learning IBM-370 Assembler language (funnier beyond expectations) and to Prof. Roberto Benzi for sharing the fun and excitement of the 'happy days' of early LBE research.

I'm deeply indebted to Dr Hudong Chen for countless inspiring discussions and to numerous other friends at EXA Corporation, Boston University and Harvard University, who made Boston a second home-town to me. In particular, I wish to thank Prof. E. Kaxiras for arranging a very enriching visiting scholarship at Harvard University.

My most sincere gratitude to Profs Steven Orszag and R. Piva, for their friendly and continued support throughout the years, and Prof. Carlo Cercignani for his crucial help in making this book possible at all. I am very grateful to Profs B. Boghosian, J. P. Boon, C. Cercignani, P. Coveney, U. Frisch and L. S. Luo for their critical reading of this book in manuscript.

I cannot close this list without mentioning my very earliest teachers in kinetic theory, Profs V. Boffi, V. Molinari and G. Spiga, for the basic facts I learned from them many years ago have been useful along the way and are still useful today. In the same vein, I wish to thank Drs K. Appert and J. Vaclavik for teaching me how beautiful theories can be turned into equally beautiful numerical methods to solve them.

I wish to thank Dott. M. Adamo for managing the the graphics content of this project, and to my colleagues G. Amati, Y. Chen, O. Filippova, O. Inamuro, I. Karlin, A. Ladd, L. S. Luo, A. Masselot, J. Yeomans, for graciously providing original postscript files of their pictures. I'm also grateful to Dr Soenke Adlung, Anja Tschoertner, R. Lawrence and J. Harris of Oxford University Press, for invaluable help all along the project.

Deep gratitude to my wife Claudia and daughter Caterina is perhaps a standard, but is nonetheless a truly meant one.

Even short books are not necessarily bug-free. This is a long one ...

My advance gratitude to compassionate readers willing to draw attention to bugs, mistakes and any other sort of infelicities using the following e-mail address: **succi@iac.rm.cnr.it**.

Rome S. S.
September 2000

CONTENTS

Part I

Theory

*This first part of the book consists of three logical sections: (1) a reminder of the basic notions of classical statistical mechanics with emphasis on kinetic theory; (2) a brief survey of the main ideas behind Lattice Gas Cellular Automata, (**LGCA**); (3) the basic theory of the Lattice Boltzmann Equation (**LBE**).*

Chapter 1 is meant for those who did not receive any specific training on the microscopic side of fluid dynamics. Those who did can safely jump ahead to Chapter 2. This chapter provides a minimal, but hopefully self-contained, introduction to the forebear of LBE, namely the lattice gas cellular automata method. In spite of being its brainchild, LBE could be discussed independently of LGCA, but we feel this option, besides foregoing the chronological development, would also needlessly deprive LBE of part of its richness.

Detailed accounts on LGCA can be found in recent beautiful monographs by Rothman and Zaleski, Chopard and Droz, and Rivet and Boon.

Finally, the third section presents the main elements of LBE theory. This is the theoretical heart of the book and should not be skipped ... on pain of wasting the money spent on this book!

1

KINETIC THEORY

Kinetic theory is the branch of statistical physics dealing with the dynamics of non-equilibrium processes and their relaxation to thermodynamic equilibrium. The Boltzmann equation, established by Ludwig Boltzmann in 1872, is its cornerstone [1]. Originally developed in the framework of dilute gas systems, the Boltzmann equation has spread its wings across many areas of modern statistical physics, including electron transport in semiconductors, neutron transport, quantum liquids, to cite but a few [2, 3]. The aim of this chapter is to provide a minimal, and yet hopefully self-contained introduction to the Boltzmann equation in the context of classical statistical mechanics.

1.1 Atomistic dynamics

Let us consider a collection of N molecules moving in a box of volume V at temperature T and mutually interacting via a two-body intermolecular potential $\Phi(\vec{r})$, $\vec{r}$ being the interparticle separation. If the linear size s of the molecules, basically the effective range of the interaction potential, is much smaller than their *mean interparticle separation* $d = (V/N)^{1/3}$, the molecules can to a good approximation be treated like *point-like structureless* particles. To the extent where the de Broglie length $\lambda = \hbar/p$ of these particles is much smaller than any other relevant length-scale (an assumption that is made throughout this book with the exception of Chapter 13), their dynamics being governed by the classical Newton equations

$$\frac{\mathrm{d}\vec{x}_i}{\mathrm{d}t} = \frac{\vec{p}_i}{m}, \tag{1.1}$$

$$\frac{\mathrm{d}\vec{p}_i}{\mathrm{d}t} = \vec{F}_i, \quad i = 1, \ldots, N, \tag{1.2}$$

where $\vec{x}_i$ is the position coordinate of the i-th particle, $\vec{p}_i \equiv m\vec{v}_i$ its linear momentum and $\vec{F}_i$ is the force experienced by the i-th particle as a result of intermolecular interactions and possibly external fields (gravity, electric field, etc.). Upon specifying initial and boundary conditions, eqns (1.1) and (1.2) can in principle be solved in time to yield a fully exhaustive knowledge of the state of the system, namely a set of $6N$ functions of time $[\vec{x}_i(t), \vec{p}_i(t)]$, $i = 1, \ldots, N$. This program is totally unviable and, fortunately, needless as well. Unviability stems from two main reasons: first, N is generally of the order of the Avogadro number $Av \sim 10^{23}$, far too big for any foreseeable computer. Second, even if we could store it, tracking so much information for sufficiently long times would be utopia since any tiny uncertainty in the initial conditions would blow up in the long run because of dynamical instability of phase-space. By dynamical instability, we refer to the fact that any uncertainty δ_0 in the initial positions and/or momenta grows exponentially in time as $\delta(t) = \delta_0 e^{\lambda t}$. The coefficient λ, known as the *Lyapunov exponent*, is a measure of the temporal horizon of deterministic behavior of the N-body system, in that at times greater than λ^{-1}, the growth of uncertainty is such as to prevent any deterministic prediction of the state of the system [4, 5]. It is estimated that a centimeter cube of argon in standard conditions (300 K, 1 Atm) produces as much as 10^{29} digits of information per second. This means that in order to keep an exact record of the state of the system over a 1 s lifespan, we need a number with nothing less than 10^{29} digits [6]! Fortunately, we manage to survive with less than that, the reason being that ... we are much bigger than the molecules our body is made of. The physical observables we are interested in, say the fluid pressure, temperature, visible flow, originate from a statistical average over a large number of individual molecular histories. A rigorous definition of what is meant by statistical average is nothing trivial, but here we shall be content with the intuitive notion of spatial average over a thermodynamic volume, namely a region of space sufficiently small with respect to the global dimensions of the macroscopic domain, and yet large enough to contain a statistically meaningful sample of molecules. It is easy to write down some typical numbers. The density of air in standard conditions is about $n_L = 2.687 \times 10^{19}$ molecules/cm^3 (Loschmidt number). Hence a millimeter cube of air hosts of the order of 10^{16} molecules, corresponding to a statistical error of about ten parts per billion. It appears wise therefore to approach the collective behavior of our ensemble of molecules from a statistical point of view. This can be done at various levels of complexity, but here we shall consider the simplest one: the *one-body kinetic* level.

At this level, the chief question is:

Q1: *What is the probability of finding a molecule around position $\vec{x}$ at time t with momentum $\vec{p}$?*

Call $f(\vec{x}, \vec{p}, t)$ the probability density (more often simply distribution function), the quantity $\Delta n = f \Delta\vec{x}\, \Delta\vec{p}$, where $\Delta\vec{x}$, $\Delta\vec{p}$ are finite cubelets centered about $\vec{x}$ and $\vec{p}$ in the so-called *phase-space* $\mu = [(\vec{x}, \vec{p}) : \vec{x}, \vec{p} \in \mathbb{R}^3]$, represents the probable number of molecules with the aforementioned property. The distribution function $f(\vec{x}, \vec{p}, t)$ is the pivotal object of kinetic theory. In 1872, the great Austrian scientist Ludwig Boltzmann was able to derive an equation describing the evolution of f in terms of microdynamic interactions. This is the celebrated Boltzmann Equation (BE), one of the greatest achievements of theoretical physics in the past century. The BE represents the first quantitative effort to attack the grand issue of why physics goes 'one-way only' on a macroscopic scale while the underlying microdynamics is apparently perfectly reversible.[2] We

[2] We shall stick to the common tenet that microscopic equations, either classical or quantum, are invariant under time reversal.

shall not be concerned in this book with such 'dizzy' issues, but rather focus on the BE as a mathematical tool to investigate, analytically or numerically, the properties of fluids far from equilibrium.

The kinetic equation for the one-body distribution function reads as follows:

$$D_t f \equiv \left[\partial_t + \frac{\vec{p}}{m} \cdot \partial_{\vec{x}} + \vec{F} \cdot \partial_{\vec{p}}\right] f(\vec{x}, \vec{p}, t) = C_{12}, \tag{1.3}$$

where the left-hand side represents the streaming motion of the molecules along the trajectories associated with the force field $\vec{F}$ (straight lines if $\vec{F} = 0$) and C_{12} represents the effect of intermolecular (two-body) collisions taking molecules in/out the streaming trajectory.

The collision operator involves the *two-body* distribution function f_{12} expressing the probability of finding a molecule, say 1, around $\vec{x}_1$ with speed $\vec{v}_1$ *and* and a second molecule, say 2, around $\vec{x}_2$ with speed $\vec{v}_2$, both at time t. It is not difficult to write down the dynamic equation for f_{12}, the only trouble being that this equation calls into play the three-body distribution function f_{123}, which in turn depends on f_{1234} and so on down an endless line known as BBGKY hierarchy, after Bogoliubov, Born, Green, Kirkwood and Yvon [7].

To close eqn (1.3), Boltzmann made a few stringent assumptions on the nature of the physical system: a *dilute* gas of *point-like, structureless* molecules interacting via a *short-range* two-body potential.

Under such conditions, intermolecular interactions can be described solely in terms of localized binary collisions, with molecules spending most of their lifespan on free trajectories (in the absence of external fields) perfectly unaware of each other. Within this picture, the collision term splits into gain and loss components:

$$C_{12} \equiv G - L = \int (f_{1'2'} - f_{12})\, g\sigma(g, \Omega)\, \mathrm{d}\Omega\, \mathrm{d}\vec{p}_2 \tag{1.4}$$

corresponding to direct/inverse collisions taking molecules out/in the volume element $\mathrm{d}\vec{v}_1$, respectively (see Fig. 1.1).

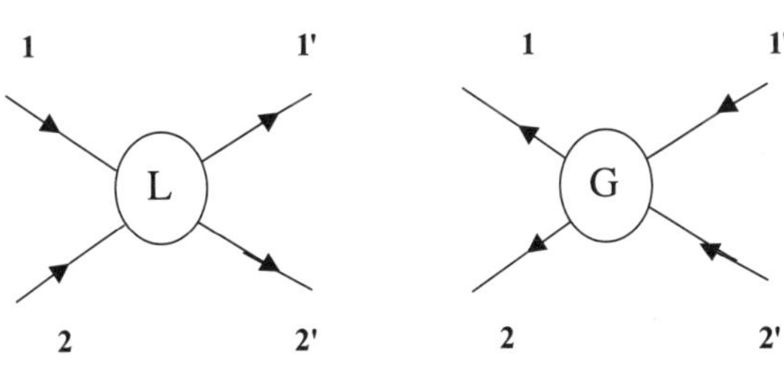

FIG. 1.1. Symbolic diagram of direct and inverse collisions. Inverse collisions (gain) place particles in state 1 while direct collisions (lose) take them away.

The link to underlying molecular dynamics is channeled into the differential cross section σ [2], expressing the number of molecules with relative speed $\vec{g} = g\vec{\Omega}$ around the solid angle $\vec{\Omega}$ (see Fig. 1.2).

Next comes the famous Boltzmann's closure assumption:

$$f_{12} = f_1 f_2. \tag{1.5}$$

This closure is tantamount to assuming no correlations between molecules entering a collision (*molecular chaos or Stosszahlansatz*). This assumption is fairly plausible for a dilute gas with short-range interactions in which molecules spend most of the time traveling in free space, only to meet occasionally for very short-lived interactions. Note that molecules are assumed to have correlations only prior to the collision, whereas after they are strongly correlated by virtue of mass, momentum and energy conservation!

Within this picture, the probability for two molecules that met at time $t - \tau$ to meet again at time t with the *same* momenta $\vec{p}_1$ and $\vec{p}_2$ is exponentially small with τ.

More precisely, this probability scales like $e^{-\tau/\tau_{int}}$ where τ_{int} is the duration of a collisional event. Since in Boltzmann's theory $\tau_{int} \sim s/v$ (v is a typical particle speed and s a typical effective molecular diameter) is negligibly small, so is the (auto)correlation function at time τ. The situation is obviously completely

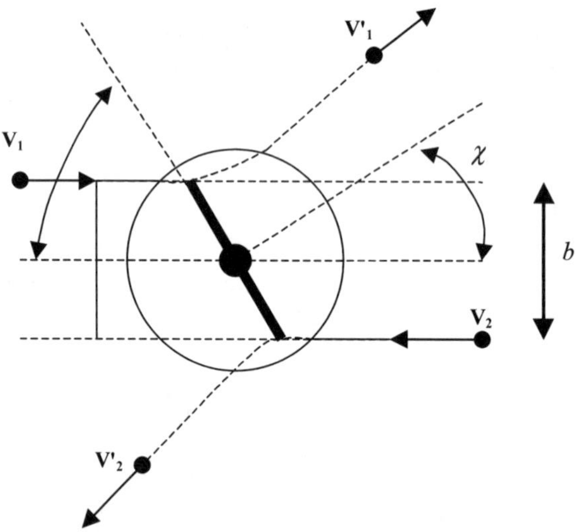

FIG. 1.2. Scattering angle associated with a binary collision. The collision takes place in the plane defined by the interparticle separation $\vec{x}_1 - \vec{x}_2$ and the relative speed $\vec{g} = \vec{v}_1 - \vec{v}_2$. The solid angle $\vec{\Omega}$ is defined by the scattering angle χ in the collisional plane and by the azimuthal angle ϕ around the collisional plane (not shown).

different in a liquid, where, due to the much higher density, the molecules are constantly interacting with one another.

Violations of Boltzmann's molecular chaos can occur even in relatively dilute fluids, due to nonlinear fluid–particle interactions. The most notable examples are the famous long-time tails first detected by Alder and Wainwright [8], where molecular correlations exhibit anomalous persistence due to self-sustained vortices generated by the molecular motion itself.[3]

Summarizing, the Boltzmann equation takes the following form:

$$\left[\partial_t + \vec{v} \cdot \partial_{\vec{x}} + \vec{F}^{\text{ext}} \cdot \partial_{\vec{p}}\right] f = \int \left(f_{1'} f_{2'} - f_1 f_2\right) g\sigma\left(g, \Omega\right) \mathrm{d}\Omega\, \mathrm{d}\vec{p}_2. \tag{1.6}$$

The left-hand side is a faithful mirror of reversible Newtonian single-particle dynamics, while the right-hand side describes intermolecular interactions under the Stosszahlansatz approximation.

1.2 Relaxation to local equilibrium

Central to the purpose of deriving hydrodynamic equations from the Boltzmann equation is the notion of *local equilibrium*. Mathematically, this is defined as a local distribution function f^{e}, such that gain and losses are in exact balance so that the collision term is annihilated:

$$C\left(f^{\mathrm{e}}, f^{\mathrm{e}}\right) = 0, \tag{1.7}$$

where superscript 'e' denotes 'equilibrium'.

This leads to the so-called *detailed balance* condition:

$$f_1' f_2' = f_1 f_2. \tag{1.8}$$

Of course, detailed balance does not mean that molecules sit idle, but rather that any direct/inverse collision is dynamically balanced by an inverse/direct partner.

Taking logarithms, the eqn (1.8) yields:

$$\ln f_1' + \ln f_2' = \ln f_1 + \ln f_2. \tag{1.9}$$

This shows that the quantity $\ln f$ is an *additive collision invariant*; that is a microscopic property which does not change under the effect of collisions. The immediate consequence is that, at thermodynamic equilibrium, $\ln f$ must be a function of dynamic collision invariants $I(v) \equiv [1, m\vec{v}, mv^2/2]$ (number, momentum, energy conservation) alone. This yields:

$$\ln f = A + B_a v_a + \frac{1}{2} C v^2, \tag{1.10}$$

where A, B_a, C are five Lagrangian multipliers carrying the functional dependence on the conjugate hydrodynamic fields ρ, u_a, E, namely density, momentum

[3] This is amazingly reminiscent of the mechanisms invoked by Aristoteles to explain the motion of arrows in air!

and energy. These Lagrangian parameters are computed by imposing conservation of these quantities:

$$m \int f \, d\vec{v} = \rho, \tag{1.11}$$

$$m \int f v_a \, d\vec{v} = \rho u_a, \quad a = 1, 2, 3, \tag{1.12}$$

$$m \int f \frac{v^2}{2} \, d\vec{v} = \rho e, \tag{1.13}$$

where ρ is the fluid density (mass per unit volume), u_a is the macroscopic flow speed and ρe the energy density.[4] Elementary quadrature of Gaussian integrals delivers the celebrated Maxwell–Boltzmann equilibrium distribution. In D spatial dimensions:

$$f^{\mathrm{e}} = \rho \left(2\pi v_T^2\right)^{-D/2} \mathrm{e}^{-c^2/2v_T^2}, \tag{1.14}$$

where c is the magnitude of the so-called *peculiar speed*

$$c_a = v_a - u_a, \quad a = 1, \ldots, D, \tag{1.15}$$

namely the relative speed of the molecules with respect to the fluid.

The quantity

$$v_T = \sqrt{\frac{K_B T}{m}} \tag{1.16}$$

is the thermal speed associated with the fluid temperature T.

Note that at this stage, the flow speed and temperature are constant in space, for otherwise the streaming operator would not be annihilated. This defines *global*, uniform, equilibria.

However, this need not be so for the collision operator. In fact, the eqn (1.14) is fulfilled even if the hydrodynamic fields in the Maxwellian are allowed space and time variations. This characterizes *local* equilibria.

It should be noted that Maxwellian equilibria inherit two basic symmetries of Newtonian mechanics [9], namely:

Translational invariance

$$t' = t - \tau, \tag{1.17}$$

$$x'_a = x_a - V_a \tau, \tag{1.18}$$

where τ and V_a are arbitrary constants.

This reflects invariance under Galilean transformations.

Rotational invariance

$$x'_a = \sum_{b=1}^{D} R_{ab} x_b, \tag{1.19}$$

where R_{ab} is a symmetric, unitary (norm-preserving) rotation matrix.

[4] Hereafter the italic subscript a denotes Cartesian components of a vector, so that the notation v_a is an equivalent substitute for $\vec{v}$ and is used interchangeably throughout the text.

Rotational invariance, which clearly applies to the particle speeds as well, is ensured by the fact that the peculiar speed $v_a - u_a$ appears through its magnitude alone, so that any sense of direction is erased. These symmetries are built-in in continuum kinetic theory, but it would be a gross mistake to take them for granted also when space–time and momentum are made discrete. Actually, this is the leading theme of subject of discrete kinetic theory, and of lattice field theories in general (although in lattice field theory the type of (gauge) symmetry to be preserved is less intuitive than just roto-translations and the Galilean group).

The most relevant hydrodynamic probe of rotational symmetry is the *momentum flux tensor*:

$$P_{ab} = m \int f v_a v_b \, \mathrm{d}\vec{v}, \tag{1.20}$$

a quantity which plays a pivotal role in lattice gas cellular automata and lattice Boltzmann theories.

1.3 *H*-theorem

We have noticed that at the level of *global* uniform equilibria both mean flow speed u and flow temperature T must be constant throughout the physical domain. This is a restriction stemming from the streaming operator only. Indeed, the collision operator does not place any restriction on the space dependence of the flow speed and temperature. This is of course related to the vanishingly small extent of the collisional region between two molecules which permits us to write down a purely local expression for the collision operator.

Whenever such a spatial dependence is allowed in the local thermohydrodynamic fields, one speaks of *local* thermodynamic equilibria, namely 'metastable' states describing a situation where thermodynamic equilibrium is only achieved on a local scale, global equilibration via gradient smoothing being still to come along on a longer time-scale.

This connects to the length scales l_μ, the particle *mean free path*, the mean distance traveled by molecules between two subsequent collisions, and l_M the typical scale of variation of macroscopic fields.

Local equilibration takes places on a time-scale $\tau_\mu \sim l_\mu / v_T$, whereas global equilibration requires much longer times, of the order of several $\tau_M = l_M / v_T$. This identifies the *transport* regime in which hydrodynamic quantities diffuse and advect along macroscopic distances within the fluid domain. It is of course a regime of great practical relevance since most real-life devices (to be of any use at all) must typically work far from thermodynamic equilibrium. Transport is naturally associated with the idea of *dissipation*, hence to the elusive concept of *irreversibility*, the subtle thread behind the second law of thermodynamics. One of the most profound contributions of Boltzmann to statistical mechanics rests with his discovery of a quantitative measure of irreversibility. We refer to the celebrated H-function and the attendant H-theorem. Boltzmann showed that the functional (H-function)

$$H(t) = -\int f \ln f \, d\vec{v} \, d\vec{x} \tag{1.21}$$

is a monotonically increasing function of time, regardless of the underlying microscopic potential. In equations:

$$\frac{dH}{dt} \geqslant 0, \tag{1.22}$$

the equality sign holding at global equilibrium, when the evolutionary potential of the system is exhausted and the *entropy* of the system is maximal. The role of $H(t)$ as a evolutionary potential (the 'time-arrow') is adamant and its intellectual magic hardly escaped.

Yet, it is difficult to think of a more debated and controversial issue in theoretical physics. We will not delve here into the details of the various paradoxes which were raised against the H-theorem as a bridge between micro and macrodynamics. Beautiful thorough accounts can be found in the literature [10, 11]. Nor shall we comment on the fact that Boltzmann derived his theorem without demonstrating that his equation, a complicated integro-differential initial-value problem, does indeed have solutions. While leaving rigor behind somehow, the H-theorem showed for the first time the way to a grand-unification of two fundamental and hitherto disconnected domains of science: mechanics and thermodynamics. But again, this is beyond the scope of this work.

Let us now discuss one of the *practical* aspects of the Boltzmann equation, namely the operational capability to compute transport parameters, diffusivity, viscosity, thermal conductivity, characterizing the approach to equilibrium on a macroscopic scale.

1.4 Length scales and transport phenomena

Qualitatively, the approach to equilibrium is controlled by the following time-scales [12]:

- $\tau_{\text{int}} \sim s/v$: *duration of a collisional event.*
- $\tau_\mu \sim l_\mu/v$: *mean flight-time between two successive collisions.*
- $\tau_h \sim \min\left[l_M/u, l_M^2/\nu\right]$: *minimum hydrodynamic (convective, diffusive) time-scale.*

Here l_M is a typical macroscopic scale, u the flow speed and ν the kinematic viscosity of the fluid.

The three time-scales identify a corresponding sequence of dynamical steps, namely:

1. *Many-body regime* ($0 < t < \tau_{\text{int}}$):
   Fast relaxation of the many-body distribution function to the single particle distribution function: $f_{12\cdots N} \to f_1$. This stage is especially fast for dilute systems where multibody collisions are very rare.

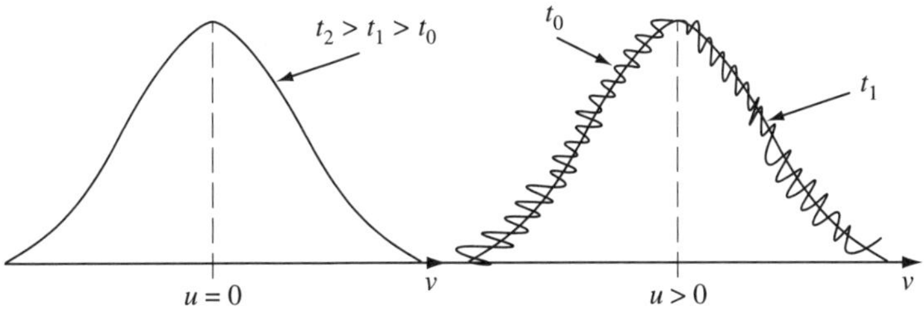

FIG. 1.3. Dynamical stages of the equilibration: $f \rightarrow$ local Maxwellian $\rightarrow$ global Maxwellian (left).

2. *Kinetic regime* ($\tau_{\text{int}} < t < \tau_\mu$):
   Relaxation to a local Maxwellian with smooth space–time-dependent flow speed and temperature: $f_1 \rightarrow M(v; u, T)$.
3. *Hydrodynamic regime* ($\tau_\mu < t < \tau_h$):
   Slow drift of the local Maxwellian with space–time-dependent flow speed and temperature to a global Maxwellian with constant speed and temperature: $M(v; u, T) \rightarrow M(v; u_0, T_0)$.

Within Boltzmann theory, step 1 is erased since collisions are treated as if they would proceed instantly. Phase 2, local equilibration, is collision-driven and brings any initial distribution function into a universal (Maxwellian) shape where space–time dependence is entirely carried on by hydrodynamic fields. Finally, phase 3 describes macroscopic transport across the system on longer hydrodynamic scales of practical interest.

1.5 From kinetic theory to fluid dynamics: the Chapman–Enskog procedure

Fluid dynamics can be seen as the mean field picture emerging from a perturbative treatment of the kinetic equations. The perturbation parameter is the *Knudsen number*, namely the ratio between the molecular mean free path and the shortest scale at which macroscopic variations can be appreciated:

$$Kn = \frac{l_{\text{mfp}}}{l_M}. \tag{1.23}$$

The Chapman–Enskog procedure is based on a double expansion in the smallness parameter of both dependent $f(x, v, t)$ and independent space–time (x, t) variables. The former reads simply ($\epsilon \equiv Kn$):

$$f = f^0 + \epsilon f^1, \tag{1.24}$$

where superscript 0 denotes local equilibrium and superscript 1 departure from this local equilibrium. The expansion of space–time variables lies at the heart of the so-called *multiscale* methods. The basic idea of multiscale analysis is to represent space and time variables in terms of a hierarchy of (slow/fast) scales, such that each variable is $O(1)$ at its own relevant scale. This greatly helps the theoretical analysis as well as the stability of numerical calculations. The time-scale hierarchy discussed previously suggests the following multiscale representation of space–time variables

$$x = \epsilon^{-1} x_1, \quad t = \epsilon^{-1} t_1 + \epsilon^{-2} t_2. \tag{1.25}$$

In this representation, x_1, t_1 describe the linear (sound wave) regime, whereas t_2 is in charge of the long-term dynamics. The multiscale representation (1.25) induces a corresponding representation of the differential operators:

$$\partial_x = \epsilon \partial_{x_1}, \tag{1.26}$$

$$\partial_t = \epsilon \partial_{t_1} + \epsilon^2 \partial_{t_2}. \tag{1.27}$$

From this expression it is apparent that in order to achieve order-by-order consistency in a Knudsen series expansions, second-order space derivatives need to be retained into the streaming operator

$$D_t \sim \epsilon \partial_{t_1} + \epsilon^2 \partial_{t_2} + \epsilon v_a \partial_{x_{1a}} + \frac{1}{2} \epsilon^2 v_a v_b \partial_{x_{1a}} \partial_{x_{1b}}. \tag{1.28}$$

Similarly, the Chapman–Enskog representation of the collision operator is

$$C[f] \sim C[f^0] + \epsilon C' f^1, \tag{1.29}$$

where the prime denotes (functional) derivative with respect to f. By recalling that the zeroth-order (ground-state) local equilibrium annihilates the collision operator, we obtain simply:

$$C[f] \sim \epsilon C' f^1. \tag{1.30}$$

These expansions provide the starting point for all order-by-order subsequent expansions. Macroscopic equations at different scales emerge naturally from the mathematical requirement that the operatorial coefficients of the expansions be identically zero term-by-term. Symbolically, from mass and momentum conservation, one obtains, respectively:

$$\epsilon \hat{M}_1 + \epsilon^2 \hat{M}_2 = 0, \tag{1.31}$$

$$\epsilon \hat{J}_1 + \epsilon^2 \hat{J}_2 = 0, \tag{1.32}$$

where $\hat{M}_i$, $\hat{J}_i$, $i = 1, 2$, are differential 1-forms whose detailed expression is given in Frisch *et al.* [13]. At order ϵ, one obtains $\hat{M}_1 = 0$ and $\hat{J}_1 = 0$,

$$\partial_{t_1} \rho + \partial_{a1} J_a = 0, \tag{1.33}$$

$$\partial_{t_1} J_a + \partial_{1b} \int v_a v_b f^0 \, \mathrm{d}\vec{v} = 0. \tag{1.34}$$

where $J_a = \rho u_a$ is the mass current density.

The velocity space integral in the above equation can be performed analytically to yield $\rho u_a u_b + \rho T \delta_{ab}$, so that eqns (1.33) and (1.34) are recognized to yield the Euler equations of inviscid flows (no dissipation).

The second-order equations are a bit more cumbersome, because they involve both levels, equilibrium and non-equilibrium, of the multiscale expansion, together with mixed space–time derivatives. Fortunately, these cross terms can be disposed of by using the condition $\hat{M}_1 = 0$ and $\hat{J}_1 = 0$, as well as the conservation relations (1.11). The final result is relatively simple:

$$\partial_{t_2} \rho = 0, \tag{1.35}$$

$$\partial_{t_2} J_a + \partial_{1b} m \int v_a v_b \left(f^1 + \partial_c v_c f^0\right) \mathrm{d}\vec{v} = 0. \tag{1.36}$$

Once these expansions are carried out, the circle is closed by reorganizing subscale derivatives into the originally continuous form. After some algebra, one finally derives the Navier–Stokes equations:

$$\partial_t \rho + \partial_a J_a = 0, \tag{1.37}$$

$$\partial_t J_a + \partial_b \left(\rho u_a u_b + P \delta_{ab}\right) = \partial_b \tau_{ab}, \tag{1.38}$$

where

$$P = \rho T \tag{1.39}$$

is the ideal gas pressure and the dissipative contribution τ_{ab} will be specified shortly. It is worth mentioning that the Chapman–Enskog perturbative treatment of the Boltzmann equation is generally *not* convergent, in the sense that higher-order (Burnett and superBurnett) equations arising from third/fourth-order truncations of the Knudsen expansion are notoriously exposed to severe numerical instabilities. This is because the smallness parameters prefactor the higher-order derivatives (streaming operator) so that singular layers may arise which cannot be captured by any finite order expansion. A particularly clear discussion of these pathologies can be found in [2].

We shall not touch any further on this thorny issue and will stay content instead with the second-order Navier–Stokes equations, which, although overly difficult to solve from the mathematical standpoint, are nonetheless physically sound and robust.

1.6 The Navier–Stokes equations

The end result of the various relaxation stages taking from the N-body dynamics all the way up to hydrodynamics are the Navier–Stokes equations [14]. For a isothermal fluid, these read:

$$\partial_t \rho + \partial_a \rho u_a = 0, \tag{1.40}$$

$$\partial_t \rho u_a + \partial_b T_{ab} = \partial_b \tau_{ab}, \tag{1.41}$$

where ρ is the fluid density and u_a the flow field. The tensorial quantities are defined as follows:

$$T_{ab} \equiv \rho u_a u_b + P\delta_{ab}, \tag{1.42}$$

$$\tau_{ab} \equiv \mu \left[\partial_a u_b + \partial_b u_a - \frac{2}{3}(\partial_c u_c)\,\delta_{ab}\right], \tag{1.43}$$

where $P = \rho T$ is the fluid pressure and μ is the dynamic viscosity.

The tensor T_{ab} is a faithful mirror of non-dissipative Newtonian dynamics, whereas τ_{ab} represents dissipative effects associated with the relaxation to local equilibria. In fact, by splitting the distribution function into equilibrium and non-equilibrium components, it can be shown that these two tensors represent the corresponding equilibrium and non-equilibrium components of the momentum flux tensor P_{ab}, respectively:

$$T_{ab} = m \int f^{\mathrm{e}} v_a v_b \, \mathrm{d}\vec{v}, \tag{1.44}$$

$$\tau_{ab} = m \int f^{\mathrm{ne}} v_a v_b \, \mathrm{d}\vec{v}. \tag{1.45}$$

The first equation in (1.40) is just mass conservation of a continuum fluid in the absence of any mass sources/sinks (continuity equation). The second is nothing but Newton's law $\vec{F} = m\vec{a}$ as applied to a finite volumelet of continuum fluid, supplemented with the (plausible, but *not* rigorous) assumption that strain of this volumelet is proportional, via the transport parameter μ, to the applied stress. From a kinetic theory standpoint, this assumption is tantamount to postulating weak departures from local equilibria. For all its apparent innocence, such an assumption is still eluding a rigorous mathematical proof [15], which means that the Navier–Stokes equations must still be regarded as phenomenological in character. Of course, this does not take a single bit away from their paramount role in fluid dynamics.

The physics of fluid flows is mainly controlled by two dimensionless parameters, the *Mach* number and the *Reynolds* number:

$$Ma = \frac{u}{c_s}, \tag{1.46}$$

$$Re = \frac{uL}{\nu}, \tag{1.47}$$

where c_s is the sound speed and $\nu = \mu/\rho$ is the kinematic viscosity. The Mach number measures the relative strength of inertial terms $\partial_b u_a u_b$ (fluid carrying around its own momentum) versus pressure gradients, whereas the Reynolds number is a measure of inertial versus dissipative effects. With reference to quasi-incompressible fluids (the main concern of lattice gas and lattice Boltzmann applications), the major difficulties standing in the way of fluid dynamicists is that the Reynolds number is typically *very high*, easily in the order of 10^6 for a standard cruising car. The name of the game is *turbulence*, one of the greatest challenges left with classical physics [16]. As we shall detail in the sequel to this book, a flow at a given Reynolds Re behaves like an extended dynamical system with approximately $Re^{9/4}$ degrees of freedom, which means that the flow around the aforementioned car involves over 10^{13} degrees of freedom! This figure alone says it all about the keen hunger for efficient numerical tools to tackle the physics of fluids.

1.7 Bhatnagar–Gross–Krook model equation

Before closing this succinct survey of kinetic theory it is worth mentioning that in order to facilitate numerical and analytical solutions of the Boltzmann equation, the complicated nonlinear integral collision operator is often replaced by simpler expressions aimed at relinquishing most of the mathematical difficulty without spoiling the basic physics. This class of simplified Boltzmann equations is epitomized by the so-called BGK (Bhatnagar–Gross–Krook) collision operator [17]:

$$C_{\mathrm{BGK}}(f) = -\frac{f - f^{\mathrm{e}}}{\tau}. \tag{1.48}$$

Here $f \equiv f_1$ and f^{e} is a local equilibrium parametrized by the local conserved quantities, density ρ, speed $\vec{u}$ and temperature T, while τ is a typical time-scale associated with collisional relaxation to the local equilibrium.

In principle, the relaxation time τ is a complicated functional of the distribution function f. The drastic simplification associated with BGK is the assumption of a constant value for this relaxation scale, which is equivalent to lumping the whole spectrum of relaxation scales into a single value! The major simplification associated with the BGK equation is that the non-locality in momentum space of the collision operator is channeled into a functional dependence on the conserved hydrodynamic variables, which serve as 'order parameters' for the particle distribution function. At a first glance, it might look as if the nonlinearity has been wiped out in the process, because *explicit* particle–particle quadratic coupling no longer appears. However, this is (obviously) not so; in fact, BGK is even *more* nonlinear than the Boltzmann equation itself because the local equilibrium depends exponentially on on the fluid speed and (inverse) temperature, both being linear function(als) of the particle distribution. Notwithstanding this strong hidden nonlinearity, the BGK equation proves very useful for analytical manipulations and sometimes even permits attainment of analytical solutions [2]. As we shall see, it plays a major role in the theory of the lattice Boltzmann equation.

1.8 Exercises

1. Consider a molecule of argon in a cubic box one centimeter in size and standard conditions. Assuming a Lyapunov exponent of the order of the inverse intercollision time, estimate the inaccuracy δ in the initial conditions such that the the position of the molecule is completely undetermined ($\delta = 1\,\mathrm{cm}$) after one millisecond.

2. Prove the Maxwellian expression (1.14).
3. Compute the gas entropy corresponding to a local Maxwellian at temperature T.
4. What percentile fraction of molecules move faster than $2v_T$ in a local Maxwellian?
5. Solve the homogeneous ($\partial_x f = 0$) BGK equation with the initial condition $f(v, t = 0) = f_0(v)$ and discuss the approach to local equilibrium.

2

LATTICE GAS CELLULAR AUTOMATA

Digital arithmetic is to scientific computing as collisions of molecules are to fluid mechanics: crucial to the implementation, but irrelevant to most scientific questions.
L. N. Trefehen, 'Maxims about numerical mathematics, computer science and life', *SIAM News*, Jan/Feb, 1998.

In 1986, Uriel Frisch, Brosl Hasslacher and Yves Pomeau produced their wonderful bombshell: a simple cellular automaton obeying nothing but conservation laws at a microscopic level was able to reproduce the complexity of real fluid flows [18]. This discovery caused great excitement in the fluid dynamics community. The prospects were promising: a round-off free, intrinsically parallel computational paradigm for fluid flows, perhaps, even more, the analogue of the Ising model for turbulence! However, a few serious problems were quickly recognized and addressed with great intensity in the following years.

The Lattice Boltzmann Equation (LBE) developed in the wake of the Lattice Gas Cellular Automata (LGCA) method and was generated precisely in response to its initial drawbacks [19, 20]. Shortly after its inception, LBE evolved into a self-standing research subject and (as of today) it could be discussed independently with no need to refer to LGCA at all [21].

This presentation route, if viable, would not do justice to the chronological development of the subject, nor would it help the didactical side of the matter. We shall therefore devote this chapter to a synthetic presentation of the basic ideas behind LGCA. Thorough accounts of this subject can be found in recent monographs [22], to which the interested reader is referred for an in-depth treatment of the subject.

2.1 Fluids in Gridland: the Frisch–Hasslacher–Pomeau automaton

Let us begin by considering a regular lattice with hexagonal symmetry such that each lattice site is surrounded by six neighbors identified by six connecting vectors $\vec{c}_i \equiv c_{ia}$, $i = 1, \ldots, 6$, the index $a = 1, 2$ scanning the spatial dimensions (see Fig. 2.1).

Each lattice sites hosts up to six particles with the following prescriptions:

- All particles have the same mass $m = 1$.
- Particles can move only along one of the six directions defined by the discrete displacements c_{ia}.

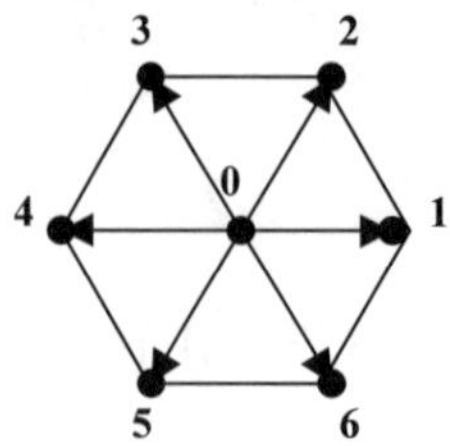

FIG. 2.1. The FHP hexagonal lattice.

- In a time-cycle (made one for convenience) the particles hop to the nearest neighbor pointed by the corresponding discrete vector c_{ia}. Both longer and shorter jumps are forbidden, which means all lattice particles have the same energy.
- No two particles sitting on the same site can move along the same direction c_{ia} (*exclusion principle*).

These prescriptions identify a very stylized gas analogue, whose dynamics represents a sort of cartoon of real-molecule Newtonian dynamics. In a real gas, molecules move along any direction (*isotropy*) whereas here they are confined to a six-barred cage.

Also, real molecules can move at virtually any (subluminal) speed, whereas here only six monochromatic beams are allowed. Finally, the exclusion principle sounds a bit weird for it applies to classical particles whose momentum and position can be specified simultaneously with arbitrary accuracy. Amazingly enough, this apparently poor cartoon of true molecular dynamics has all it takes to simulate realistic hydrodynamics!

With the prescription kit given above, the state of the system at each lattice site is unambiguously specified in terms of a plain *yes/no* option saying whether or not a particle sits on the given site. That is all we need to know.

This dichotomic, *tertium non datur*, situation is readily coded with a single binary digit (bit) per site and direction so that the entire state of the lattice gas is specified by $6N$ bits, N being the number of lattice sites. Borrowing the parlance of statistical mechanics, we introduce an *occupation number* n_i, such that

$$n_i(\vec{x}, t) = 0 \quad \text{particle absence at site } \vec{x} \text{ and time } t, \tag{2.1}$$

$$n_i(\vec{x}, t) = 1 \quad \text{particle presence at site } \vec{x} \text{ and time } t. \tag{2.2}$$

The collection of occupation numbers $n_i(\vec{x}, t)$ over the entire lattice defines a $6N$-dimensional time-dependent Boolean field whose evolution takes place in a Boolean phase-space consisting of 2^{6N} discrete states. This Boolean field takes the intriguing name of *Cellular Automaton* (CA) to emphasize the idea that not only space and time, but also the dependent variables (matter) take on discrete (Boolean) values.

The fine-grain microdynamics of this Boolean field can *not* be expected to reproduce the true molecular dynamics to any reasonable degree of microscopic accuracy. However, as has been known since Gibbs, many different microscopic systems can give rise to the same macroscopic dynamics, and it can therefore be hoped that the macroscopic dynamics of the lattice Boolean field would replicate real-life hydrodynamic motion even if its microdynamics does not.

We shall dub elementary Boolean excitations with such a property simply as *fluons*.

2.2 Fluons in action: LGCA microdynamic evolution

So much for the definitions. Let us now prescribe the evolution rules of our CA. Since we aim at hydrodynamics, we should cater for two basic mechanisms:

- *Free-streaming*;
- *Collisions*.

Free-streaming consists of simple particle transfers from site to site according to set of discrete speeds c_{ia}. Thus, a particle sitting at site $\vec{x}$ at time t with speed c_{ia} will move to site $\vec{x}+c_{ia}$ at time $t+1$.

In equations:

$$n_i\left(\vec{x}+c_{ia},t+1\right)=n_i\left(\vec{x},t\right). \tag{2.3}$$

This defines the discrete free-streaming operator Δ_i as

$$\Delta_i n_i \equiv n_i\left(\vec{x}+c_{ia},t+1\right)-n_i\left(\vec{x},t\right). \tag{2.4}$$

This equation is a direct transcription of the Boltzmann free-streaming operator $D_t \equiv \partial_t + v_a\partial_a$ to a discrete lattice in which space–time are discretized according to the synchronous 'light-cone' rule:

$$\Delta x_{ia} = c_{ia}\Delta t. \tag{2.5}$$

The relation between the discrete and continuum streaming operators reads

$$\Delta_i = \mathrm{e}^{D_i} - 1, \tag{2.6}$$

where $D_i \equiv \partial_t + c_{ia}\partial_a$ is the generator of space–time translations along the i-th direction (sum over repeated indices is implied). Manifestly the speed magnitude

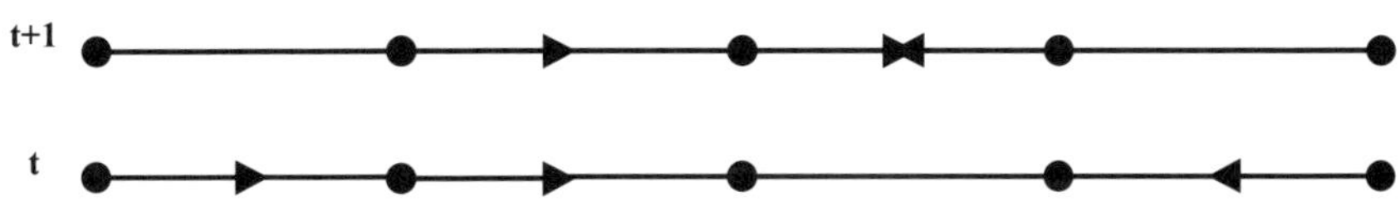

FIG. 2.2. Free-streaming in a discrete one-dimensional lattice.

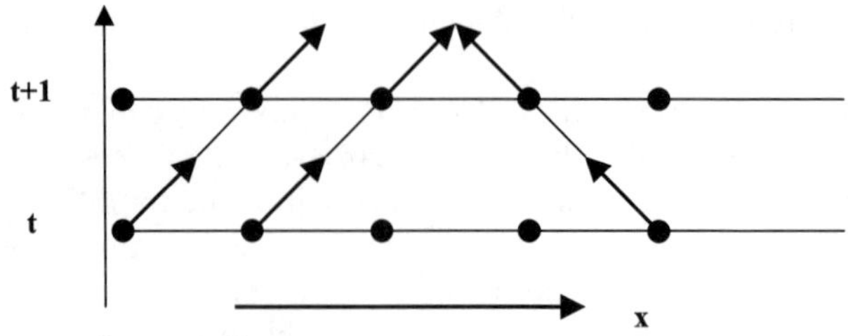

FIG. 2.3. Free-streaming along the light-cones of x–t space–time.

c plays the role of the 'light speed' in the discrete world, in that no signal can propagate faster than c in the lattice. Once on the same site, particles interact and reshuffle their momenta so as to exchange mass and momentum among the different directions allowed by the lattice (see Fig. 2.4).

This mimics the real-life collisions taking place in a real gas, with the crude restriction that all pre- and post-collisional momenta are forced to 'live' on the lattice. As compared with Continuum Kinetic Theory (CKT), the LGCA introduces a very radical cut of degrees of freedom in momentum space: just one speed magnitude (all discrete speeds share the same magnitude $c = 1$, hence the same energy) and only six different propagation angles. Not bad for an originally set of ∞^6 degrees of freedom!

Space–time is also discretized (see eqn (2.5)), but this is common to all dynamical systems destined to computer simulation. At this stage, it is still hard to believe that such a stylized system can display all the complexities of fluid phenomena. And yet it does!

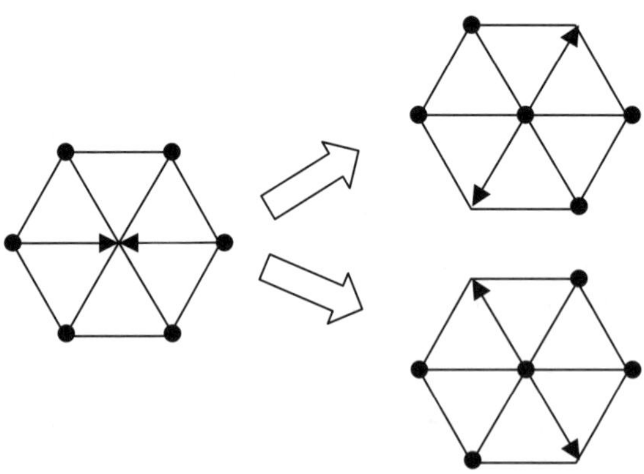

FIG. 2.4. A FHP collision with two equivalent outcomes.

The reader acquainted with modern statistical mechanics surely smells the sweet scent of *universality*: for all its simplicity, the FHP automaton may belong to the same class of universality of a real fluid. The name of this magic is *symmetry and conservation.*

Let us dig a bit deeper into this matter.

Consider the FHP collision depicted in Fig. 2.4.

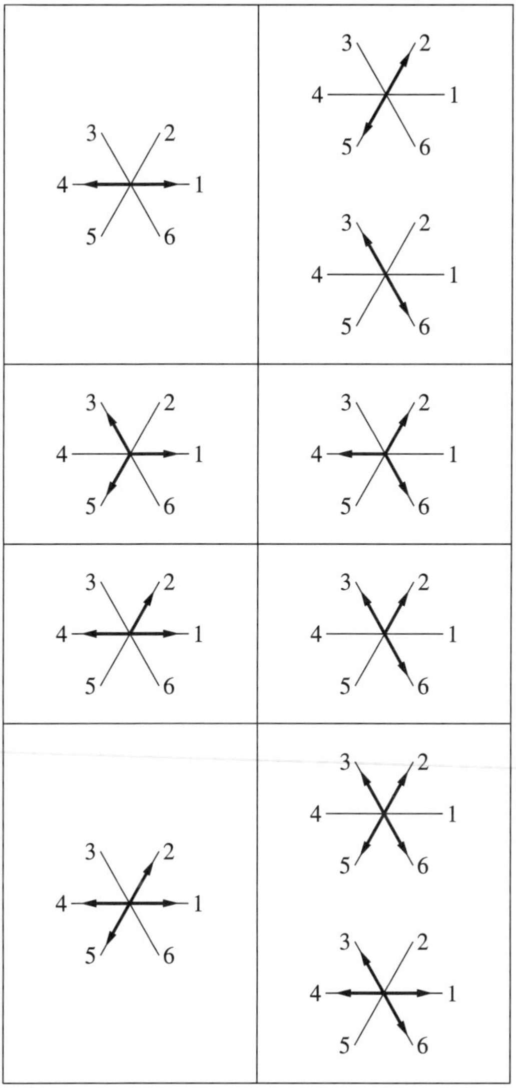

FIG. 2.5. Table of all possible FHP collisions. (Courtesy of L. S. Luo.)

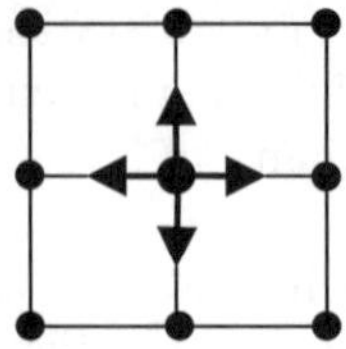

FIG. 2.6. The square grid of the HPP automaton.

Albeit stylized, this collision shares two crucial features with a real molecular collision:

- It conserves particle number (2 before, 2 after).
- It conserves total momentum (0 before, 0 after).

These properties are a '*conditio sine qua non*' to achieve hydrodynamic behavior once a sufficiently large group of particles is considered; this is what sets a fluid apart from a 'wild bunch' of particles. It is a necessary but not sufficient condition, though. One may ask why not make things even simpler and consider for instance just a four-state automaton like the one depicted in Fig. 2.6.

This is the so-called Hardy–Pomeau–de Pazzis (HPP) cellular automaton, predating FHP by more than 10 years [23]. Like FHP, the HPP automaton can also secure conservation laws. However, it fails to achieve a further basic symmetry of the Navier–Stokes equations, namely *rotational invariance.*

Making abstraction of dissipative terms for simplicity, the hydrodynamic probe of isotropy is the tensor $T_{ab} \equiv \rho u_a u_b + P\delta_{ab}$.

For a 2D Navier–Stokes fluid, this reads:

$$T_{xx} = P + \rho u^2, \tag{2.7}$$

$$T_{xy} = T_{yx} = \rho uv, \tag{2.8}$$

$$T_{yy} = P + \rho v^2, \tag{2.9}$$

where P is the fluid pressure and u, v the Cartesian components of the flow field. This is an isotropic tensor because its components are invariant under arbitrary rotations of the reference frame.

With HPP, we would obtain instead

$$T_{xx} = P + \rho(u^2 - v^2), \tag{2.10}$$

$$T_{xy} = T_{yx} = 0, \tag{2.11}$$

$$T_{yy} = P + \rho(v^2 - u^2). \tag{2.12}$$

Besides missing isotropy, this 'Squareland' tensor has little to do with real fluids altogether.[5] Thus, the four HPP particles do not qualify for the status

[5] As personally experienced by this author, it produces beautiful *square* vortices (sic!).

of *fluons*; that is, elementary excitations whose collective dynamics reproduces macroscopic hydrodynamics, while the six FHP particles do.

It is important to realize that this lack of isotropy cannot be cured by going to finer space–time resolutions, no matter how finer. In mathematical language, this is because the continuous group of rotations $SO(1)$ is compact. In two dimensions, the group of discrete rotations Z_6 (rotations by multiple of $2\pi/6$) can substitute for the continuous group $SO(1)$ while Z_4 cannot. More precisely, the lattice must generally provide enough symmetry to ensure the following tensorial identity (spatial indices are summed upon):

$$\left[\sum_{i=1}^{b} c_{ia}c_{ib}\left(c_{ic}c_{id} - \frac{c^2}{D}\delta_{cd}\right)\right] u_c u_d = u_a u_b \tag{2.13}$$

for any choice of the dyadic $u_a u_b$. Here D is the space dimensionality and b is the number of discrete speeds (not to be confused with the subscript b denoting spatial indexing).

This is a very stringent condition that weeds out most discrete lattices. As it stands, the constraint (2.13) comes a bit out of the blue. Intuitively, however, the point is that as opposed to, say, a scalar field, fourth-order tensor fields are demanding probes of the lattice since they sense more details of the space–time 'fabric' (they sense multiple directions at a time). As a result, it is more difficult to cheat on them, and have them believe that a square and a circle are the same thing, that simply does not work! This is not surprising, after all. More surprising (and pleasing) is that a simple hexagon does serve the purpose.[6] Even after this blend of definitions of isotropy and hand-waving arguments, the inquiring reader may still ask why fourth-order tensors and not some other, say second or sixth?

To make the picture truly compelling, a little plunge into the mathematics of discrete kinetic theory cannot be helped [13]. A partial justification, if not explanation, will nonetheless be given later in this chapter. Let us for the time being suspend criticism and accept for a fact that a hexagon is 'good enough' to replace a circle, whereas a square is not.

Now, back to the collision operator.

Symbolically, its effect on the occupation numbers is a change from n_i to n'_i on the same site

$$n'_i - n_i = C_i\left(\underline{n}\right), \tag{2.14}$$

where $\underline{n} \equiv [n_1, n_2, \ldots, n_b\rangle$ denotes the set of occupation numbers at a given lattice site.

To formalize the expression of C_i it proves expedient to label phase-space via a bit-string $\underline{s} = [s_1, s_2, \ldots, s_b\rangle$ spanning the set of all possible (2^b) states at a given lattice site. For instance, numbering discrete speeds 1–6 counterclockwise starting from rightward propagation, $c_{1x} = 1$, $c_{1y} = 0$, the pre- and post-collisional states read $\underline{s} = [100100\rangle$ and $\underline{s}' = [010010\rangle$, respectively.

[6] Said by Uriel Frisch: 'The symmetry gods are benevolent'.

It is natural to define a *transition matrix* $A(\underline{s}, \underline{s}')$ flagging all permissible collisions from source state $\underline{s}$ to destination state $\underline{s}'$, $\underline{s} \to \underline{s}'$ as follows:

$$A(\underline{s}, \underline{s}') = 1 \quad \text{collision allowed,} \tag{2.15}$$

$$A(\underline{s}, \underline{s}') = 0 \quad \text{collision forbidden.} \tag{2.16}$$

'Allowed' here means compliant with conservation laws (see Fig. 2.5).

The transition matrix obeys the *semi-detailed balance* condition:

$$\sum_{\underline{s}} A(\underline{s}, \underline{s}') = 1, \tag{2.17}$$

meaning that every destination state necessarily comes from a source state within the phase-space of the automaton. This condition does not imply a one-to-one source–destination relationship, as it is the case of detailed balance:

$$A(\underline{s}, \underline{s}') = A(\underline{s}', \underline{s}) . \tag{2.18}$$

The latter ensures micro-reversibility, while the former does not. Indeed, it is easily shown that a given pre-collisional FHP input state can land into more than one (actually, two) post-collisional output state compliant with conservation laws (head-on collisions can equally well rotate particle pairs $\pi/6$ left or right). Consequently, unlike HPP, the time evolution of the FHP automaton is no longer deterministic. In practice, the resulting lack of chiral invariance is easily disposed of by choosing either collision with equal probabilities.

Next, let us define the probability to have $\underline{s}$ as input state with occupation number $\underline{n}$:

$$\mathrm{P}(\underline{s}, \underline{n}) = \prod_{i=1}^{b} n_i^{s_i} \, \bar{n}_i^{\bar{s}_i} . \tag{2.19}$$

Here the overbar denotes complement to one, i.e., $\bar{n}_i \equiv 1 - n_i$ as befits to fermionic degrees of freedom.

Let us clarify with an example. The probability of occupying state $\underline{s} = [100100\rangle$ as an input string is given by $\mathrm{P}[100100\rangle = n_1 \bar{n}_2 \bar{n}_3 n_4 \bar{n}_5 \bar{n}_6$.

This quantity is manifestly always zero, except when a particle with speed $\vec{c}_1$ AND a particle with speed $\vec{c}_4$ are sitting simultaneously on the node. In passing, we note that 'particle absence' is to all intents tantamount to 'hole presence', echoing the particle–hole symmetry of fermionic matter.

With these preparations, the collision operator can be formally recast in the traditional gain minus loss form

$$C_i = \sum_{\underline{s}\,\underline{s}'} (s_i' - s_i) \, \mathrm{P}(\underline{s}, \underline{n}) \, A(\underline{s}, \underline{s}') . \tag{2.20}$$

The reader is encouraged to check that C_i is a 'trit', namely a 3-state variable taking values -1 (annihilation), 0 (no action), +1 (generation) (see Fig. 2.7).

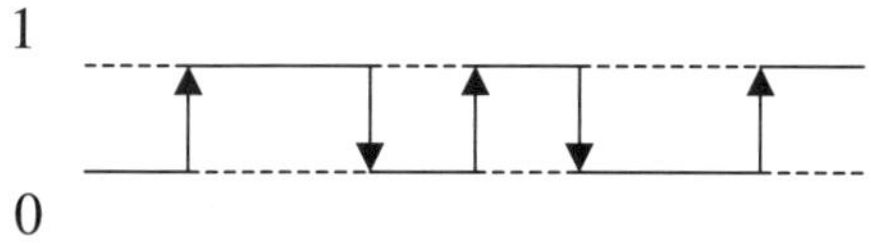

FIG. 2.7. The railway analogy of Boolean computing. Only $0 \to 1$, $1 \to 0$ transitions can occur, leaving the system on either upper or lower rails. No drifts allowed.

The trit associated with the head-on collision depicted in Fig. 2.4 is easily computed: $C_i = [-1, 1, 0, -1, +1, 0\rangle$. It is also readily checked that the sum over all discrete speeds yields identically zero, in fulfilment of the requirement of mass conservation. This is formally traced back to the identity:

$$n_i = \sum_{\underline{s}'\underline{s}} s_i \mathrm{P}\left(\underline{s}, \underline{n}\right) A\left(\underline{s}, \underline{s}'\right), \tag{2.21}$$

which is easily checked by direct computation.

Owing to its definition, the collision operator C_i obeys the remarkable property of preserving the Boolean nature of the occupation numbers. This can be verified directly by noting that the Boolean-breaking occurrences, $n_i = 0, C_i = -1$ or $n_i = 1, C_i = 1$, can never take place. They are automatically ruled out by the fact that if $n_i = 0$, the collision operator cannot subtract particles to the input state, and conversely, if $n_i = 1$ it cannot add them to the pre-collisional state.

This closely evokes the well-known properties of fermionic annihilation/generation operators as described by the language of second quantization.

To sum up, the final LGCA update rule reads as follows:

$$\Delta_i n_i = C_i, \tag{2.22}$$

or, which is the same:

$$n_i\left(\vec{x} + \vec{c}_i, t + 1\right) = n_i'\left(\vec{x}, t\right), \tag{2.23}$$

where all quantities have been defined previously.

The eqns (2.22) and (2.23) represent the microdynamic equation for the Boolean lattice gas, the analogue of Newton equations for real molecules.

As already pointed out, this formulation evokes closely the second quantization formalism, and it casts fluid dynamics into a many-body language probably more familiar to condensed matter and quantum field theorists than fluid dynamicists.

This equation constitutes the starting point of a lattice BBGKY hierarchy, ending up with the Navier–Stokes equations. At each level, one formulates a lattice counterpart of the various approximations pertaining to the four levels of the hierarchy (see Fig. 2.8).

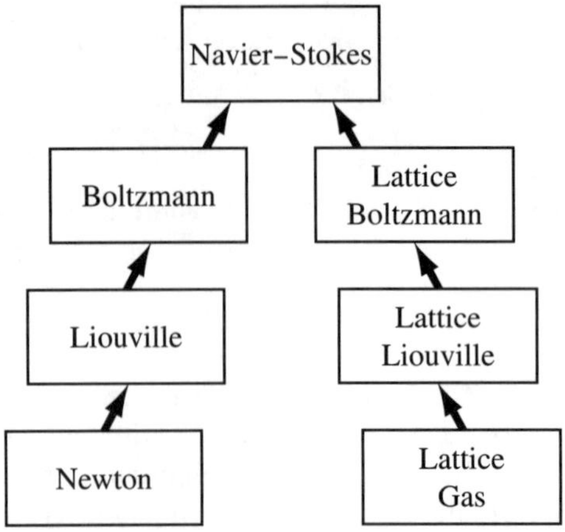

FIG. 2.8. The BBGKY and lattice BBGKY hierarchies. The lack of microscopic detail of the lattice branch becomes less and less important as one walks upward along the hierarchy.

The remarkable point is that, notwithstanding the drastic reduction of microscopic degrees of freedom, the lattice Navier–Stokes equation can be made basically to coincide with its continuum counterpart. In the LGCA jargon, the lattice is 'erased' from the macroscopic dynamics.

The guiding lights of this clever procedure are the fundamental conservation laws of classical mechanics [13].

In Boolean terms:

- *Mass and momentum conservation*:

$$\sum_i C_i = 0, \tag{2.24}$$

$$\sum_i c_{ia} C_i = 0. \tag{2.25}$$

- *Angular momentum conservation (rotational invariance)*:

$$\sum_i c_{ia} c_{ib} c_{ic} c_{id} = \frac{bc^4}{D\,(D+2)} \left(\delta_{ac}\delta_{bd} + \delta_{ad}\delta_{bc} - \frac{2}{D}\delta_{ab}\delta_{cd} \right). \tag{2.26}$$

As it stands, the microdynamic eqn (2.22) represents the dynamics of a $6N$-body system of hopping spins ('spin fluid') with local mass and momentum conservation, a very rich and interesting model of non-equilibrium statistical mechanics. That the large-scale dynamics of this model does indeed reproduce the fluid dynamic equations remains still to be proved. We shall only sketch the main ideas, directing the reader fond of full details to the excellent monographs available in the recent literature [22].

2.3 From LGCA to Navier–Stokes

The standard *bottom-up* procedure[7] taking us from many-body particle dynamics (Newton–Hamilton) all the way up to continuum fluid-like equations proceeds through the three formal steps familiar from classical statistical mechanics:

1. *From Newton–Hamilton to Liouville*
   Upon ergodicity assumption[8] the deterministic Newton–Hamilton equations of motion are replaced by the probabilistic Liouville equation for the N-body distribution function $f_N(\vec{x}_1, \vec{v}_1, \ldots, \vec{x}_N, \vec{v}_N; t)$ expressing the probability of finding particle 1 around $\vec{x}_1$ with speed $\vec{v}_1$, *and* particle 2 around $\vec{x}_2$ with speed $\vec{v}_2$, *and* particle N around $\vec{x}_N$ with speed $\vec{v}_N$, at the same time t.

   Symbolically: $\underline{z}(t) \to f_N(\underline{z}, t)$, where $\underline{z} \equiv [\vec{x}_1, \ldots, \vec{v}_N]$.

2. *From Liouville to Boltzmann*
   Upon diluteness assumption, high-order distribution functions are expressed in terms of the low-order ones by integration over many-body phase-space coordinates. This procedure ends up at the lowest level (Boltzmann) dealing with the single-body distribution function. The hierarchy is not closed because the distribution function at a given level depends on the upper-lying one. To close the hierarchy, a truncation in the diluteness parameter $(s/d)^3$, s being the effective molecular diameter and d the mean intermolecular distance, is used.

   Symbolically: $f_N(\underline{z}, t) \to f(\vec{z}_1, t)$.

3. *From Boltzmann to Navier–Stokes*
   Upon integration over momentum space degrees of freedom, a set of partial differential equations for the space–time dependent moments of the Boltzmann distribution function is obtained.
   Upon retaining a finite, say $m = 1, \ldots, M$, number of moments, the set of moment equations takes the form of an M-dimensional *generalized continuity equation*

   $$\partial_t \rho_m + \partial_a J_{am} = C_m, \tag{2.27}$$

   where

   $$\rho_m = \int f \phi_m(\vec{v})\, \mathrm{d}\vec{v}, \tag{2.28}$$

   $$J_{am} = \int f v_a \phi_m(\vec{v})\, \mathrm{d}\vec{v} \tag{2.29}$$

[7] By 'bottom' we imply the most fundamental, microscopic level, the idea being that upper levels 'emerge' to the status of macroscopic observables from coarse-graining of the lower levels.

[8] Loosely speaking, ergodicity means that the probability of visiting a given region of phase-space is proportional to the volume of that region. This permits replacement of the phase-space averages with time averages.

are the generalized densities and currents, respectively, and

$$C_m = \int C\,[f]\,\phi_m\,(\vec{v})\,\mathrm{d}\vec{v} \tag{2.30}$$

represents the effect of intermolecular interactions.
Finally, $\{\phi_m\}$ is a complete set of basis functions in momentum space, typically Hermite polynomials.
The most relevant hydrodynamic moments are:

Density:	$\rho = m \int f \,\mathrm{d}\vec{v},$
Speed:	$u_a = m \int f v_a \,\mathrm{d}\vec{v}/\rho,$
Momentum flux tensor:	$P_{ab} = m \int f v_a v_b \,\mathrm{d}\vec{v}.$

The system (2.27) is not closed either since the time derivative of a given generalized density involves the divergence of the corresponding generalized flux lying one step higher in the hierarchy.
The hierarchy is closed by splitting up the one particle distribution into a local equilibrium and non-equilibrium components:

$$f = f^{\mathrm{e}} + f^{\mathrm{ne}},$$

with the assumption that $f^{\mathrm{ne}} \sim O(Kn)f^{\mathrm{e}}$, Kn being the Knudsen number, namely the ratio of the particle mean free path to a typical macroscopic length-scale (*Chapman–Enskog procedure*).
To leading order $O(Kn)$, the power expansion in Knudsen number of the Boltzmann equation delivers the Euler equations of inviscid flows, whereas at the next order $O(Kn^2)$ the dissipative Navier–Stokes equations are obtained.

Exactly the same steps are involved in the process of deriving the lattice Navier–Stokes equations from the lattice BBGKY, with the notable caveat that lattice discreteness needs to be handled with great care since continuum symmetries (Galilean invariance, roto-translations, parity) are always at risk of being broken by the lattice discreteness.

The most dangerous effects bear upon the discrete lattice equilibria.

2.3.1 *Discrete local equilibria*

Since LGCA microdynamics obeys an exclusion principle, lattice equilibria are expressed by a Fermi–Dirac distribution

$$f_i^{\mathrm{eq}} = \frac{\rho/b}{1 + \mathrm{e}^{\Phi_i}}, \tag{2.31}$$

where Φ_i is a linear combination of the collisional invariants, mass and momentum. Restricting our attention to isothermal ideal fluids, where energy and mass are the same, we obtain:

$$\Phi_i = A + Bc_{ia}u_a, \tag{2.32}$$

where A and B are free Lagrangian parameters to be adjusted in order to secure mass and momentum conservation.

As usual, the goal is to express the Lagrange multipliers A, B as a function of fluid density ρ and momentum u_a, by using the conservation relations (2.24) and (2.25). This task is easy in the continuum, much less so in a discrete momentum space. Let us see why.

By plugging the Fermi–Dirac expression into the conservation relations (2.24) and (2.25), we obtain two nonlinear equations for the Lagrangian parameters A, B as a function of the fluid density ρ and speed u_a. Actually, ρ can always be scaled out since it is just a multiplicative prefactor (we are considering incompressible fluids). The dependence on u_a is non-trivial, in the sense that it is generally *not* possible to come up with a closed, analytical expression of the form $A = A(u_a), B = B(u_a)$.

The point is by no means a mere technicality!

What we would like to obtain is something in the form $\Phi = A + B_a(c_{ia} - u_a)$ to ensure Galilean invariance. The exponential form (2.31) reflects the minimization of the Boltzmann entropy

$$H = -\int \tilde{f} \ln \tilde{f} \, d\vec{v} \tag{2.33}$$

($\tilde{f} \equiv f/(1-f)$ for fermions), which is strictly related to Stosszahlansatz assumption. The subsequent possibility to express the Lagrangian parameters in a closed form is in turn strictly tied up to the continuity of momentum space.

In a discrete momentum world, the exponential function plays no special role (if it did, the plain replacement $v_a = c_{ia}$ would work!), and we may expect it to be replaced by some other functional form, perhaps suitable polynomials.

In the absence of a closed analytical solution for local equilibria, it appears quite natural to seek approximate solutions obtained by expanding the exponential in powers of the flow field, or more precisely, of the Mach number $Ma = u/c_s$. Conservation constraints can then be imposed order-by-order in the perturbation series and since the ultimate target, the Navier–Stokes equations, exhibit a quadratic nonlinearity (the convective term $\partial_b u_a u_b$), it is legitimate to expect that a second-order expansion should meet the need.

This is indeed the case, modulo some anomalies that will be commented on shortly.

The specific expression of LGCA equilibria is:

$$f_i^{\text{eq}} = \frac{\rho}{b}\left[1 + u_i + Gq_i\right], \tag{2.34}$$

where

$$u_i = \frac{c_{ia}u_a}{c_s^2}, \tag{2.35}$$

$$q_i = \frac{Q_{iab}u_a u_b}{2c_s^4} \tag{2.36}$$

and

$$Q_{iab} = c_{ia}c_{ib} - c_s^2\delta_{ab} \tag{2.37}$$

is the projector along the i-th discrete direction.

Here

$$c_s = \frac{c}{\sqrt{D}} \tag{2.38}$$

is the lattice sound speed.

The breaking of Galilean invariance reflects into the anomalous prefactor

$$G(\rho) = \frac{1-2d}{1-d} \neq 1, \tag{2.39}$$

where

$$d = \frac{\rho}{b} \tag{2.40}$$

is the *reduced fluid density.*

The isotropy constraint (2.13) unveils now part of his mysterious nature: plugging discrete quadratic equilibria (2.34) into the definition of the (equilibrium) momentum flux tensor (see Chapter 1), $P^{\mathrm{e}}_{ab} = \int f^{\mathrm{e}} v_a v_b \, \mathrm{d}\vec{v}$, reveals that (2.13) is just the condition to match the tensorial equality:

$$\int f^{\mathrm{e}} v_a v_b \, \mathrm{d}\vec{v} = \rho\left(u_a u_b + c_s^2 \delta_{ab}\right) \tag{2.41}$$

discussed in Chapter 1.

Let us now take a short break from theory and spend a few comments on the practical use of lattice gas cellular automata as a tool for computer simulation.

2.4 Practical implementation

The main computational assets of the LGCA approach to fluid dynamics are:

- Exact computing (round-off freedom);
- Virtually unlimited parallelism.

The Boolean nature of the LGCA update rule implies that the corresponding algorithm can be implemented in pure Boolean logic, without ever needing floating-point computing. This is very remarkable, since it offers a chance to sidestep a number of headaches associated with the floating-point representation of real numbers, primarily round-off errors. Among others, the Boolean representation eases out the infamous problem of numerical drifts plaguing long-time simulations both in fluid dynamics and fundamental studies in statistical mechanics.

To see this in more detail, let us consider the head-on collision of Fig. 2.4. The collision is encoded by the following logic statement:

'If there is a particle in state 1 AND a hole in state 2 AND a hole in state 3 AND a particle in state 4 AND a hole in state 5 AND a hole in state 6, then the collision occurs. Otherwise nothing happens.'

All we need to put this plain logical statement into practice are the elementary Boolean operations AND (logical exclusion), OR (logical inclusion), NOT (negation) and XOR (exclusive OR). The effect of these operations is usually represented by the so-called 'truth table' (1 = TRUE and 0 = FALSE) which we report below for simplicity.

The statement 'the collision occurs' corresponds to setting up a collision mask flagging collisional configurations $M = 1$ and $M = 0$ all others:

$$M = n_1.\mathrm{AND}.(\bar{n}_2).\mathrm{AND}.(\bar{n}_3).\mathrm{AND}.(n_4).\mathrm{AND}.(\bar{n}_5).\mathrm{AND}.(\bar{n}_6), \tag{2.42}$$

where the overbar means negation.

Once the collision mask M is set up, the post-collisional state

$$n_i' = n_i + C_i \tag{2.43}$$

is obtained by simply 'XORing' the pre-collisional state with the collisional mask.

$$n_i' = M.\mathrm{XOR}.n_i. \tag{2.44}$$

This simple procedure is applied simultaneously in lock-step mode to all lattice sites in the typical 'blindfold' fashion so dear to vector computers. More importantly, each site is updated independently of all others, thus making the scheme ideal for parallel processing. Aptness to massively parallel processing is probably worth a few additional words of comment.

Roughly speaking, parallel computing works on the time-honored Roman principle *divide et impera* (divide and conquer). To solve a large problem, first break it into small parts, then solve each part independently, and finally glue all the parts together to produce the global solution [24]. By doing so, a collection of, say, P processors would ideally solve a given problem at a fraction $1/P$ of the cost on a single computer.

Apparently, nothing could be easier, but actual practice often tells another story because many sources of inefficiency, both practical and conceptual, may hamper the practical realization of this general principle.

Among others, a chief quantity to consider is the communication/computation ratio. Generally speaking, a numerical algorithm is good for parallel computing when the amount of communication between the different processors

TABLE 2.1. Truth table of the basic Boolean operators.

A	B	A.AND.B	A.OR.B	A.XOR.B	.NOT.A
0	0	0	0	0	1
0	1	0	1	1	
1	0	0	1	1	0
1	1	1	1	0	

FIG. 2.9. One (unhappy) overwhelmed worker. (Courtesy of A. Masselot.)

needed to coordinate their interaction can be made negligible in comparison with the in-processor computations (see Appendix C). This is quite intuitive, because if communication is scanty, all processors can proceed (almost) independently without wasting too much time in exchanging information (the cell-phone syndrome).

LGCA are fairly well positioned in this respect since the collision step, by far the most time-consuming operation, is completely local and requires no communication at all among the processors. This low communicativity makes LGCA ideal candidates to massively parallel processing.

The elegance and the practical appeal of round-off-free parallel computing justify the blue-sky scenarios envisaged in the late 80s when LGCA held promise of simulating turbulence on parallel machines at unprecedented resolutions.

A closer look at the method revealed a number of flaws that ultimately ground it almost to a halt for fully developed turbulence (at least in the academic environment) in the early 90s. Basically, it was realized that LGCA are plagued by a number of anomalies, namely broken symmetries which cannot be restored even in the limit of zero lattice spacing [25]. Ingenious cures were found, but still the final picture did not live up to the (real high) initial expectations, at least for

FIG. 2.10. Many (happy?) workers, sharing the workload. (Courtesy of A. Masselot.)

the simulation of fully developed turbulent flows.

These anomalies surface neatly only once the task of deriving quantitative hydrodynamics out of the LGCA Boolean dynamics is undertaken.[9] This is a serious technical task, which proceeds along the lines of standard statistical mechanics, with the additional burden of 'slalomizing' through the technical catches littered by space–time–momentum discreteness.

2.5 Lattice gas diseases and how to cure them

The formal derivation of the 'Navier–Stokes look-alike' equations from LGCA microdynamics highlights two basic diseases related to the lattice discreteness, namely:

A1. *Lack of Galilean invariance.*

A2. *Anomalous velocity dependence of the fluid pressure.*

Additional drawbacks coming along with phase-space discretization are:

A3. *Statistical noise.*

A4. *High viscosity (low Reynolds number).*

A5. *Exponential complexity.*

A6. *Spurious invariants.*

Anomalies A1, A2 trace back to the fact that a finite number of speeds does not allow a *continuum* family of (Fermi–Dirac) equilibria parameterized by the flow speed $\vec{u}$. As a result, hydrodynamic equilibria can only be defined via a perturbative expansion in the flow field, and this expansion cannot match exactly the form of the Navier–Stokes inertial and pressure tensors. It delivers instead the following:

$$\sum_i f_i^{\mathrm{e}} c_{ia} c_{ib} = \rho \left[g u_a u_b + c_s^2 \left(1 - g M^2\right) \delta_{ab} \right], \tag{2.45}$$

where

$$g = \frac{D}{D+2} \frac{1-2d}{1-d} \tag{2.46}$$

is the Galilean breaking factor.

The quadratic term in this expression is responsible for the advection term $u \nabla u$ in the Navier–Stokes equation, whereas the remaining term is associated with fluid pressure. Both terms are anomalous, and in a truly devious way. To make advection Galilean invariant we need $g = 1$. First, there is *no* value of the fluid density for which this can occur. Second, the value $g = 1$ just maximizes the anomalous contribution gM^2 to the fluid pressure. This spells true frustration.

[9] As opposed to the 'wallpaper' prejudice looming on LGCA in the early days. According to this prejudice, CA are very good at pointing out colorful (wallpaper) analogies, much less so at nailing down quantitative hard numbers. It is only fair to say that this kind of criticism did not strike home with LGCA. Shortly after the seminal FHP paper, Frisch, d'Humières, Lallemand, Hasslacher, Pomeau and Rivet hastened to produce a thorough quantitative description of the way the large-scale LGCA dynamics yields real-life hydrodynamic equations [13].

Admirably enough, pioneers in the field responded quite promptly to this disheartening state of affairs.

It was rapidly recognized that for quasi-incompressible flows (low Mach number $u/c_s \ll 1$), the pressure anomaly becomes negligibly small, the density is virtually a constant, and so is the Galilean breaking factor g. Under these conditions, a plain rescaling of time $t \to t/g$ and viscosity $\nu \to \nu/g$ makes it possible to reobtain anomaly-free Navier–Stokes equations!

2.5.1 *Statistical noise*

Like any particle method, LGCA are exposed to a fair amount of statistical fluctuations. This problem was quickly recognized by Orszag and Yakhot [26] and shortly later demonstrated by actual numerical simulations [27] against pseudospectral methods for the case of two-dimensional turbulence. The problem was that, like any honest N-body Boolean system, the automaton was still crunching a lot of needless many-body details. Very noticeable efforts, mainly in Nice, Paris and Shell Research, brought the 2nd and 3rd generation FHPs to the point of competitiveness with conventional methods. Besides clever theory, computer hardware has some say too. Comparing LCGA and floating-point methods on what we usually call general purpose computers is subtly unfair, simply because what we call a general purpose computer is de-facto a special purpose computer for floating-point calculations! (Think of the following nightmare: computational fluid dynamics with floating-point operations in software ...!).

Why not do the same for LGCA, namely build a LGCA-chip performing, say, a collision (in fact, very many) in just a single clock cycle? On such a computer LGCA would probably win hands down over 'traditional' methods.

Some prototypes in this breed were indeed built at MIT and at the Ecole Normale in Paris [28,29]. Discussing why these prototypes did not make it into a fully fledged LGCA machine would be too long a story, that certainly has to do with the mind-boggling escalation of computer power afforded by 'off the shelf' commercial floating-point processors. But, again, that is another story.

Before moving on to the next item, it is important to mention that there is a positive side to noise, too. As shown by Grosfils *et al.* [30] the statistical noise intrinsic to LGCA dynamics has many *quantitative* features in common with true noise in actual thermodynamic systems. This puts LGCA in a privileged position to address problems related to modern statistical (micro)hydrodynamics.

2.5.2 *Low Reynolds number*

The maximum Reynolds number achievable by a LGCA simulation is controlled by the minimum mean free path one can reach in the lattice, namely the maximum number of collisions per unit time that the automaton is able to support. More precisely, it is not just the number of collisions which matters, but also their quality factor, namely the amount of momentum they are able to transfer across the different directions of the discrete lattice. Phase-space discreteness plays a two-faced role here. On the one side, the limited number of discrete speeds combined with compliance to mass–momentum conservation, puts heavy restrictions on the number of possible collisions. For instance, even in FHP III, a low-viscosity variant of FHP with rest particles (7 states per site), only 22 out of 128 configurations are liable to collisions, which means that only about twenty per cent of phase-space is collisionally active. This is to be contrasted with continuum kinetic theory in which virtually all of phase-space is collisionally active. This lowly twenty per cent efficiency is partially rescued by the streaming step. Here lattice discreteness is on our side because the discrete speeds force particles to meet at lattice sites with no chance for 'near miss' events (a sort of 'highway' effect).

One may thus hope to achieve good collisionality, namely a mean free path smaller than the lattice spacing a. Indeed, that is exactly what happens.

The figure of merit of LGCA collisionality is the dimensionless Reynolds number R^*, defined as

$$R^* = \frac{g c_s a}{\nu}. \tag{2.47}$$

With this definition, the Reynolds number becomes

$$Re = R^* \, Ma \, N, \tag{2.48}$$

where N is the number of lattice site per linear dimension ($L = Na$) and Ma is the Mach number.

By recalling that fluid viscosity ν is basically the mean free path l_μ times the particle speed c, we obtain $R^* = ga/l_\mu$. This definition shows that R^* is essentially the number of mean free paths in a lattice spacing a. Since the computational power required to simulate a three-dimensional flow of size L scales roughly like L^4 (three spatial dimensions times the temporal span of the evolution), it is clear that raising R^* by a factor of two, buys us a factor of eight in memory occupation and sixteen in computer power. This explains the intense efforts spent on the optimization of this parameter in the late 80s.

The earliest FHP model featured $R^* = 0.4$, which means a collision every two and a half streaming hops. This also means that hydrodynamics cannot be attained at the scale of a single lattice spacing, and consequently that heavy coarse-graining is needed to extract sensible hydrodynamics out of the noisy

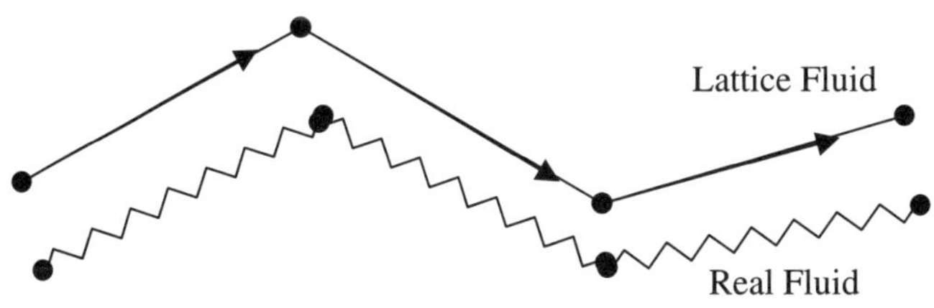

FIG. 2.11. True and LGCA mean free path.

LGCA signal. Once again, we are steered in the direction of trading spatial degrees of freedom for visibility of coherent signals. This contrasts with conventional Computational Fluid Dynamics (CFD) where the problem simply does not exist and the lattice pitch, a, can be freely chosen in the close vicinity of the smallest hydrodynamic scale. Since, by definition, hydrodynamics works in the regime of low Knudsen numbers $Kn \ll 1$, it follows that $l_\mu/a = Kn \ll 1$.

Therefore, conventional computational fluid dynamics works at $R^* \sim 1/Kn \sim 100$ or more. In other words, the situation is just the opposite as for early LGCA: many mean free paths in a lattice spacing. The impact on the Reynolds number is obvious; consider just simple figures: a 1000 cubed grid with a flow at $Ma = 0.1$ and $R^* = 0.2$, yields a lowly $Re = 0.1 \times 0.2 \times 1000 = 20$! We are two orders of magnitude below conventional CFD.

On the other hand, since the computational work grows with the 4-th power of R^*, it is clear that even minor improvements on this parameter bring a huge pay-off in terms of computational efficiency. This explains the intense research efforts spent in the late 80s to enhance LGCA collisionality. This activity yielded a remarkable top value of $R^* = 13$; that is, almost the hardly sought for two orders of magnitude.

The gap was basically bridged even on 'general purpose' machines: undeniably a very remarkable deed.

The large number of speeds (24) required by the algorithm, however, made it rather impractical even on most powerful computers. This takes us to the next issue, namely exponential complexity of the collision rule.

2.5.3 *Exponential complexity*

A weird feature of LGCA is that there is no three-dimensional crystal ensuring the isotropy of fourth-order tensors, as required to recover isotropy of the Navier–Stokes equations. For a (short) while, this seemed another unforgiving 'no go' to three-dimensional LGCA hydrodynamics. A clever way out of this problem was found by d'Humières, Lallemand and Frisch in 1989 [31]. These authors realized that a suitable lattice can be found, provided one is willing to add an extra space dimension to the LGCA world. In particular, they showed that the *four*-dimensional Face Centered HyperCube (FCHC) has the correct properties. The FCHC consists of all neighbors of a given site (center) generated by the speeds $c_i = [\pm 1, \pm 1, 0, 0]$ and permutations thereof. This yields 24 speeds, all sharing the same magnitude $c_i^2 = 2$.

The three-dimensional projection of FCHC is shown in Fig. 2.12, from which we see that the 24 speeds are generated as follows:

- *6 nearest neighbors (nn) (connecting the center C to the centers of the six faces of the cube).*
- *12 next-to-nearest neighbors (nnn) connecting C to the centers of the edges of the cube.*
- *The remaining six speeds are accounted for by counting a factor of two (degeneracy) to the nn connections as they possess a nonzero speed along*

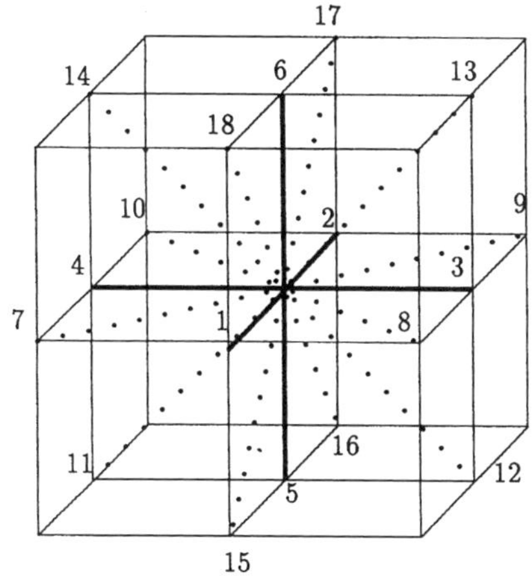

FIG. 2.12. The face centered hypercubic lattice. Thick solid lines represent degenerate directions along the fourth dimension ($c_4 = \pm 1$).

the fourth dimension.

Thus, *nn* links can be thought as of carrying two particles instead of one.

The use of extra dimensions to remove anomalies is not so foreign to field theory; also the famous ten/eleven dimensions of superstring theories come from a quest for cancellations of unwanted anomalies. But let us come back to the point. The FCHC restores the missing symmetry, namely isotropy, but not for nothing. In fact, the number of Boolean operations needed to exhaustively represent the collisional states is a $O(1)$ fraction of the single site phase-space of the automaton. As a result, a b-bit automaton generates a computational complexity of order $O(2^b)$.

In 2D $b = 6, 7$ and the collision operator is nearly a matter of 'back of the envelope' calculation. In 3D, however, $b = 24$, and the situation is much less favorable. Some actual numbers help in putting the picture straight.

With $b = 24$, we must make allowance for $2^{24} = 16.128$ millions of Boolean operations. Even assuming a closed, analytical, Boolean expression for such a 'monster' operator can be found, its practical value would be spoiled by the huge amount of computer time required to process it.

Because of these daunting difficulties, researchers turned to the use of pre-hard-wired look-up tables storing the post-collisional state $\underline{s}'$ as a function of the pre-collisional one $\underline{s}$. Interestingly enough, the compilation of these table look-ups requires a computer program on its own taking a few CRAY-2 hours Central Processing Unit (CPU) time!

Once the table is computed, the simulation code needs only to read-off the post-collisional state $\underline{s}'$ corresponding to each pre-collisional state $\underline{s}$. The conceptual simplicity of this strategy should not deceive the inexperienced reader.

From a practical point of view, reading off the output state given the input one, is a potentially costly operation. First, a matter of size: the mapping $A(\underline{s}, \underline{s}')$ generates a table look-up consisting of 24×2^{24} bits, namely 48 Mbytes (1 byte = 8 bits) of storage. What is worse, this huge piece of storage is to be accessed almost randomly because there is no way of storing the state $\underline{s}(\vec{x})$ at site $\vec{x}$ and those of its neighbors $\underline{s}(\vec{x} + \vec{c}_i)$ in contiguous memory locations.

As a result, reading off the table look-up may imply relatively long and expensive memory searches.

Pioneering simulations with FCHC–LGCA were performed by Rivet, Frisch, and Henon in Nice, who used a highly tuned version for the CRAY-2 machine to study transitional flows past a flat plate [32]. By then, R^* was raised by more than a factor of ten with respect to the original FHP, $R^*_{\mathrm{FCHC}} \sim 7$. Subsequently, P. Rem and J. Somers, then at Shell Research, devised ingenious strategies to work with reduced look-up tables (64 Kbytes instead of 48 Mbytes), by sacrificing approximately a factor of five in R^* [33].

2.5.4 *Spurious invariants*

In the absence of external sources/sinks continuum fluids conserve mass, momentum and energy: nothing less, nothing more either. Discrete fluids, as a result of leaving in a cage, may exhibit additional conserved quantities that draw their existence solely from the lattice discreteness. Being pure artifacts, they are currently referred to as *spurious invariants.* The effect of these spurious invariants is to shrink the accessible portion of phase-space (lack of ergodicity) with undesirable effects on the coarse-graining procedures leading to hydrodynamic equations. Systematic procedures to detect these invariants have been devised by a number of authors [34].

Although staggered invariants may be regarded as subtler pathologies as compared to the anomalies discussed in this chapter, they must be watched carefully in the simulations, especially in the vicinity of generic boundaries non-aligned with the lattice grid lines where spurious modes are more likely to be excited and propagate inside the fluid region. Fortunately, no serious damage seems to occur for simple boundaries aligned with the flow.

2.6 Summary

In spite of the remarkable progress achieved in the late 80s, leading to substantial improvements over the initial state of affairs, the LGCA algorithms remained nonetheless rather heavy and stiff with respect to changes in the collision rules. On the other hand, these inconveniences were not compensated by any dramatic advantage in terms of accessible Reynolds numbers. As a result, the interest in LGCA as a tool for high Reynolds flow simulations leveled off in the early 90s. This was the time the lattice Boltzmann method took off.

2.7 Exercises

1. List all possible collisions of HPP.

2. Construct the collision mask for a triple collision $[101010\rangle$ going into $[010101\rangle$.
3. Write down the collision mask and the trit operator for the above collisions.
4. Estimate the signal-to-noise ratio in a FHP block with 16×16 Boolean sites per fluid variable.
5. Suppose to simulate a 3D flow at $Re = 10\,000$ on a 100 cube grid. What R^* would you plan?

3

LATTICE BOLTZMANN MODELS WITH UNDERLYING BOOLEAN MICRODYNAMICS

The earliest Lattice Boltzmann Equation (LBE) was generated in response to one of the main drawbacks of LGCA; that is, statistical noise. Shortly after, it became apparent that all other anomalies plaguing LGCA could also be naturally disposed of by the LBE approach. As a result, LBE rapidly evolved into a self-standing research subject bearing an increasingly fainter relation to its LGCA ancestor. Better said, LBE development was steered in the direction of retaining the best of LGCA while giving away needless constraints and ensuing anomalies. The effort met with significant success, this is a fact. Still, this success did not come entirely for nothing, because one of the most heralded assets of LGCA, round-off freedom, had to be surrendered in the process. In this chapter we shall take a walk into the Jurassic's of LBE, namely the earliest lattice Boltzmann model that grew up out of underlying LGCA.

3.1 Nonlinear LBE

The earliest LBE was first proposed by G. McNamara and G. Zanetti in 1988, with the explicit intent of sidestepping the statistical noise problem plaguing its LGCA ancestor [19]. The basic idea is simple: just replace the Boolean occupation numbers n_i with the corresponding ensemble-averaged populations

$$f_i = \langle n_i \rangle, \tag{3.1}$$

where brackets stand for suitable ensemble averaging. The change in perspective is exactly the same as in Continuum Kinetic Theory (CKT); instead of tracking single Boolean molecules we content ourselves with the time history of a collective population representing a 'cloud' of microscopic degrees of freedom.

Formally, it is convenient to split up the average and fluctuating part of the Boolean occupation numbers as follows:

$$n_i = f_i + g_i, \tag{3.2}$$

where the fluctuation g_i averages to zero by definition ($\langle g_i \rangle = 0$).

By plugging (3.2) into the microdynamic Boolean eqns (2.23) given in the previous chapter, we obtain

$$\Delta_i f_i = C_i\left(\underline{f}\right) + G_i, \tag{3.3}$$

where G_i collects all contributions from interparticle correlations. It is important to remark that the left-hand side of eqn (3.3) is a direct transcription of the corresponding Boolean counterpart, with the plain replacement

FIG. 3.1. Cluster diagram $f_{12} = f_1 f_2 + g_{12}$. The springy line denotes correlations.

$$n_i \to f_i. \tag{3.4}$$

This relates to the linearity of the streaming operator. This is not the case for the multibody collision operator which, being a b-th order polynomial, introduces particle–particle correlations up to order b. These are collected in the term G_i.

A simple example for the 4-state HPP automaton yields

$$\langle n_1 \overline{n}_2 n_3 \overline{n}_4 \rangle = f_1 \overline{f}_2 f_3 \overline{f}_4 + P\left(f_1 \overline{f}_2 G_{34}\right) + G_{1234}, \tag{3.5}$$

where P denotes all possible permutations, and $\bar{f}_i = 1 - f_i$ are the hole populations. We note that the correlation chain of the lattice BBGKY hierarchy and the attendant Mayer cluster expansions are significantly more cumbersome than is usual in kinetic theory because of the multibody nature of the lattice collision operator. Nonetheless, the fact that the kinetic space is finitely dimensional permits interesting simplifications in lattice ring theory [35, 36].

In line with CKT, one makes the 'Stosszahlansatz' assumption of no correlations between particles entering a collision, namely:

$$G_i = 0. \tag{3.6}$$

By weeding out all multiparticle correlations we are left with a closed, nonlinear, finite difference equation for the one particle distribution f_i:

$$\boxed{\Delta_i f_i = C_i\left(f_1, \ldots, f_b\right).} \tag{3.7}$$

This is the earliest Lattice Boltzmann Equation (LBE).

In order to underline the nonlinear nature of the collision operator, we shall hereafter refer to it as *nonlinear LBE.*

It is interesting to point out that nonlinear LBE is a *déjà vu* from LGCA theory, where it was derived as an intermediate stopover in the journey from LGCA to Navier–Stokes. However, in the LGCA treatment LBE was just regarded as an intermediate *conceptual* step, with no hint at the possibility that it might constitute a valuable tool for fluid dynamics computations on its own.

McNamara and Zanetti's merit was precisely to point out a new way to interpret and use a known equation! In a way, they reinvented the LBE by looking at it from a different angle.

Having assessed that the nonlinear LBE is a direct transcription of LGCA microdynamics with the plain replacement $n_i \to f_i$, the trade-off is clear: noise is erased because f_i is *by definition* an averaged, smooth, quantity. On the other hand, the price to pay for this smoothness is twofold: on the physical side, the physics of particle correlations ('non-Boltzmann effects') is lost. This means that fundamental issues such as the breakdown of molecular chaos, long time tails, mode–mode coupling and related points, become off-limits.

Second, and more practical, round-off freedom falls apart since, being real numbers, the f_i are no longer amenable to exact Boolean algebra. The ground-breaking idea of 'exact' computing dissolves and with it perhaps also the intriguing idea of 'Ising model for turbulence'.

Even if it is fair to concede that nonlinear LBE is much less revolutionary in scope than LGCA, we are still left with some reasons for excitement.

First, the LBE can be used, *easily* and profitably on any type of computer, today.

Second, it can accommodate additional of mesoscopic physics not so easily accessible to continuum-based models. Thus, leaving the fundamental world of many-body systems, does not mean landing in a culturally flat landscape either.

McNamara and Zanetti demonstrated nonlinear LBE by means of a shear-wave decay numerical experiment with the six-state FHP scheme. By proving that the fluid viscosity is in an excellent accord with the theoretical expectation given by LGCA theory, they showed that nonlinear LBE does indeed correctly capture (linear) transport phenomena. However, they did not perform any large-scale flow computations to test nonlinear LBE on physically richer situations.

At this stage it should be noted that, apart from statistical noise, all other 'dark sides' of LGCA are still left with nonlinear LBE.

In particular, the unviability to 3D computations is just as bad as for LGCA, due to the overwhelming quest for compute power (floating-point instead of Boolean) raised by the collision operator.

This is precisely the problem taken up and solved by the next member of the LBE family, introduced by Higuera and Jimenez (HJ) basically at same time as McNamara and Zanetti [19,20,37]. Before going into the details of the HJ theory, a few more comments on nonlinear LBE are in order.

As already noticed, nonlinear LBE does away with statistical noise but still retains a multibody collision operator, which is where it draws its complexity reduction from. This is in a marked contrast with the Boltzmann equation in CKT, which deals with two-body collisions since higher-order collisions are ruled by the combined assumption of low-density and short-range interactions, the typical scenario of rarefied gas dynamics.

Such an assumption is *not* made here: The potential, although short-ranged, does involve multiple particles, a situation evoking a solid-state-like dense medium rather than a rarefied gas.

The question is: what are these higher-order collisions for? Do we really need them for the purpose of describing large-scale hydrodynamics?

A minute's thought reveals the answer: yes, we do need them *to make the mean free path small*, actually much smaller than purely binary collisions would allow.

Thus multibody collisions help to cushion the detrimental effect of discrete momentum space (the cage) on the mean free path, namely momentum diffusivity. Of course we can always trade efficiency (R^* as defined in the previous chapter) for complexity; in other words, we can reduce the complexity of the collision operator by compensating the raise of diffusivity by a larger number of sites according to the relation

$$Re = R^* \, Ma \, N, \tag{3.8}$$

where N is the number of sites per direction.

This is pure 'VLSI' (Very Large Size Integration) line of thought: do something simple, the simplest possible, on each site and replicate it on as many sites as possible.

However, since the computational work goes like the fourth power of R^*, we know that there is wide scope for making R^* as large as possible without stressing the size of the spatial grid.

Another LGCA asset retained by the nonlinear collision operator is the existence of an H-theorem, with the attendant nonlinear stability of the numerical scheme.

In other words, the fact of sticking to the underlying microdynamics implies not only the same mean free path, but also the same physical behavior *vis-à-vis* the second principle of thermodynamics. All this, of course, modulo the numerical drifts possibly introduced in the numerical scheme by round-off errors.

3.1.1 *Lattice quantum fluids*

Before quitting nonlinear LBE, we wish to point out a formal analogy with the quantum Boltzmann equation for quasi-particles in quantum liquids, i.e. elementary fermionic excitations whose de Broglie wavelength $\lambda_B = \hbar/p$ is much smaller than the mean free path. The quantum nature surfaces only via the the explicit presence of holes $\bar{f}_i \equiv (1 - f_i)$ in the collision operator, as dictated by the exclusion principle. This is again a legacy of LGCA, and, as we shall see shortly, it has no reason to survive in a framework where the final goal is only classical hydrodynamics. It is natural to ask whether other quantum (Bose–Einstein) or classical (Maxwell–Boltzmann) or maybe fancier fractional statistics might find some use here.

In fact, one can easily generalize the collision operator by defining a conjugate hole function $\tilde{f}_i$ as follows:

$$\tilde{f}_i = (1 + \gamma f_i), \tag{3.9}$$

where $\gamma = -1, 0, 1$ corresponds to fermionic, classical and bosonic degrees of freedom, respectively.

The various statistics reflect into the expression of the Galilean breakdown factor, which takes the form

$$g \sim \frac{b + 2\gamma\rho}{b + \gamma\rho}. \tag{3.10}$$

This expression sets a pointer in favor of classical statistics ($\gamma = 0$) for two reasons. First, in this case $g = 1/2$ independently of ρ and can be safely rescaled out at any density regime.

As a further bonus, Maxwell–Boltzmann statistics halves the complexity of the collision operator since the conjugate populations all become unit constants. Summarizing, once classical hydrodynamics is the mission, there is no reason to retain quantum statistics. That said, it would be interesting to learn whether fermionic/bosonic LBEs could be used as numerical tools for quantum fluids. To the best of the author's knowledge, this issue has not been pursued as yet.

3.2 The quasilinear LBE

The problem of the unviability of nonlinear LBE to three-dimensional computations did not last long; indeed it was circumvented by Higuera and Jimenez basically at the same time as the McNamara–Zanetti paper.

The spirit is always the same; since the final aim is macroscopic hydrodynamics, one tries to spot the unnecessary degrees of freedom and work out systematic procedures to get rid of them.

One of the crucial steps in the Chapman–Enskog treatment leading from kinetic theory to hydrodynamics, is the low-Knudsen, small mean free path assumption. In our discrete language this reads:

$$f_i = f_i^{\mathrm{e}} + f_i^{\mathrm{ne}}, \tag{3.11}$$

where superscripts e and ne stand for equilibrium and non-equilibrium, respectively, and the ne component is of order $O(k)$ with respect to the equilibrium one, k being the Knudsen number of the flow.

Being aware that, at some point, the discrete speeds force a low Mach number expansion into the lattice Chapman–Enskog treatment, Higuera and Jimenez resolve to perform this expansion right away on top of (3.11). This yields:

$$f_i = f_i^{\mathrm{e}0} + f_i^{\mathrm{e}1} + f_i^{\mathrm{e}2} + f_i^{\mathrm{ne}} + O\left(kMa^2\right), \tag{3.12}$$

where superscripts 0, 1, 2 refer to the order of the Mach number expansion. With specific reference to the equilibria introduced in Chapter 2, these perturbations read as follows:

$$f_i^{\mathrm{e}0} = \frac{\rho_0}{b}, \tag{3.13}$$

$$f_i^{\mathrm{e}1} = \frac{\rho_0 c_{ia} u_a}{b c_s^2}, \tag{3.14}$$

$$f_i^{\mathrm{e}2} = \frac{\rho_0 Q_{iab} u_a u_b}{2 b c_s^4}, \tag{3.15}$$

where ρ_0 is the uniform, global, density.

Next, expand the collision operator around *global* equilibria f_i^{e0}, as follows:

$$C_i(f) = C_i^0 + C_{ij}^0 \phi_j + \frac{1}{2} C_{ijk}^0 \phi_j \phi_k, \tag{3.16}$$

where $\phi_i = f_i - f_i^{e0}$, $C_{ij} = \partial C_i / \partial f_j$, $C_{ijk} = \partial^2 C_i / \partial f_j \partial f_k$, and superscript 0 denotes evaluation at $f_i = f_i^{e0}$.

Recalling that $C_i^0 = 0$ because global equilibria annihilate the collision operator and discarding terms higher than $O(kMa^2)$, we obtain

$$C_i(f) = C_{ij}^0 f_j^1 + \frac{1}{2} C_{ijk}^0 f_j^1 f_k^1 + C_{ij}^0 f_j^{ne}. \tag{3.17}$$

Now specialize (3.17) to the case of a *local* equilibrium $f_i = f_i^e$, namely $f_i^{ne} \equiv 0$, to obtain

$$C_{ij}^0 f_j^1 + \frac{1}{2} C_{ijk}^0 f_j^1 f_k^1 = 0. \tag{3.18}$$

This permits expression of the Hessian term as a function of the Jacobian, which is mostly welcome since it leaves us with the simple quasilinear expression

$$C_i = C_{ij}^0 f_j^{ne} \equiv C_{ij}^0 \left(f_j - f_j^e\right). \tag{3.19}$$

This finally leads to the *quasilinear LBE* in the form proposed by Higuera and Jimenez

$$\boxed{\Delta_i f_i = A_{ij} \left(f_j - f_j^e\right),} \tag{3.20}$$

where the notation $A_{ij} \equiv C_{ij}^0$ stresses the link with the transition matrix $A(\underline{s}, \underline{s}')$ governing the underlying LGCA microdynamics.

Several remarks are in order.

First, we note that (3.20) looks like a linear (quasilinear) equation but it is definitely not: since it describes the nonlinear Navier–Stokes dynamics it simply cannot be linear!

The (quadratic) nonlinearity is hidden in the local equilibrium term f_i^e via the quadratic dependence on the flow field $\rho u_a = \sum_i f_i c_{ia}$. This is a typical *mean-field* coupling where the slow hydrodynamic degrees of freedom couple back into the kinetic dynamics by dictating the local form the non-equilibrium distribution function must relax to.

Second, we note that the scattering matrix A_{ij} is still one-to-one related to the underlying LGCA dynamics since it is the second-order derivative (Hessian) of the nonlinear collision operator C_i as derived from LGCA Boolean dynamics.

The distinctive feature here is that since this Hessian is evaluated at the global uniform equilibrium f_i^{e0}, *the scattering matrix C_{ij}^0 can be computed analytically from the LGCA transition matrix $A(\underline{s}, \underline{s}')$ once and for all.*

In equations, it can be shown that the exact relation is:

$$A_{ij} = \sum_{\underline{s}\,\underline{s}'} (s'_i - s_i)\, A(\underline{s}, \underline{s}')\, (s'_j - s_j)\, d^p (1-d)^q\,, \tag{3.21}$$

where $p = \sum_i^b s_i$ is the number of particles in state s and q is its complement to b, $q = b - p + 1$, namely the number of holes.

Once the transition matrix $A(\underline{s}, \underline{s}')$ is prescribed, this explicit formula allows precomputation of A_{ij} once and for all at the outset. The resulting reduction of complexity is dramatic: C_i entails $O(2^b)$ operations whereas the scattering matrix only requires $O(b^2)$. Everybody agrees that flipping base and exponent when b is large ($b = 24$) makes a great deal of a difference!

Actually, it is precisely this nice property that paved the way to fully three-dimensional LBE simulations. Which is why, in this author's opinion, this is one of the most significant breakthroughs in LBE theory, probably the one that mostly contributed to making the LBE practical and competitive 'out there' in the world of computational fluid dynamics.

Having highlighted the most important advantages associated with quasi-linear LBE, we now move on to inspect its basic properties.

3.3 The scattering matrix A_{ij}

A few important properties of the scattering matrix are easily identified. The matrix is:

- *Cyclic and isotropic: A_{ij} depends on $|i - j|$ only.*
- *Symmetric: just a consequence of isotropy.*
- *Negative-definite.*

Isotropy and symmetry stem from the fact that the scattering between populations f_i and f_j only depends on $\mu_{ij} = |\vec{c}_i \cdot \vec{c}_j / c^2|$, the absolute value of the cosine of the scattering angle between directions i and j. This reduces the number of independent matrix elements from $O(b^2)$ to $O(b)$ only. Owing to the periodicity of the cosine, it is easily seen that the matrix A is cyclic; that is, it repeats itself in cycles of a few elements. This is a very important property, as it permits analysis of its spectrum in full depth *on purely analytical means.*

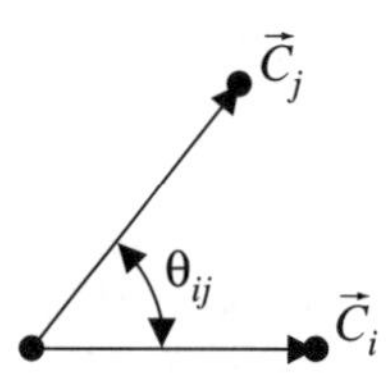

FIG. 3.2. The scattering angle θ_{ij}.

As an example, consider the 6×6 scattering matrix of the FHP model. There are only four possible scattering angles, $\theta = [0, \pi/6, \pi/3, \pi]$, with which we associate the four independent matrix elements a, b, c, d. With this notation, the FHP scattering matrix is explicitly written down as follows:

$$A_{ij} = \begin{pmatrix} a & b & c & d & c & b \\ b & a & b & c & d & c \\ c & b & a & b & c & d \\ d & c & b & a & b & c \\ c & d & c & b & a & b \\ b & c & d & c & b & a \end{pmatrix}. \tag{3.22}$$

Negative-definiteness hinges on the spectral properties of A_{ij}. First, we focus on the all-important mass and momentum conservation principles.

The requirement of mass and momentum conservation translate into the following sum rules:

$$\sum_i A_{ij} = 0, \quad j = 1, \ldots, b \tag{3.23}$$

$$\sum_i c_{ia} A_{ij} = 0, \quad j = 1, \ldots, b, \quad a = 1, \ldots, D, \tag{3.24}$$

where use has been made of the symmetry $A_{ij} = A_{ji}$. These simply mean that $D+1$ eigenvalues are zero, and describe the collisional invariant subspace $\mathcal{H}$ of the b-dimensional kinetic space $\mathcal{K}$ spanned by the discrete populations f_i, $i = 1, \ldots, b$.

Next come the leading nonzero eigenvalues associated with quasihydrodynamic, transport, modes. It can be checked by direct computation that these eigenvalues are associated to the kinetic projector $Q_{iab} = c_{ia}c_{ib} - c_s^2\delta_{ab}$ and yield the expression of the kinematic viscosity

$$\nu = -\frac{D}{D+2} c_s^2 \left(\frac{1}{\lambda} + \frac{1}{2} \right), \tag{3.25}$$

well known from the theory of lattice gas cellular automata.

Finally, the purely kinetic modes, whose eigenvalues must lie in the strip $]-2, 0[$ for reasons of numerical stability. Once it is proved that all eigenvalues lie in this strip, the negative-definiteness property immediately follows:

$$\sum_{ij} V_i A_{ij} V_j = \sum_{k=1}^{b} \lambda_k ||E_k||^2 < 0, \tag{3.26}$$

where V_i is a generic vector in kinetic space and $||E_k||$ is the norm of the k-th eigenvector of the scattering matrix.

Conservation laws guarantee compliance with the first principle of thermodynamics, whereas negative-definiteness ensures fulfilment of the second principle.

It is worth noting that the sum rules (3.23) enforce mass and momentum conservation even if the local equilibrium does *not* carry exactly the same amount of mass and momentum of the actual distribution f_i. Actually, (3.23) implies that *any* weak departure from local equilibrium is reabsorbed in a time lapse of the inverse nonzero eigenvalues of the scattering matrix.

Viewed from this angle, the quasilinear LBE is most conveniently regarded as a *multiple-time relaxation* method. As a such, in principle, it can generate other sorts of PDEs besides Navier–Stokes, depending on the form of the non-equilibrium departure as a function of the slow fields $\rho(x,t)$ and $u(x,t)$. *The quasilinear LBE wipes out the three-dimensional barrier, but what about the high-diffusivity/low-Reynolds limitation?*

Equation (3.25) stands firm to say that this limitation is still with us.

More precisely the momentum diffusivity of the quasilinear LBE gas is still given by eqn (3.25), where λ is the leading nonzero eigenvalue of the scattering matrix. At a first glance one might come to the (wrong) conclusion that since A_{ij} has a simple two-body structure, the associated mean free path should be higher than for nonlinear LBE. Happily enough, this is not true; we should not let ourselves be fooled by the two-body notation. Here, 'effect of population i on population j' does *not* mean two-body collisions, but *all* collisions, including many-body ones, involving populations f_i and f_j!

The much coveted reduction of degrees of freedom *does not result from weeding out multibody collisions* but rather from clustering their effects into the factor $d^p(1-d)^q$ in eqn (3.21). This factor only depends on the number of particles in a given configuration, regardless of the specific configuration itself.

Clearly this clustering is made possible only by the fact of evaluating the collision operator at the uniform equilibrium, because only then all directions collapse into a single value $f_i^{\mathrm{e}} = \rho_0/b$. This is the key to the magic which allows precomputation of the scattering matrix once and for all independently of the flow configuration.

As usual, no free lunch, this magic has its price, specifically nonlinear stability (H-theorem) must be surrendered. For the simple (linear) case of a uniform equilibrium, $f_i^{\mathrm{e}} = \rho_0/b$, quasilinear LBE has a straightforward H-theorem

$$-\frac{\mathrm{d}H}{\mathrm{d}t} = -\sum_{ij} \phi_i A_{ij} \phi_j > 0, \tag{3.27}$$

where

$$H = \sum_i \phi_i^2 \tag{3.28}$$

and $\phi_i = f_i - f_i^{\mathrm{e}0}$ is the departure from global equilibrium.

As already noted, the H-theorem is reconduced to the negative-definiteness of the scattering matrix A. Under general flow conditions, and notably when the quadratic term $f_i^{\mathrm{e}2}$ is included, general statements on the existence of a H-theorem become a fairly non-trivial matter.

This means that quasilinear LBE is potentially exposed to nonlinear numerical instabilities, typically observed whenever the local flow field u_a and/or its spatial gradients become 'too large' for the hosting grid.

Again, one more nice property of the LGCA ancestor falling apart, or, better said, hard to keep standing.

Is the gain worth the pain?

Surely it is, for otherwise three-dimensional simulations would be totally out of reach. Nonetheless, the lesson is that there is still room for improvement, as we shall see in the next chapter.

3.4 Numerical experiments

The first numerical simulations using a two-dimensional quasilinear LBE were performed by Higuera [37] and Higuera and Succi [38] for the case of moderate Reynolds number flow past a cylinder.

These simulations compare successfully with existing literature at a competitive cost in terms of computer resources.[10] Soon after, a number of three-dimensional simulations of flow in porous media were performed [39], which, although still restricted to moderately low Reynolds numbers, proved the viability of the method versus existing floating-point techniques.

Somewhat ironically, the Boolean lattice gas paradigm had generated a floating-point spin-off!

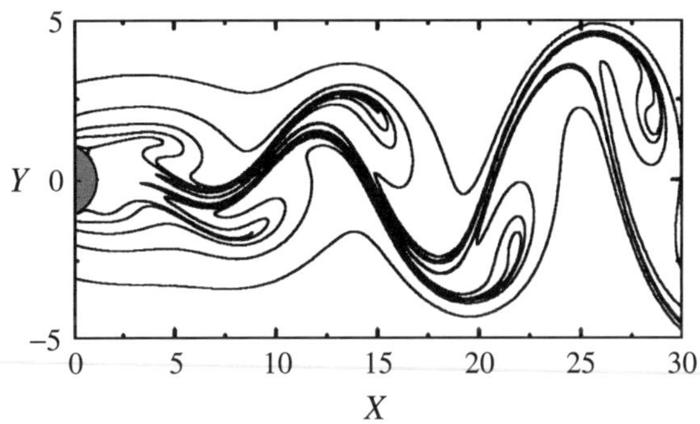

FIG. 3.3. Streaklines of a moderate Reynolds flow past a cylinder (after [38]).

[10] Practical hint: Note that, by symmetry arguments, C^0_{ij} annihilates both ϕ^1_i and ϕ^2_i, so that (3.19) further simplifies to

$$C_i = C^0_{ij} f^{\mathrm{ne}}_j \equiv C^0_{ij} \left(f_j - f^{\mathrm{e}2}_i\right). \tag{3.29}$$

This reduces the number of floating operations needed to encode the collision operator.

Second practical hint: In computational practice it is advisable to work with reduced populations $\tilde{f}_i \equiv (f_i - f^{\mathrm{e}0}_i)/f^{\mathrm{e}0}_i$, in such a way that full machine precision can be directed to departures from uniform equilibria, rather than to uniform equilibria themselves!

3.5 Exercises

1. Write down the explicit form of nonlinear LBE for the FHP scheme.
2. Write down the explicit form of quasilinear LBE for the FHP scheme.
3. Prove that the scattering matrix is negative-definite.
4. Prove the particle–hole duality of quasilinear LBE.
5. Try to prove the H-theorem with quadratic terms included in the local equilibria. Analyze the mathematical terms which prevent you from succeeding in full generality.

4

LATTICE BOLTZMANN MODELS WITHOUT UNDERLYING BOOLEAN MICRODYNAMICS

For some players luck itself is an art.
The Color of Money

The work by McNamara and Zanetti and Higuera and Jimenez described in the previous chapter showed how to circumvent two major stumbling blocks of the LGCA approach, namely statistical noise and exponential complexity of the collision rule. It is then natural to wonder whether other inconveniences of LGCA can also be wiped out by further refinements of the LBE approach. The answer is a plain yes; by further departing from LGCA, it is possible to eliminate basically all LGCA idiosyncratic aspects. In particular, it is possible to break the high-viscosity barrier set by the LGCA collision rules and take LBE into the territory of high-Reynolds flows. This program was accomplished by the so-called LBE with enhanced collisions, the main focus of the present chapter.

4.1 LBE with enhanced collisions

Now that two major drawbacks of LGCA models, statistical noise and practical inviability to three-dimensional simulations, have been lifted, we can turn our attention to the third major limitation of LGCA: high momentum diffusivity, read low Reynolds flows. This further step was taken in short sequence by Higuera, Succi and Benzi [40], just a few months after quasilinear LBE was successfully demonstrated by the numerical simulations described in the previous chapter [37, 38, 41].

The main observation is that the high diffusivity constraint is again a needless limitation inherited from LGCA via the one-to-one correspondence between the transition matrix $A(\underline{s}, \underline{s}')$ and the scattering matrix A_{ij}. In other words, neither nonlinear LBE nor quasilinear LBE are allowed to produce a single collision that the subjacent LGCA would not permit. This tyrannic dependence is encoded in the one-to-one relation between the transition matrix $A(\underline{s}, \underline{s}')$ and the scattering matrix A_{ij} given in the previous chapter. On the other hand, since quasilinear LBE is known to yield the Navier–Stokes equations, why not take quasilinear LBE itself as a *self-standing* mathematical model of fluid behavior, regardless of its microscopic Boolean origin?

This stance of 'independence' is perfectly legal, provided one secures that the scattering matrix (a set of free parameters at this stage) is chosen with the correct symmetries ensuring the Navier–Stokes equation as a macroscopic limit.

The right properties of the scattering matrix are easily guessed from what we learned with quasilinear LBE: it is simply the array of conservation properties listed in the previous chapter, securing compliance with the first and second principles of thermodynamics:

- *Conservativeness (sum-rules);*
- *Isotropy and symmetry;*
- *Negative-definiteness.*

The basic idea is that *the spectral properties of A_{ij}, controlling the transport coefficients, are no longer dictated by any underlying microscopic theory, but they are handled as free parameters of the theory instead.* Consequently, they can be freely tuned in such a way as to achieve minimum viscosity, hence maximum Reynolds numbers. This idea proves exceedingly fruitful for all subsequent developments of LBE methods.

Technically, the name of the game is simply 'spectral decomposition': once the spectrum of eigenvalues and corresponding set of eigenvectors is known, any matrix A can be reconstructed via the spectral decomposition

$$A_{ij} = \sum_{k=1}^{b} \lambda_k P_{ij}^k, \tag{4.1}$$

where P_{ij}^k is the projector along the k-th kinetic eigenvector E_i^k. We are in fairly good shape, because in a discrete velocity space the linear eigenvalue problem associated with the spectrum of the scattering matrix can be solved exactly! This conveys a concrete feel for what 'minimal' kinetic theory means and the fringe benefits it brings.

Standard analysis shows that if the set of eigenvectors $\underline{E} = [E_i^1, \ldots, E_i^b]$ is orthonormal, the projectors P_{ij}^k are expressed by the dyadic matrix $E_i^k E_j^k$, whose matrix element (i, j) is given by the product of the i-th and j-th element of the k-th column eigenvector. Once the formal constraint (brackets denote sum over Boolean states with weight $d^p(1-d)^{b-p+1}$ as defined in the previous chapter)

$$A_{ij} = \left\langle \left(s_i' - s_i\right) A\left(\underline{s}, \underline{s}'\right) \left(s_j' - s_j\right) \right\rangle \tag{4.2}$$

is relaxed, nothing could be easier than minimizing the viscosity, since the relation between the leading nonzero eigenvalue and the momentum diffusivity is still given by the basic relation obtained by the Chapman–Enskog analysis of LGCA microdynamics, namely:

$$\nu = c_s^2 \frac{D}{D+2} \left(-\frac{1}{\lambda} - \frac{1}{2} \right). \tag{4.3}$$

This expression prompts a greedy question: can we achieve *zero* viscosity by simply choosing $\lambda = -2$?

Should this be the case, we would have a real ground-breaking tool at our fingertips, the dream of every turbulence researcher!

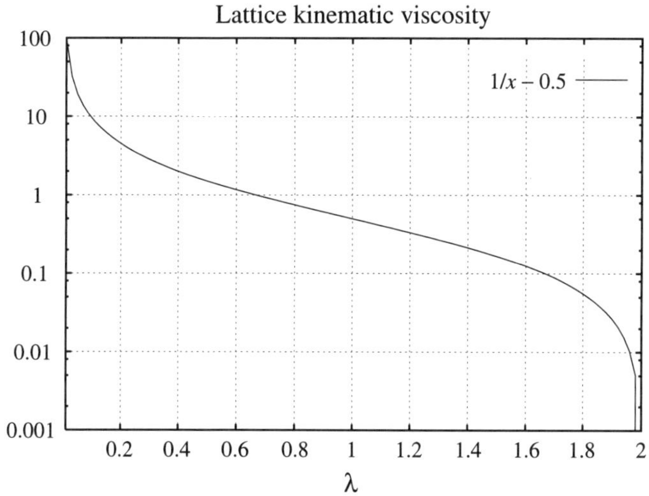

FIG. 4.1. Lattice viscosity as a function of $|\lambda|$ (the prefactor $c_s^2 D/(D+2)$ made unit).

Unfortunately, the answer is a down-to-earth 'no', as we shall detail shortly.

Before doing so, let us discuss the spectrum of the quasilinear LBE collision operator in some detail.

As anticipated, the spectral analysis of the scattering collision operator can be taken to the very end on purely analytical means [21].

For the sake of conciseness, we shall refer to the FCHC lattice, not because this is the simplest case, but because it represented the workhorse of LGCA and LBE simulations for some years.

Our task is to analyze the 24×24 matrix A_{ij} generated by the FCHC discrete speeds.

Many properties of the spectrum $\sigma[A]$ are known beforehand.

First, direct inspection of the cosines $\cos\theta_{ij}$ between directions $\vec{c}_i$ and $\vec{c}_j$, shows that there are only *five* independent scattering channels/elements, denoted by a, b, c, d, e and corresponding to $\theta = 0,\ \pi/3,\ \pi/2,\ 2\pi/3,\ \pi$, respectively (don't forget we are in $D = 4$).

This means that the spectrum consists of only five distinct eigenvalues

$$\sigma(A) = [\lambda_k,\ k = 1, \ldots, 5], \tag{4.4}$$

each with multiplicity μ_k, such that $\sum_{k=1}^{5} \mu_k = 24$.

These eigenvalues are real since the matrix is symmetric, and in addition they must lie within the strip $]-2, 0[$ for reasons of numerical stability.

Mass and momentum conservation enforce the following equalities:

$$a + 8b + 6c + 8d + e = 0, \tag{4.5}$$
$$a + 4b - 4c - e = 0, \tag{4.6}$$

which leave us with only *three* distinct nonzero eigenvalues.

This completes the reconstruction of the conservative part of the spectrum. Now comes the dissipative/transport component. At this stage we turn our attention to the eigenvectors $\underline{E}^k$. Four eigenvectors are immediately spotted from mass and momentum conservation

$$E_i^0 = 1, \tag{4.7}$$
$$E_i^a = c_{ia}, \quad a = 1, \ldots, 4. \tag{4.8}$$

The remaining ones can be constructed by the well-known Gram–Schmidt procedure. A quicker alternative is to recognize [42,43] the *circulant* property of the scattering matrix, which allows us to write down the analytical expression of the spectrum. We shall however proceed through a different route because we feel it conveys more physical insight into the problem.

For the sake of clarity, we report here the explicit computation. We seek a set of eigenvectors orthogonal to $\underline{E}^0$, $\underline{E}^1$. A sensible guess is $E_i^{ab} = c_{ia}c_{ib} - K_{ab}$, where K_{ab} is a yet unknown tensor to be computed by imposing orthogonality to the previous ones, namely:

$$\sum_i (c_{ia}c_{ib} - K_{ab})\, 1_i = 0, \tag{4.9}$$
$$\sum_i (c_{ia}c_{ib} - K_{ab})\, c_{ic} = 0. \tag{4.10}$$

The second of these equations is automatically satisfied by the parity of the lattice (all odd tensors vanish identically by parity), whereas the first one yields

$$K_{ab} = \sum_i c_{ia}c_{ib} = \frac{bc^2}{D}\delta_{ab}. \tag{4.11}$$

This delivers a set of $5 \times 4/2 - 1 = 9$ distinct eigenvectors

$$E_i^k = c_{ia}c_{ib} - c_s^2\delta_{ab}, \quad k = 6, \ldots, 14, \tag{4.12}$$
$$\lambda = \lambda_1, \quad \mu_1 = 9, \tag{4.13}$$

where $c_s^2 = c^2/D$ is the sound speed, as usual.

On the right-hand side we recognize the all-important kinetic projector $Q_{iab} = c_{ia}c_{ib} - c_s^2\delta_{ab}$.[11]

[11] There is a manifest link with tensor Hermite polynomials: See for instance [3, p. 205].

One can proceed further with the diagonalization procedure and derive the remaining eigenvalues and eigenvectors

$$E_i^{15} = \chi_i', \qquad E_i^{16} = \chi_i'', \tag{4.14}$$
$$\lambda = \lambda_2, \qquad \mu_2 = 2, \tag{4.15}$$

$$E_i^k = \chi_i' c_{ia}, \qquad E_i^k = \chi_i'' c_{ia}, \quad k = 17, \ldots, 24, \tag{4.16}$$
$$\lambda = \lambda_3, \qquad \mu_3 = 8, \tag{4.17}$$

where the weights χ_i', χ_i'' will be discussed shortly.

The final relation between the matrix elements and the nonzero eigenvalues is:

$$\lambda_1 = a - 2c + d, \tag{4.18}$$
$$\lambda_2 = \frac{3}{2}(a - d), \tag{4.19}$$
$$\lambda_3 = \frac{3}{2}(a + 6c + d), \tag{4.20}$$

with multiplicities $\mu_1 = 9$, $\mu_2 = 2$, $\mu_3 = 8$, respectively.

The spectral analysis of the scattering matrix is summarized in the following eigenvalue–multiplicity–eigenvector–conjugate fields synoptic table:

$\lambda = 0,$	$\mu = 5,$	$[1_i, c_{ia}],$	$[\rho, J_a],$
$\lambda_1,$	$\mu = 9,$	$[Q_{iab}],$	$[S_{ab}],$
$\lambda_2,$	$\mu = 2,$	$[\chi_i', \chi_i''],$	$[\rho', \rho''],$
$\lambda_3,$	$\mu = 8,$	$[\chi' c_{ia}, \chi'' c_{ia}],$	$[J_a', J_a''].$

This completes the spectral analysis of the FCHC scattering matrix A_{ij}.

It is now instructive to examine the physical meaning of the conjugate macroscopic fields $\Psi_k(x,t) = \sum_i f_i(x,t) E_i^k$.

4.2 Hydrodynamic and ghost fields

In line with the material presented in previous chapters, we shall define conjugate macroscopic fields as the projection of the distribution function f_i over the the set of eigenvectors $\underline{E}^k$ in momentum space (underline stands for vector in momentum space $\underline{E}^k \equiv E_i^k$). For simplicity, we shall refer to the three-dimensional projection, so that we shall deal with a 18×18 matrix. The remaining 6 quenched populations are reabsorbed into a set of effective weights:

$$\underline{w} = [(2,2,2,2,2,2 \mid 1,1,1,1 \mid 1,1,1,1 \mid 1,1,1,1)], \tag{4.21}$$

where the bars group the following sublattices:

- 6 direct links along x, y, z, respectively;

- 4 cross links along xy;
- 4 cross links along xz;
- 4 cross links along yz.

With this definition:

$$\Psi_k(x,t) = \sum_{i=1}^{18} w_i E_i^k f_i. \tag{4.22}$$

The hydrodynamic sector of the spectrum is well known, and generates mass and current densities associated with conserved quantities and null eigenvalues:

$$\rho \equiv \Psi_0(x,t) = \sum_{i=1}^{18} w_i f_i, \tag{4.23}$$

$$J_a \equiv \Psi_a(x,t) = \sum_{i=1}^{18} w_i f_i c_{ia}, \quad a = 1,2,3. \tag{4.24}$$

The transport sector of the spectrum defines the slowest non-equilibrium decaying quantities. The corresponding macro-field is the deviatoric part of the momentum flux tensor, namely:

$$S_{ab} \equiv \Psi_{ab}(x,t) = \sum_{i=1}^{18} w_i f_i Q_{iab}, \quad a,b = 1,2,3, \tag{4.25}$$

with the obvious relation:

$$P_{ab} \equiv \sum_{i=1}^{18} w_i f_i c_{ia} c_{ib} = S_{ab} + P\delta_{ab}, \tag{4.26}$$

where $P = \rho c_s^2$ is the fluid pressure.

Being symmetric and traceless, the tensor S_{ab} generates $3+3+2+1=9$ independent fields associated with the leading nonzero eigenvalue of the matrix A.

Next comes the 'fast', mesoscopic component of the spectrum; that is, the modes that never manage to make it from the microscopic to the macroscopic level. They were dubbed 'ghost fields', to stress the idea that they live 'behind the scenes', but cannot be erased from the dynamics for pain of disrupting essential symmetries.

By sheer definition:

$$\rho' = \sum_{i=1}^{18} \chi_i' f_i, \tag{4.27}$$

$$\rho'' = \sum_{i=1}^{18} \chi_i'' f_i, \tag{4.28}$$

where

$$\underline{\chi}' = [(1,1,1,1,-2,-2) \mid (-2,-2,-2,-2) \mid (1,1,1,1) \mid (1,1,1,1)], \quad (4.29)$$
$$\underline{\chi}'' = [(1,1,-1,-1,0,0) \mid (0,0,0,0) \mid (1,1,1,1) \mid (1,1,1,1)]. \quad (4.30)$$

From these definitions, we see that the ghost densities ρ' and ρ'' correspond to a pair of staggered densities: $\rho' = \rho^x + \rho^y - 2\rho^z - 2\rho^{xy} + \rho^{xz} + \rho^{yz}$ and $\rho'' = \rho^x - \rho^y + \rho^{xz} + \rho^{yz}$, where the suffixes label the four distinct sublattices introduced previously.

Being differences rather than sums, the staggered densities carry higher-order, short-lived information corresponding to derivatives in velocity space.

The remaining ghost fields are easily identified as ghost currents

$$J'_a = \sum_{i=1}^{18} \chi'_i f_i c_{ia}, \quad (4.31)$$

$$J''_a = \sum_{i=1}^{18} \chi''_i f_i c_{ia}. \quad (4.32)$$

These relations show that the ghost fields can also be given a transparent physical interpretation.

The full set of equations for the 24 moments of the FCHC–LBE scheme can be written in full splendor as follows:

$$\partial_t \rho + \partial_a J_a = 0, \quad (4.33)$$
$$\partial_t J_a + \partial_a P_{ab} = 0, \quad (4.34)$$
$$\partial_t S_{ab} + \partial_c R_{abc} = \lambda_1 \left(S_{ab} - S^{\mathrm{e}}_{ab}\right), \quad (4.35)$$
$$\partial_t \rho' + \partial_a J'_a = \lambda_2 \rho', \quad (4.36)$$
$$\partial_t \rho'' + \partial_a J''_a = \lambda_2 \rho'', \quad (4.37)$$
$$\partial_t J'_a + \partial_b P'_{ab} = \lambda_3 J'_a, \quad (4.38)$$
$$\partial_t J''_a + \partial_b P''_{ab} = \lambda_3 J''_a. \quad (4.39)$$

The hydrodynamic and ghost sectors of the theory make contact through the triple tensor:

$$R_{abc} = \sum_{i=1}^{18} w_i f_i Q_{iab} c_{ic}. \quad (4.40)$$

Direct substitution of the hydrodynamic component

$$f_i^H \equiv \rho \left[1 + \frac{1}{c_s^2} c_{ia} u_a + \frac{1}{2c_s^4} Q_{iab} S_{ab}\right] \quad (4.41)$$

into the expression of R_{abc}, shows that the divergence of this triple tensor reorganizes sensibly into an isotropic form corresponding to the Navier–Stokes stress tensor:

$$H_{ab} \equiv \partial_c R^H_{abc} = (\partial_a J_b + \partial_b J_a) - \frac{2}{D} (\partial_c J_c) \, \delta_{ab}. \tag{4.42}$$

The same is *not* true for ghost contributions because, owing to the different weights χ'_i, χ''_i, the corresponding ghost tensor $G_{ab} \equiv \partial_c R^G_{abc}$ does not fulfil the isotropy requirements.

It should be appreciated that we have made use of the freedom on the choice of the hydrodynamic component of f_i in order to fine-tune the coefficients of expansion (4.41) in such a way as to achieve the desired anomaly-free result.

The system of partial differential equations (4.33)–(4.39) has a rich and formally elegant structure, which invites some further comments.

4.2.1 *Field-theoretical analogies*

The 24-moment FCHC equations exhibit a beautiful symmetry between hydro- and ghost fields: they turn one into another under the dual transformation

$$w_i \to \tilde{w}_i, \tag{4.43}$$

where w_i are the weights associated with the degeneracies and $\tilde{w}_i = [w_i \chi_i, w_i \chi''_i]$ are the corresponding weights for the ghost fields.

These weights can also be interpreted as elementary masses of the particles propagating along the various directions of the lattice. Note that, in tune with their physical meaning, the hydrodynamic (normalized) weights are all positive and sum to 1, whereas some of the ghost weights are forcibly negative because they must sum to zero. The physical content of the hydrodynamic and ghost sectors of the theory is very different.

Consistent with this difference, the above duality is broken by the nonzero mass (eigenvalues) of the ghost fields, which goes with the short-lived nature of these excitations. In the numerical analyst's language, hydro-fields are smoothed-out (integral) variables in momentum space, whereas the ghost fields represent the short-scale detail (derivatives).

The question is: do we really want these ghosts fields, or should we rather try to erase them from the dynamics?

Let us handle this question in two separate parts.

First and foremost, why do ghost fields appear at all?

The reason is that *they represent the excess of information from kinetic space required to formulate the Navier–Stokes equations in a fully covariant, hyperbolic form* [44]. This short-scale information, if unable to emerge from the meso to the macroscopic scale, is definitely necessary to secure the basic local symmetries which guarantee correct hydrodynamics. A formal analogy with the famous ghost fields of quantum field theory is recklessly tempting:

Here come the arguments [45, 46]:

- *Quantum Field Theory*
    1. Ghost fields are required to cancel unwanted degrees of freedom arising from the covariant treatment of massless gauge vector fields (photons, gluons).

    2. Their task is to restore unitarity. They live only inside Feynman loops, and must disappear from the ultimate physical picture.

- *Lattice Hydrodynamics*
    1. Ghost fields are required to ensure the basic *local* space–time symmetries of classical physics within a covariant formulation of hydrodynamics.
    2. Their task is to restore Galilean and rotational invariance. They are short-lived and do not surface into the macroscopic physics.

In light of this analogy, the condition (4.42) plays a role similar to the Ward identity in quantum field theory, namely anomalies cancellation [47]. In our case, the anomaly is lack of Galilean and rotational invariance, whereas quantum field theory involves more sophisticated symmetries typically related to internal degrees of freedom not (necessarily) associated with space–time transformations.

4.2.2 *Dimensional compactification*

Let us spend a few final comments on the present hydrodynamic derivation for the FCHC–LBE model. First of all, the 24-moment equations do correspond to the Navier–Stokes dynamics of a *four-dimensional fluid* (4+1 in field theory language).

Actually, the fluids we are interested in, only need three, so that it is natural to ask what to do with the extra spatial dimension.[12]

Normally, the fourth dimension is 'curled up' by imposing a zero current along the fourth dimension, $J_4 = 0$. This introduces a degeneracy in the populations with $c_{i4} = 0$, which collapses into three doublets, thus leaving only eighteen independent populations out of the original twenty-four. The corresponding numerical simulations are then performed by shrinking the fourth 'Kaluza–Klein' dimension into a single layer across which periodicity is imposed so as to erase any dependence on x_4.

On the other hand, if this degeneracy is lifted by allowing $J_4 \neq 0$, and the dimensional curling of the fourth dimension is still retained, the resulting scheme corresponds to a three-dimensional fluid carrying along the current J_4 as a passive scalar.

This is formally understood as the limit $\partial_4 \to 0$, namely no observable dependence on the hidden dimension. As a result, the 4D FCHC scheme with 18 discrete speeds and 24 populations describes a three-dimensional fluid plus a passive scalar, for nothing.

By the same token, a 2D FCHC scheme, with 9 discrete speeds and 24 populations, describes a two-dimensional fluid, plus *two* passive scalars and a 1D FCHC with three speeds describes a 1D fluid plus three passive scalars.

[12] And again, another amazing analogy with field theory, namely the famous 5-dimensional Kaluza–Klein attempt to unify gravitation and electromagnetism. By now, we know that even if the Kaluza–Klein theory fails to achieve the goal, the idea of pursuing unification in higher-dimensional spaces has proved overly productive in modern developments of quantum field theory and quantum gravity [47].

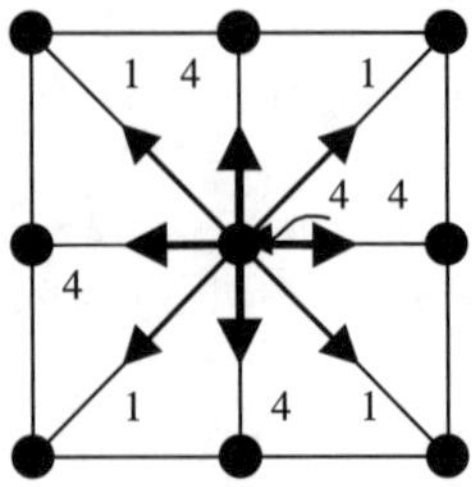

FIG. 4.2. The 2D projection of FCHC and the corresponding weights.

Such a peculiar property can be leveraged to develop fast-chemistry combustion, magnetohydrodynamics and thermal flows, as will be partly detailed in Part III.

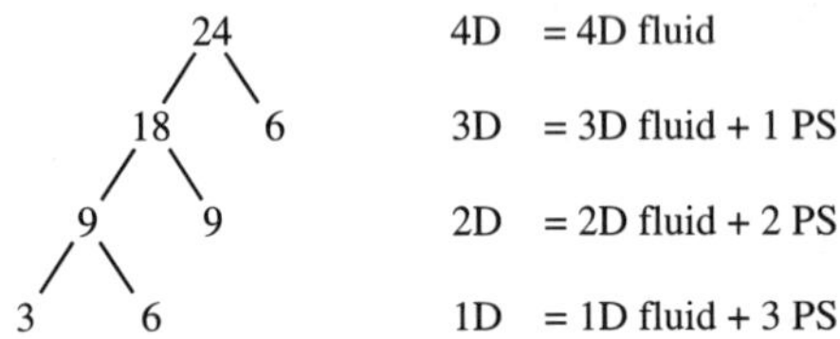

SKETCH 4.1. The hierarchical tree of the FCHC lattice. The rightmost figure counts the number of degenerate populations produced by projection.

4.2.3 *Removing ghost fields*

Since ghosts are no hydrodynamic observables, it is natural to ask how to decouple them from hydrodynamics.

A simple, if approximate, recipe is readily available: Take

$$\lambda_2 = \lambda_3 = -1, \tag{4.44}$$

so as to damp out ghost excitations in a single time-step.

The effect of this move is easily appreciated from the homogeneous version of eqn (4.33):

$$\rho'(t+1) - \rho'(t) = -\rho'(t). \tag{4.45}$$

This yields $\rho'(t+1) = 0$, which means that the ghost density at time $t+1$ is erased from the dynamics. In the non-homogeneous case, however, ghosts are revived at each time-step by the divergence of the ghost currents, which in turn are excited by the ghost momentum flux tensor.

$$\rho'(t+1) - \rho'(t) = -\rho'(t) + \partial_a J'_a(t). \tag{4.46}$$

It easily seen that the same recipe guarantees that these terms cannot live longer than a single time-step, so that the prescription (4.44) gives rise to a sort of low amplitude sawtooth oscillations, whose amplitude should remain small unless pathological initial or boundary conditions are met (remember the staggered invariants discussed earlier in this book).

Once the basics of ghost dynamics are grasped, we can move one step further and ask ourselves whether we can make any good use of them, for instance put them at work to 'improve' hydrodynamic behavior.

Here 'improve hydrodynamic behavior' essentially stands for sub-grid-scale modeling, namely generate an effective viscosity corresponding to the effect of small, unresolved scales on the resolved ones. Such attempts have been made in the past [48] but, to the best of the author's knowledge, they have not been demonstrated numerically. Since these models are rather technical, we direct the interested reader to the original reference and take as a final stance that ghosts should be kept at a minimum. Of course, the brave reader is encouraged to challenge this conservative statement and find his own way to a smart use of ghost fields.

4.3 The route to Navier–Stokes: adiabatic assumption

The set of eqns (4.33)–(4.39) is a 24-moment *hyperbolic superset of the Navier–Stokes equations.* It contains *exactly* the same amount of information as the LBE, only recast (projected) onto a different basis set. As for any discrete kinetic model, closure is automatically provided by the finite dimensionality of the kinetic space $\mathcal{K}$. To recover the Navier–Stokes dynamics we must necessarily break the hyperbolic character of this system and cast it into a dissipative form where time and space derivatives are no longer in balance.

The route is *adiabatic assumption* on the slow, non-conserved modes, in our case the shear tensor S_{ab}. By adiabatic assumption, we mean the shear tensor relaxing to its local equilibrium value on a time-scale much shorter than any typical hydrodynamic scale. Mathematically, this is tantamount to dropping the time derivative in the evolution equation for the shear tensor, eqn (4.35), and obtaining a simple algebraic solution

$$S_{ab} \simeq S_{ab}^{\mathrm{e}} + \frac{H_{ab}}{\lambda_1}, \tag{4.47}$$

where ghosts have been phased out via the condition (4.44).

By plugging (4.47) into (4.33), we obtain

$$\partial_t J_a + \partial_b \left(S_{ab}^{\mathrm{e}} + \frac{H_{ab}}{\lambda_1} \right) = 0. \tag{4.48}$$

All we need at this stage is simply to recall that the equilibrium shear tensor reads

$$S_{ab}^{\mathrm{e}} = \rho u_a u_b \tag{4.49}$$

and realize that eqn (4.48) is nothing but the Navier–Stokes equation with a kinematic viscosity $\nu \sim -1/\lambda_1$.

This is of course a very cheap treatment of the hydrodynamic limit, in the sense that the subtleties of the multiscale Chapman–Enskog approach have been swept under the carpet. We recall that in the multiscale approach not only the dependent variables, f_i but also space–time variables are expanded in multiple scales driven by a smallness parameter, typically the Knudsen number. This implies that different equations are obtained at different orderings of this smallness parameter.

From LGCA theory, it is known that being cavalier about multiscale considerations does not spoil the qualitative structure of the macroscopic equations, but it does introduce quantitative errors in the transport coefficients. In particular, it is known that the Taylor expansion of the finite difference streaming operator $\Delta_i f_i \equiv f_i(\vec{x} + c_{ia}, t+1) - f_i(\vec{x}, t)$ to second-order in the lattice spacing, brings about a factor $-1/2$ in front of the second-order spatial derivative.

It is fortunate that such an exquisitely numerical effect can be likened to a *physically negative* viscosity, known as *propagation viscosity*.

The final, quantitative expression therefore reads

$$\nu = -c_s^2 \frac{D}{D+2} \left(\frac{1}{\lambda_1} + \frac{1}{2} \right), \tag{4.50}$$

which is a known fact from LGCA theory.

This brings us back to the issue of zero viscosity.

4.4 The mirage of zero viscosity

One of the crucial results of the spectral analysis is the expression relating the fluid viscosity to the leading nonzero eigenvalue of the scattering matrix A_{ij}. This equation was known long before LBE, but it is the change in perspective which makes the whole difference. Here the eigenvalue λ_1 is no longer imposed by any underlying microdynamics, but becomes a free handle that can be fine-tuned to minimize fluid viscosity. In principle, zero viscosity could be achieved by selecting $\lambda_1 = -2$, which corresponds to the collisional diffusivity coming in perfect balance with the negative propagation viscosity $\nu_p = -1/2$.

As usual, when two quantities become serious about exactly balancing each other, new effects need be taken into account.

This is the thorny issue of nonlinear stability, namely the effect of large macroscopic gradients on the numerical stability of LBE once the viscosity becomes really small, i.e., so small that the grid Reynolds number $u/\nu > 1$ (in lattice units). This is the so-called *sub-grid* scale regime, whose distinctive feature is the potential existence of dynamic scales shorter than the lattice pitch.

This important issue is examined apart in a separate chapter. What we wish to point out here is that LBE actually bridges the gap with conventional CFD, in that *the minimal fluid viscosity is dictated only by the mesh resolution and not by intrinsic limitations of the LGCA collisionality (lattice mean free path).*

In other words, the issue of 'low collisionality' is lifted, and one can go as low in viscosity as the lattice permits.[13]

It is natural to ask why this is so, given the fact that the structure of the lattice, and in particular the number of discrete speeds, has not been touched at all!

The point is that by freely adjusting the scattering matrix A_{ij}, *and* the discrete equilibrium f_i^e, we implicitly take into account *all* possible collisions involving the discrete speeds $\vec{c}_i$ and $\vec{c}_j$ even though only a (very) limited set is retained in the actual evolution scheme. We thus have a sort of set of virtual collisions with a continuous set of speeds even if, in the end, only a fraction of these discrete speeds keeps being tracked all along.

Once this picture is realized, the issue of Galilean invariance dissolves on its own, because the local equilibrium can be chosen at will (within hydrodynamic constraints) without being tied down to specific statistical distributions of continuum physics.

In particular, the following expression

$$f_i^e = \frac{\rho}{b}\left(1 + \frac{c_{ia}u_a}{c_s^2} + \frac{Q_{iab}u_a u_b}{2c_s^4}\right), \tag{4.51}$$

which has already been used in this chapter to obtain the correct form of the Navier–Stokes stress tensor, is manifestly compliant with all hydrodynamic constraints and Galilean invariant as well.

In conclusion, the self-standing LBE contains two quasi-free dynamical 'knobs':

1. *The scattering matrix* A_{ij} ;
2. *The discrete local equilibrium* f_i^e.

By tuning the associated set of parameters, one can flexibly generate a whole class of nonlinear partial hyperbolic equations, including of course Navier–Stokes. Even though little has been changed in the formal LBE apparatus, this standpoint dissolves two issues in one swoop, low collisionality and Galilean invariance, thus making full contact with standard computational fluid dynamics.

Moreover, by capitalizing on the freedom to play with the scattering matrix and the local equilibria, the present LBE paves the way for a host of generalized hydrodynamic applications, such as magnetohydrodynamics, thermal convection and many other situations involving *field–fluid* interactions [49, 50].

4.5 Numerical experiments

The LBE scheme with enhanced collisions was demonstrated shortly after quasi-linear LBE for a 2D transitional flow past a thin plate at moderate Reynolds

[13] For a while it was even hoped that one could go *below* the grid ('sub-grid modeling' in CFD jargon), but this exciting perspective was abandoned as it was unsupported by numerical evidence.

number [40]. Subsequently, high Reynolds 2D LBE turbulence was tested against spectral methods [51]. Details of these simulations are reported in Part II. These simulations proved that LBE with enhanced collisions performs fairly competitively against spectral methods even on moderate resolutions (512^2). In terms of the quality factor R^*, these simulations showed that one can safely go as high as $R^* \sim 30$, by choosing $\lambda_1 \sim -1.9$, corresponding to a lattice viscosity $\nu \sim 0.01$, basically the smallest value compatible with the lattice resolution. The calculation is simple: take $U = 0.1$, $N = 1000$ lattice sites and $\nu = 0.01$, all in lattice units, the resulting Reynolds is $Re = 10^4$, more or less the state of the art in the direct numerical simulation of fluid turbulence. *We hasten to emphasize that, having swept the H-theorem under the carpet (at no point have we discussed compliance with the H-theorem in the nonlinear flow regime), it is by no means obvious a priori that LBE can be taken to the lowest viscosities allowed by the grid resolution without incurring kinetic numerical instabilities!*

This is once again a lucky break, since the theoretical justification for such friendly behavior has started to become apparent only in the recent past [52].

Besides turbulence, LBE with enhanced collisions has been used for the earliest calculations of three-dimensional single and multiphase flows in porous media [53], Rayleigh–Bénard turbulence and others [49]. Its prime time as a tool of choice for LBE simulations did not last more than a few years, though, the time it took to realize that the LBE formalism still had a further nugget in store. This takes us to the last, and probably ultimate, 'release' of the LBE family, the lattice Bhatnagar–Gross–Krook version discussed in the next chapter.

4.6 Exercises

1. Verify by direct computation that the FCHC only admits five distinct scattering angles (don't forget the fourth dimension!).
2. Write down the scattering matrix for the 1D FCHC.
3. Compute the eigenvectors of the scattering matrix using the Gram–Schmidt procedure.
4. How many ghost fields does one have in 2D FCHC and 1D FCHC?
5. Compare the weights of 1D FCHC with numerical quadrature formulae.

5

LATTICE BHATNAGAR–GROSS–KROOK

If it doesn't look easy it is that we have not tried hard enough yet.
F. Astaire

In the previous chapter, we have learned how to cut down the complexity of the LBE collision process by formulating a discrete scattering version of the collision operator. This scattering version is de-facto a multi-relaxation model equation in which, owing to the discreteness of velocity space, it becomes possible to analyze the spectrum of the scattering operator in full depth. This analysis delivers an exact expression for the fluid viscosity via the single eigenvalue associated with the slowest non-conserved quantity, namely the momentum flux tensor. In light of this, it is natural to wonder whether the scattering operator can be simplified further and brought down to a single-time relaxation form. The idea is again patterned after continuum kinetic theory, in particular after the celebrated Bhatnagar–Gross–Krook (BGK) model Boltzmann introduced as early as 1954. Its transcription to the discrete lattice world proves exceedingly fruitful, and it leads to a hydrodynamic-compliant kinetic equation of great simplicity and efficiency. In this chapter we shall describe the Lattice Bhatnagar–Gross–Krook scheme, LBGK for short, the ultimate LBE model in terms of simplicity and effectiveness.

5.1 Single-time relaxation

The main lesson taught by the self-standing LBE of the previous chapter is that the scattering matrix and local equilibria can be regarded as free parameters of the theory, to be tuned to our best purposes within the limits set by the conservation laws and numerical stability. Within this picture, the viscosity of the LB fluid is entirely controlled by a single parameter, namely the leading nonzero eigenvalue of the scattering matrix A_{ij}. The remaining eigenvalues are then set so as to minimize the interference of non-hydrodynamic modes (ghosts) with the dynamics of macroscopic observables. This spawns a very natural question: since transport is related to a single nonzero eigenvalue, why not simplify things further by choosing a one parameter scattering matrix? The point, raised almost simultaneously by a number of authors [54], is indeed well taken. Such a formal change is tantamount to choosing the scattering matrix in a diagonal form

$$A_{ij} \to -\omega\delta_{ij}, \tag{5.1}$$

where the parameter $\omega > 0$ is the inverse relaxation time to local equilibrium, a constant here and throughout.

This means moving from a multi- to a single-time relaxation scheme in which *all* modes relax on the same time-scale $\tau = 1/\omega$.

Upon such a change, the diagonal LBE reads as follows:

$$\boxed{\Delta_i f_i = -\omega\,(f_i - f_i^{\mathrm{e}})\,.} \tag{5.2}$$

Given the direct link with the time-honored Bhatnagar–Gross–Krook model Boltzmann equations in continuum kinetic theory [55], the eqn (5.2) has appropriately been named *Lattice BGK, LBGK* for short.

It is difficult to imagine something simpler than this equation to recover the Navier–Stokes equations. Still a few points need some clarification.

As already noted, the diagonalization of the collision operator implies that all modes decay at the same rate. This is undesirable for ghost fields and downright inadmissible for conserved quantities.

Leaving aside ghosts for a while, let us consider the hydrodynamic fields. The fix is simple. Since the conservation laws can no longer be encoded at the level of the scattering matrix (no way a diagonal matrix can fulfil the sum-rules eqns (3.23) and (3.24)), they must necessarily be enforced on the the local equilibrium, precisely as in continuum kinetic theory.

As a result, at variance with quasilinear LBE, we shall explicitly require local equilibria to carry the same density and momentum as the actual distribution function:

$$\sum_i f_i^{\mathrm{e}} = \sum_i f_i \quad = \rho, \tag{5.3}$$

$$\sum_i f_i^{\mathrm{e}} c_{ia} = \sum_i f_i c_{ia} = \rho u_a. \tag{5.4}$$

Note that these equalities need *not* be obeyed by the quasilinear LBE equilibria because sum rules of the scattering matrix were automatically taking care of conservation.

This is why quasilinear LBE local equilibria could be reduced to the second-order term in the Mach expansion, upfronted by the global density ρ_0 instead of the local one ρ. The first task with LBGK is therefore to derive explicit equilibria fulfilling (5.3) and (5.4).

5.2 LBGK equilibria

In facing the task of defining LBGK equilibria we shall aim at a greater generality than previously done with LBE. In particular we shall cater for the possibility of *multi-energy* levels, which means lifting the constraint of the same magnitude for all discrete speeds.

By introducing two distinct 'quantum numbers', $j = 0, \ldots, j_E$, labeling the energy level and $i = 1, \ldots, b_j$ labeling the momentum eigenstate within energy level j, we can write:

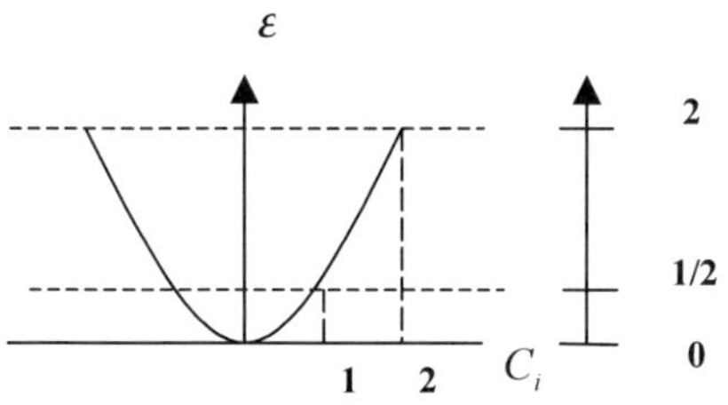

FIG. 5.1. Energy–momentum relation with three energy levels.

$$\epsilon_j = \frac{1}{2}c_{ji}^2, \quad j = 0, \ldots, j_E, \quad i = 1, \ldots, b_j. \tag{5.5}$$

With these preparation, a generic family of LBGK equilibria can be expressed as the low Mach expansion of a multi-energy Maxwellian. This reads:

$$f_I^e = \rho w_I \left(1 + \frac{c_{Ia}u_a}{c_s^2} + \frac{Q_{Iab}u_a u_b}{2c_s^4}\right), \tag{5.6}$$

where I is a shorthand for (i, j) and the weights w_I and the constant c_s (sound speed) depend on the specific choice of the discrete speeds c_{Ia}. In light of what we have learned with the FCHC lattice, the weights w_I can either be interpreted as degeneracy numbers from higher dimensions or simply as different masses of the particles moving along the different directions.

Either way, mass and momentum conservation, as well as isotropy, impose the following equalities:

$$\sum_I w_I = 1, \tag{5.7}$$

$$\sum_I w_I \, c_{Ia} = 0, \tag{5.8}$$

$$\sum_I w_I \, c_{Ia} \, c_{Ib} = P\delta_{ab}, \tag{5.9}$$

$$\sum_I w_I \, Q_{Iab} = 0, \tag{5.10}$$

$$\sum_I w_I \, c_{Ia} \, Q_{Ibc} = 0, \tag{5.11}$$

$$\sum_I w_I \, c_{Ia} \, c_{Ib} \, Q_{Ibc} = \rho u_a u_b. \tag{5.12}$$

Unlike LBE, these equations are generally under-constrained, i.e., they admit several solutions because, due to the extra freedom provided by the multi-energy lattice, there are more discrete speeds than constraints.

Optimal use of this extra freedom is key to the success of the LBGK approach.

Qian *et al.* (1992) provide a whole family of solutions, dubbed DnQm for m speed model in n dimensions.

Popular examples are D1Q3, D1Q5, D2Q9, whose diagrams are sketched in Figs 5.2 and 5.4, together with the synoptic table of their main parameters.

An important asset of the multi-energy schemes is that *there is no problem in going to three dimensions.* In fact there are a number of viable three-dimensional LBGKs, such as the D3Q15 and D3Q19 depicted in Fig. 5.4.

The D3Q15 is the crystal made up by the center of the cell, the six face-centers (nearest neighbor, *nn*) and the eight vertices (next-nearest-neighbor, *nnn*). Note that there is a spectral gap, namely no particles with speed two.

D3Q19 is a close relative of the three-dimensional projection of FCHC, with different weights though (see Table 5.1).

The reader with some background in numerical analysis probably cannot help but notice a close similarity between the LBGK weights and the numerical stencils (sometimes called 'computational molecules') of finite difference and finite element calculations. This analogy is real, and will be commented in some more detail later in this chapter.

5.3 LBGK versus LBE

The lattice BGK seemingly represents the ultimate version of LBE in terms of simplicity, elegance and efficiency. This explains why the vast majority of modern

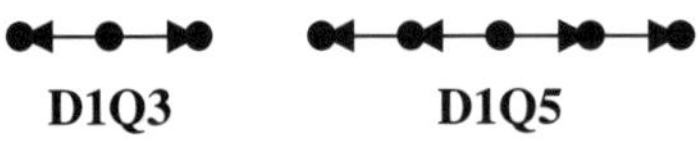

FIG. 5.2. The D1Q3 and D1Q5 lattices.

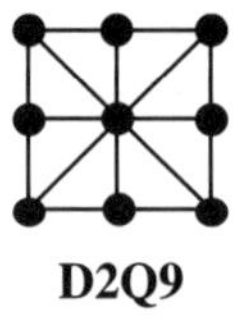

FIG. 5.3. The D2Q9 lattice.

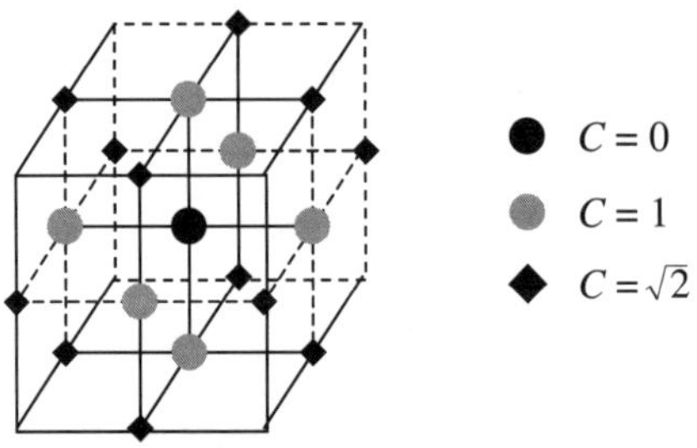

FIG. 5.4. The D3Q19 lattice.

TABLE 5.1. Main parameters of some DnQm BGK lattices.

	c_s^2	Energy	Weight
D1Q3	1/3	0	4/6
		1/2	1/6
D1Q5	1	0	6/12
		1/2	2/12
		2	1/12
D2Q9	1/3	0	16/36
		1/2	4/36
		1	1/36
D3Q15	1/3	0	16/72
		1/2	8/72
		3/2	1/72
D3Q19	1/3	0	12/36
		1/2	2/36
		1	1/36

LBE applications justly make use of it. Still, a few words of comment are in order.

First, the gain in efficiency over LBE is less dramatic than what one would naively expect from the diagonal form of the collision operator.

This is because LBE does not need the full equilibrium distribution but only the second-order term in the Mach number expansion. In addition, LBE tolerates weak density fluctuations in the equilibria since it can work with a constant, uniform density ρ_0 instead of the local value ρ. All in all, LBGK is found to provide roughly a 30 per cent improvement in performance. This is not mind-boggling, but can nonetheless make a significant difference for large-scale three-dimensional computations, as directly experienced by this author. Efficiency, besides simplicity, therefore marks a definite point in favor of LBGK.

A limitation of BGK not shared by LBE is that all modes decay at the same time rate ω^{-1}. Depending on initial and physical conditions, this may imply unphysical short-wave oscillations which need be filtered out with a consequent loss of spatial resolution. In addition this may also degrade numerical stability. Another LBE bonus not shared by LBGK is the possibility of seamless inclusion of passive scalars arising from the dimensional projection of the FCHC lattice in dimensions $D < 4$ (see previous chapter).

Such a bonus is not available with LBGK where, in principle, each additional passive scalar calls for a separate set of $2D$ populations to implement a random walk in D spatial dimensions. This is an unacceptable burden for applications, such as combustion, involving many species. Only recently, it has been realized that combined use of LBGK with the Lax–Wendroff finite difference scheme permits to use just one population per passive scalar thus bridging the gap with

LBE. It is nonetheless more complicated than the native LBE–FCHC scheme.

Finally, a single relaxation time implies that mass, momentum and heat transfer all take place at the same rate.

This is appropriate only for ideal gases.

This restriction can be partially cured by using a two-time relaxation operators [56] of the form

$$C_i = -\omega\,(f_i - f_i^{\mathrm{e}}) - \omega^*\,(f_{i^*} - f_{i^*}^{\mathrm{e}})\,, \tag{5.13}$$

where i^* denotes the parity conjugate of i ($\vec{c}_i + \vec{c}_{i^*} = 0$).

By tuning the two relaxation parameters, one can change the momentum and heat diffusivity ν and χ independently, thereby achieving a non-unit Prandtl number ν/χ. The dynamic range of realizable Prandtls remains relatively narrow, though.

All the aforementioned cautionary remarks do not change the fact that LBGK is significantly simpler than all its forebears and consequently it should be recommended as the method of choice to start off with for most LB simulations of fluid flow.

Before closing this section, we wish to mention recent 'beyond BGK' developments in the direction of reviving the scattering matrix LBE approach [57]. The idea is to devise near-optimal spectral representations of the scattering matrix which, besides minimizing viscosity, prove also to be capable of enhancing (linear) numerical stability [164].

The practical implementation of such 'beyond BGK' schemes proceeds via a double real space–momentum space representation of the discrete populations, say f_i and ψ_j, which diagonalize the streaming and collision operators, respectively (the former being the standard representation).

It can be shown that careful optimization of matrix product operations keeps the computational overhead only marginally above the one associated with 'plain' LBGK, so that the price to pay for enhanced stability seems definitely worthwhile.

5.4 Relation to continuum kinetic theory

In spite of its physics, as opposed to numerically-oriented genealogy, it is clear that LBGK can be regarded as 'yet another finite difference scheme' for the continuum BGK equation introduced in Chapter 1, [58]:

$$\partial_t f + v_a \partial_a f = -\omega\,(f - f^{\mathrm{e}})\,. \tag{5.14}$$

To see this, let us write down the generalized continuity equations corresponding to the projection of the continuum BGK equation onto a set of basis functions $\phi_k(v)$:

$$\partial_t \rho^{(k)} + \partial_a J_a^{(k)} = -\omega\left(\rho^{(k)} - \rho^{\mathrm{e},(k)}\right), \tag{5.15}$$

where $\rho^{(k)} = \int f \phi_k\,(v)\,\mathrm{d}v$ and $J_a^{(k)} = \int f v_a \phi_k\,(v)\,\mathrm{d}v$.

Now, let us evaluate these integrals by some numerical quadrature formula:

$$\rho^{(k)} = \int f \phi_k(v)\, \mathrm{d}v \sim \sum_i w_i\, \phi_k(v_i)\, f_i, \tag{5.16}$$

where v_i and w_i are the nodes and weights, respectively, of the quadrature formula.

By combining the projection (5.15) with the quadrature (5.16) the continuum BGK takes manifestly the form of a discrete BGK equation for the pivotal values $F_i \equiv f_i w_i$ [59]:

$$\partial_t F_i + \partial_a F_i v_{ia} = -\omega\left(F_i - F_i^{\mathrm{e}}\right). \tag{5.17}$$

By marching each population in time with an explicit Euler forward scheme along the characteristics $\Delta x_{ia} = v_{ia}\Delta t$, the left-hand side yields:

$$\frac{F_i\left(x + v_{ia}\Delta t, t + \Delta t\right) - F_i\left(x, t\right)}{\Delta t}, \tag{5.18}$$

which, by making $\Delta t = 1$, delivers precisely LBGK.

It is therefore apparent that every quadrature scheme gives rise to its own LBGK. It is equally apparent that not all of these quadrature schemes are congruent; that is, can tile three-dimensional space with no 'holes' in between. A list of viable choices can be found in the original paper [59].

This procedure encapsulates the LBGK formalism within the big box of numerical analysis. This is sometimes hailed as a major finding, one that confers on LBE theory a higher status, by showing an explicit formal link with 'true' kinetic equations [60].

In this author's opinion, this point is indeed potentially far-reaching, but slightly overemphasized on the basis of the present achievements (of course, future achievements may change this somewhat conservative statement).

Proving that the continuum limit of LBGK does reproduce the Navier–Stokes is good enough to qualify LBGK as a viable tool for numerical fluid dynamics, regardless of whether or not LBGK can be derived by continuum BGK. This is a very pragmatic stance, but it should not be forgotten that BGK is only a useful model, not a fundamental equation in itself.

This said, the fact remains that pinning down the exact nature of the link between LBGK and continuum kinetic theory is useful for the systematic analysis and, more importantly, for the potential derivation of novel LBGK schemes. It might well be that the major contribution from this viewpoint is a conceptual shift in devising models of complex fluid behavior, namely:

1. *Start with a continuum BGK-like kinetic equation.*
2. *Apply the Hermite–Gauss discretization procedure described above to derive a series of LBGKs.*
3. *Select the 'best' LBGK for your problem (efficiency, accuracy and the like).*

The physical modeling effort is entirely concentrated in step 1. Step 2 is automatized, which is a welcome methodological benefit, and finally step 3 relies again upon physical/computational intuition.

In an optimistic vein, one could speculate that the above modeling strategy may give rise to a new paradigm in non-equilibrium statistical mechanics, one that is based on suitable minimal kinetic equations rather than on partial differential equations (continuum approach) or truly atomistic motion (microscopic approach). That is precisely the type of revival of modern kinetic theory described in the Preface.

Whether this procedure will indeed contribute new or better LBGK schemes than the existing ones, remains to be seen. At the time of this writing, it seems that the number of applications of this viewpoint is on the rise, especially with regard to complex flows situations. The forthcoming years will surely tell us more about this fascinating subject.[14]

5.5 Relation to discrete velocity models

This chapter concludes our introduction to the basic LBE theory. Before moving on to the illustration of how this theory can be put to practice, a few remarks are in order. In particular, we wish to mention the relation between LBGK and Discrete Velocity Models (DVM) of the Boltzmann equation.

First and foremost, one should not forget that DVMs were invented long before LBE. Actually, they date back to Mark Kac's work in the 50s and, more systematically, to Broadwell and Gatignol's work in the 60s.

How does LBE relate to this work?

Surely DVM and LBE are very close relatives in mathematical terms, since they both are based on grid-bound particles moving along a set of discrete speeds.

However, as far as we can judge, they depart significantly in their practical aim. DVM is genuinely concerned with kinetic theory, the main aim being to develop stylized Boltzmann equations possibly amenable to analytical solutions, or theorem demonstration [61]. Computer simulation is also in focus, but with no particular obsession on reaching the hydrodynamic scales.

LBE is less and more ambitious at the same time.

Less ambitious, because the idea of a faithful description of kinetic phenomena is not pursued at all. More, because it challenges the Lion King in his own den, aiming as it does at capturing hydrodynamic phenomena (and beyond) more effectively than continuum models themselves! As we shall see, such a daring task often meets with significant success.

5.6 LBE genealogy

We summarize this first part of the book in the following genealogical tree of LBE:

[14]Incidentally, this procedure reveals how indebted we are to Mr Friedrich Gauss, for his numerical quadrature formulae provide a wonderful importance-sampling device of velocity space which lies at the heart of LBE theory.

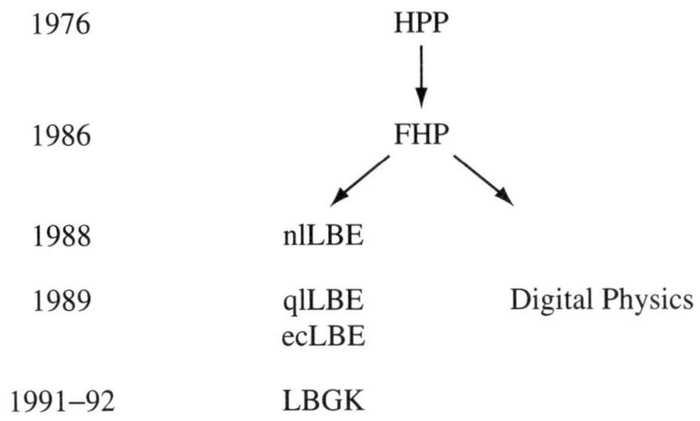

SKETCH 5.1. The genealogical tree of LBE. nl: nonlinear, ql: quasilinear, ec: enhanced collision.

From this (very schematic) picture, we infer that LBGK somehow marks the endpoint of the basic development of LBE theory. Other major theoretical progress has been made in the subsequent years, and more are certainly needed in the future to address more advanced issues (see Parts II and III), but it seems fair to say that the hard-core of the basic theory lies within the 1988–1992 developments. The right-side branch (Digital Physics) is intended to remind the reader that besides LBE, the floating-point spin-off of FHP, a less radical departure from FHP Boolean ideas kept (and still keeps) going in the form of digital physics and integer Boltzmann alternatives [62]. This second branch is mentioned only for the sake of completeness, and a serious description of this matter would warrant another book on its own.

5.7 Warm-up code

A simple warm-up LBGK code is available at the following WWW site:

```
ftp.iac.rm.cnr.it
cd incoming
mget lbe.f, lbe.inp, lbe.par
```

Enjoy!

5.8 Exercises

1. Prove that the D2Q9 equilibria fulfil the constraints (5.7)–(5.12).
2. Express the viscosity ν as a function of the relaxation parameter ω.
3. Imagine to give a different relaxation rate to each discrete speed ω_i. What do you expect to gain?
4. Compute the range of temperature associated with D1Q3, D1Q5, D2Q9.
5. Derive the LBGK generated by second-order Gaussian quadrature in eqn (5.16).

Part II

Fluid dynamics applications and advanced theory

Li Wenzi always thinks three times before he acts. Twice is enough.
Confucius

Having covered the basic theoretical grounds of the LBE method, we now move on to the discussion of some selected fluid dynamic applications, namely:

1. *Flows at moderate Reynolds number.*
2. *Low Reynolds flows in porous media.*
3. *Turbulent flows in simple geometries.*

This list is by no means exhaustive of the wide spectrum of LBE applications developed to date, but only responds to criteria of simplicity and space economy. We shall not enter application details in any depth, but stick instead to the main purpose of illustrating the basic methodological aspects of the LBE technique. Since no application can be discussed without specification of boundary conditions, Part II starts off with a chapter devoted to this subject.

6

BOUNDARY CONDITIONS

The actual dynamics of fluid flows is highly dependent on the surrounding environment. This influence is described mathematically via the prescription of suitable boundary conditions. Boundary conditions play a crucial role since they select solutions which are compatible with external constraints. Accounting for these constraints may be trivial under idealized conditions (periodic flows) but generally speaking, it constitutes a very delicate (and sometimes nerve-probing!) task. In this chapter we illustrate the most common ways to impose boundary conditions to LBE flows.

6.1 General formulation of LBE boundary conditions

Consider a fluid flowing in a bounded domain Ω confined by a surrounding boundary $\partial\Omega$. Generally speaking, the problem of formulating boundary conditions within the LBE formalism consists in finding an appropriate relation expressing the *incoming* (unknown) populations $f_i^<$ as a function of the *outgoing* (known) ones $f_i^>$. Outgoing populations at a boundary site $\vec{x}$ are defined by the condition

$$\vec{c}_i \cdot \vec{n} \equiv c_{ia} n_a > 0, \tag{6.1}$$

where n_a is the outward normal to the boundary element $\mathrm{d}S$ centered in $\vec{x}$. Incoming populations are defined by the opposite sign of the inequality. In mathematical terms, this relationship translates into a linear integral equation

$$f_i^< (\vec{x}) = \sum_y \sum_j B_{ij} (\vec{x} - \vec{y}) \, f_j^> (\vec{y}) \, , \tag{6.2}$$

where the kernel $B_{ij} (\vec{x} - \vec{y})$ of the boundary operator generally extends over a finite range of values $\vec{y}$ inside the fluid domain. This boundary operator reflects the interaction between the fluid molecules and the molecules in the solid wall. Consistent with this molecular picture, boundary conditions can be viewed as special (sometimes simpler) collisions between fluid and solid molecules.

Physical fidelity can make the boundary kernel quite complicated [63], which is generally not the idea with LBE. Instead, we look for expressions minimizing the mathematical burden without compromising the essential physics.

In particular, we shall require minimal kernels fulfilling the desired constraints on the macroscopic variables (density, speed, temperature and possibly the associated fluxes as well) at the boundary sites $\vec{x}$. This may lead to a mathematically under-determined problem, more unknowns than constraints. *This opens up an appealing opportunity to accommodate more interface physics into the formulation of the boundary conditions, but it also calls for some caution to guard against mathematical ill-posedness.*

The major element controlling the complexity of the boundary problem is whether or not the collection of boundary points lies on a surface aligned with the grid. The latter case is significantly more complex and still not entirely settled at the time of this writing. In this book we shall focus on geometrically simple applications where the former situation applies. Advanced industrial applications of the method can deal with much more complicated situations which go beyond the scope of this book.

6.2 Survey of various boundary conditions

We shall distinguish two basic classes of boundary conditions:

- *Elementary*;
- *Complex*.

By 'elementary' we imply situations in which *the physical boundary is aligned with the grid coordinates*. Elementary boundaries need not be smooth surfaces, such as straight lines or planes, for instance they could be staircase 'Legoland' approximates of fairly complex surfaces. The distinctive mark is that they *do not cut through mesh cells*.

Complex boundaries, on the contrary, can take virtually any shape, including mesh-cutting surfaces. Consequently, they can describe complex shapes more accurately. They are manifestly more difficult to implement, but the pain is often worth the gain for practical engineering applications.

In this respect LBE theory has still some way to go in the academic environment, in that advanced boundary conditions have only recently begun to be implemented [64, 65]. Not unexpectedly, commercial software leads the way in this area [66, 67].

We shall consider the following classes of boundary conditions:

- *Periodic*;
- *No-slip*;
- *Free-slip*;
- *Frictional slip*;
- *Sliding walls*;
- *Open inlet/outlets*.

In the case of simple boundaries, these conditions can be dealt with by elementary mechanical arguments involving straight single particle trajectories. Complex boundaries require instead the introduction of more sophisticated techniques such as:

- Staircasing;
- Extrapolation;
- Surface elements (*Surfels*) dynamics.

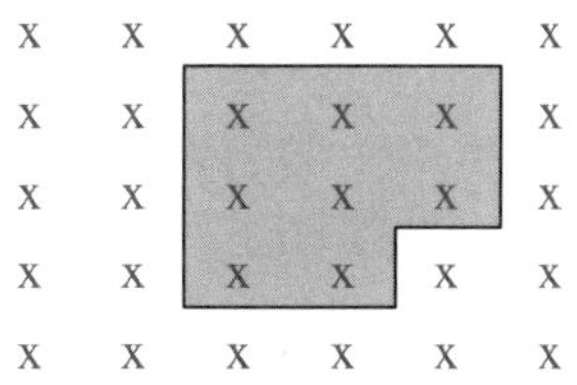

FIG. 6.1. Regular boundaries staircasing through the lattice.

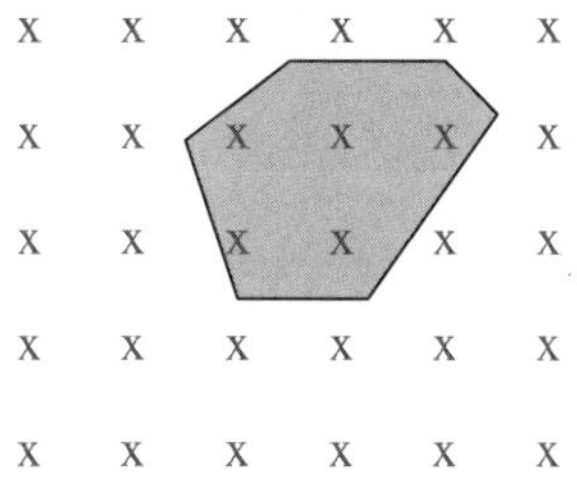

FIG. 6.2. Irregular boundaries cutting through the lattice.

All of these boundary conditions apply to the velocity field. For many steady-state flows of engineering interest it is more adequate to deal with mixed pressure and velocity boundary conditions. Since LBE is intrinsically a weakly compressible method, pressure boundaries are somewhat sensitive and will therefore be considered separately.

6.2.1 *Periodic boundary conditions*

Periodic boundary conditions are the simplest instance of boundary conditions one can think of, the joy of the programmer. They are typically intended to isolate bulk phenomena from the actual boundaries of the real physical system and consequently they are adequate for physical phenomena where surface effects play a negligible role. A typical example is homogeneous isotropic incompressible turbulence, discussed later in this book.

The practical implementation of periodic boundaries goes as follows: To fix ideas, let us refer to the nine-speed 2D FCHC scheme numbered 0–8 in counterclockwise direction starting with $\vec{c}_1 = (1, 0)$. It will be useful to keep the expressions of 2D FCHC equilibria and non-equilibria in full display:

$$f_i^{\mathrm{e}} = \rho w_i \left[1 + 2c_{ia}u_a + 2Q_{iab}u_a u_b\right], \tag{6.3}$$

$$f_i^{\mathrm{ne}} = \rho w_i \left[Q_{iab}\left(\partial_a u_b + \partial_b u_a\right)\right]. \tag{6.4}$$

Explicitly:

$$f_0^{\mathrm{e}} = 4\rho\left[1 - u^2 - v^2\right]/24, \tag{6.5}$$
$$f_1^{\mathrm{e}} = 4\rho\left[1 + 2u + u^2 - v^2\right]/24, \tag{6.6}$$
$$f_2^{\mathrm{e}} = \rho\left[1 + 2(u+v) + (u+v)^2 + 2uv\right]/24, \tag{6.7}$$
$$f_3^{\mathrm{e}} = 4\rho\left[1 + 2v - u^2 + v^2\right]/24, \tag{6.8}$$
$$f_4^{\mathrm{e}} = \rho\left[1 + 2(-u+v) + (-u+v)^2 - 2uv\right]/24, \tag{6.9}$$
$$f_5^{\mathrm{e}} = 4\rho\left[1 - 2u + u^2 - v^2\right]/24, \tag{6.10}$$
$$f_6^{\mathrm{e}} = \rho\left[1 - 2(u+v) + (u+v)^2 + 2uv\right]/24, \tag{6.11}$$
$$f_7^{\mathrm{e}} = 4\rho\left[1 - 2v - u^2 + v^2 + 2uv\right]/24, \tag{6.12}$$
$$f_8^{\mathrm{e}} = \rho\left[1 + 2(u-v) + (u-v)^2 - 2uv\right]/24 \tag{6.13}$$

and

$$f_0^{\mathrm{ne}} = 0, \tag{6.14}$$
$$f_1^{\mathrm{ne}} = 2\rho s_{xx}/24, \tag{6.15}$$
$$f_2^{\mathrm{ne}} = 2\rho s_{xy}/24, \tag{6.16}$$
$$f_3^{\mathrm{ne}} = -2\rho s_{yy}/24, \tag{6.17}$$
$$f_4^{\mathrm{ne}} = -2\rho s_{xy}/24, \tag{6.18}$$
$$f_5^{\mathrm{ne}} = 2\rho s_{xx}/24, \tag{6.19}$$
$$f_6^{\mathrm{ne}} = 2\rho s_{xy}/24, \tag{6.20}$$
$$f_7^{\mathrm{ne}} = -2\rho s_{yy}/24, \tag{6.21}$$
$$f_8^{\mathrm{ne}} = -2\rho s_{xy}/24, \tag{6.22}$$

where $2s_{xy} = \partial_x v + \partial_y u$ is the shear stress tensor and $s_{xx} = \partial_x u$, $s_{yy} = \partial_y v$ describe (volume conserving) normal compression/dilatation effects.

Consider a square box consisting of $N \times M$ lattice sites; the discrete fluid is then represented by 9 matrices $f_i(l,m)$, $l = 1,\ldots,N$, $m = 1,\ldots,M$, $i = 0,\ldots,8$.

Let us now introduce four extra layers (buffers) of sites N, S, W, E (standing for North, South, West, East):

$$\mathrm{N} = [i = 0,\ldots,N+1;\ j = M+1], \qquad \mathrm{W} = [i = 0;\ j = 0,\ldots,M+1],$$
$$\mathrm{S} = [i = 0,\ldots,N+1;\ j = 0], \qquad \mathrm{E} = [i = N+1;\ j = 0,\ldots,M+1].$$

Note that four corners are shared and may possibly require separate treatment. Periodicity along x is imposed by simply filling the W buffer with the populations of the rightmost column of the physical domain:

$$f_{\mathrm{in,W}}(\mathrm{W}) = f_{\mathrm{out,E}}(\mathrm{EF}), \tag{6.23}$$
$$f_{\mathrm{in,E}}(\mathrm{E}) = f_{\mathrm{out,W}}(\mathrm{WF}). \tag{6.24}$$

nw	n	n	n	n	n	n	n	n	n	n	ne	6
w	f	f	f	f	f	f	f	f	f	f	e	5
w	f	f	f	f	f	f	f	f	f	f	e	4
w	f	f	f	f	f	f	f	f	f	f	e	3
w	f	f	f	f	f	f	f	f	f	f	e	2
w	f	f	f	f	f	f	f	f	f	f	e	1
sw	s	s	s	s	s	s	s	s	s	s	se	0
0	1	2	3	4	5	6	7	8	9	10	11	

SKETCH 6.1. Numbering a 10×5 periodic lattice with buffers. f: fluid, w: west boundary, e: east boundary, n: north boundary, s: south boundary.

Here EF is a shorthand for the Eastward Fluid row EF $= [i = N;\ j = 0, \ldots, N+1]$.

The subscripts 'in' and 'out' denote inward and outward populations, respectively. With the speed numbering adopted here, the map is:

```
{in ,W}  = {1,2,8}
{out,W}  = {4,5,6}
{in ,E}  = {out,W}
{out,E}  = {in, W}
```

The four corners require special treatment:

```
f_{in}(NW) = f_{out}(SE)
f_{in}(SW) = f_{out}(NE)
f_{in}(NE) = f_{out}(SW)
f_{in}(SE) = f_{out}(NW)
```

Once the buffers are filled by the boundary procedure, everything is in place for the streaming step to move particles into the proper location at the next time-step.

A typical code sequence looks as follows:

1. *Boundary*;
2. *Move*;
3. *Collide*.

Technical aside The buffer policy described here is just one out of several possible strategies. One could of course dispense with buffers and copy variables directly in the physical domain. The crudest way to do this is to place 'IF' statements within the 'DO' loops. This solution is quickly programmed but certainly not elegant nor efficient on most computers since conditional statements usually break the pipeline execution on vector and superscalar computers thus resulting in a needless CPU burden. Dispensing with buffers saves a factor $1/N$ (surface

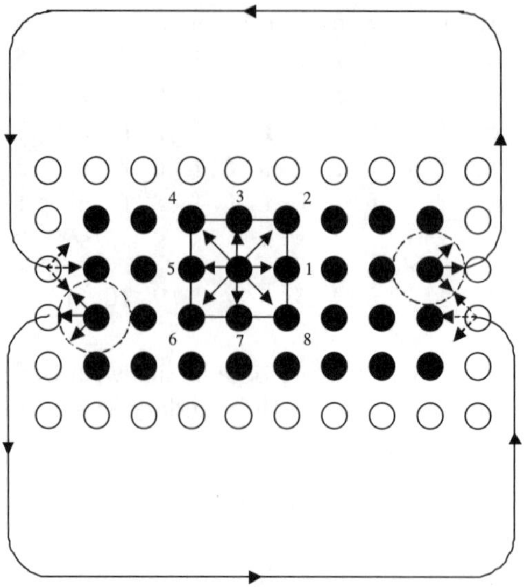

FIG. 6.3. Periodic boundaries. Particles leaving from the right fill in the left buffers and vice versa.

to volume) in memory but requires extra care in the time-ordering of the operations in order to protect the correct values of the variables prior to subsequent use. Typically one would store boundary values in temporary arrays prior to using them. Since nowadays huge memories are available, the linear size N can be chosen large enough (greater than hundred) so as to make the extra memory factor almost irrelevant to practical purposes.

6.2.2 *No-slip boundary conditions*

The next simplest boundary condition is the so-called 'no-slip' situation, namely zero fluid velocity at a given solid surface. This is physically appropriate whenever the solid wall has a sufficient rugosity to prevent any net fluid motion at the wall. Again, we shall assume the solid surface is aligned with the grid, like for instance the north/south walls in a two-dimensional rectangular channel.

Here we distinguish two types of implementations:

- On-grid;
- Mid-grid.

The on-grid condition means that the physical boundary lies exactly on a grid line, whereas the mid-grid case refers to the situation where the boundary lies in between two grid lines (still being aligned though, see Fig. 6.5).

The on-grid situation is easy; just reverse all populations sitting on a boundary node:

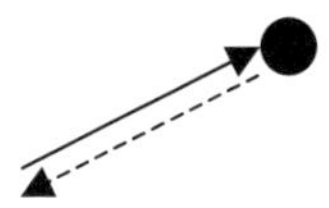

FIG. 6.4. On-grid bounce-back, no-slip boundary condition.

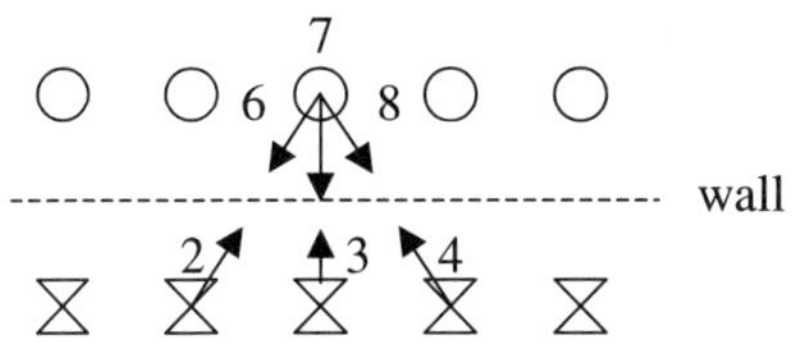

FIG. 6.5. Mid-grid bounce-back, no-slip boundary condition. Populations 6, 7, 8 on the buffer receive populations 2, 3, 4, respectively, from the mirror conjugate sites in the fluid.

$$f_{\text{in}}\,(\text{N}) = f_{\text{out}}\,(\text{N})\,, \tag{6.25}$$

$$f_{\text{in}}\,(\text{S}) = f_{\text{out}}\,(\text{S})\,, \tag{6.26}$$

where N and S denote the North and South rows of the domain, respectively.

In terms of the boundary kernel (6.2), the bounce-back kernel (on the top wall) is the 3×3 matrix

$$\begin{bmatrix} f_6\,(x,y) \\ f_7\,(x,y) \\ f_8\,(x,y) \end{bmatrix} = \begin{pmatrix} 1 & 0 & 0 \\ 0 & 1 & 0 \\ 0 & 0 & 1 \end{pmatrix} \begin{bmatrix} f_2\,(x,y) \\ f_3\,(x,y) \\ f_4\,(x,y) \end{bmatrix}. \tag{6.27}$$

This complete reflection guarantees that both tangential and normal to the wall fluid speed components, identically vanish. In fact:

$$J_x = (f_1 + f_2 + f_8) - (f_4 + f_5 + f_6) = (f_1 - f_5) = 0, \tag{6.28}$$

$$J_y = (f_2 + f_3 + f_4) - (f_6 + f_7 + f_8) = 0. \tag{6.29}$$

The first equality is ensured by initializing $f_1 = f_5$ at $t = 0$, these values not being altered by the dynamics at any subsequent time-step. The second equality is guaranteed by the complete reflection (6.25) and (6.26). All this is very simple, and yet sufficient to set up concrete LBE simulations of fluid flows in simple bounded domains.

The on-grid bounce-back is generally credited to be only first-order accurate because of the one-sided character of the streaming operator at the boundary. Mid-line reflection buys second-order accuracy at the price of a modest complication:

$$f_{\text{in}}(\text{N}) = f_{\text{out}}(\text{NF}), \tag{6.30}$$

$$f_{\text{in}}(\text{S}) = f_{\text{out}}(\text{SF}), \tag{6.31}$$

where N and S denote a generic site on the top and bottom walls, respectively, whereas NF, SF is the *set* of fluid sites connected to N, S, respectively (*domain of influence*), somehow the range of the boundary potential discussed previously. Since particles meet at mid-time $t + 1/2$ between the N, S and NF, SF rows, this implies a fully centered space–time scheme with resulting second-order accuracy.[15]

```
0 1 2 3 4 5 6 7

n n n N n n n n
----------------------------
f f F F F f f f
```

SKETCH 6.2. The north site n = 3 interacts with the fluid sites 2, 3, 4 (marked F) in the lower-lying fluid row. The wall (- - -) is in between.

The explicit scheme reads:

$$\begin{bmatrix} f_6(x,y) \\ f_7(x,y) \\ f_8(x,y) \end{bmatrix} = \begin{pmatrix} 1 & 0 & 0 \\ 0 & 1 & 0 \\ 0 & 0 & 1 \end{pmatrix} \begin{bmatrix} f_2(x-1,y-1) \\ f_3(x,y-1) \\ f_4(x+1,y-1) \end{bmatrix}, \tag{6.32}$$

where $(x, y) \equiv \text{N}$ is the generic site on the North wall. The south wall is obviously handled in the same way.

We remark that this weakly non-local formulation requires some care in handling the corners.

For instance in the leftmost site $x = 1$, the left neighbor falls within the inlet buffer, and must therefore be drawn therefrom.

6.2.3 *Free-slip boundary conditions*

Free-slip boundary conditions apply to the case of smooth boundaries with negligible friction exerted upon the flowing gas or liquid.

In this case, the tangential motion of the fluid flow on the wall is free and no momentum is to be exchanged with the wall along the tangential component.

The simplest on-grid implementation consists of an elastic scattering on the wall

[15]The reader should be cautioned that according to some authors this distinction between mid-grid and on-grid accuracy is not very meaningful, since on-grid can also be made second-order accurate. The only true difference between the two methods is the time it takes to attain a zero speed on the wall.

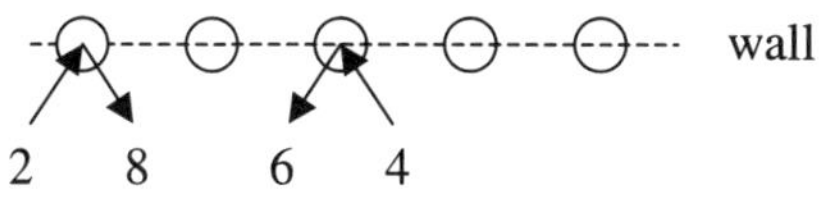

FIG. 6.6. On-site specular reflection. Population 2 goes into 8, and 4 into 6.

$$f_{\text{in}}\,(\text{N}) = f_{\text{out}}\,(\text{N})\,, \tag{6.33}$$

where N is again the North wall layer.

In terms of our cherished boundary matrix,

$$\begin{bmatrix} f_6 \\ f_7 \\ f_8 \end{bmatrix} = \begin{pmatrix} 0 & 0 & 1 \\ 0 & 1 & 0 \\ 1 & 0 & 0 \end{pmatrix} \begin{bmatrix} f_2 \\ f_3 \\ f_4 \end{bmatrix}, \tag{6.34}$$

all evaluated at (x, y).

This implies no tangential momentum transfer to the wall, as required for a free-slip fluid motion. The condition $f_3 = f_7$ implies no normal-to-wall speed. This is easily checked by simple algebra:

$$\delta J_x \equiv J'_x - J_x = (f_8 - f_2) - (f_6 - f_4) = 0, \tag{6.35}$$

where δ refers to the change due to the boundary collision (prime means after boundary collision).

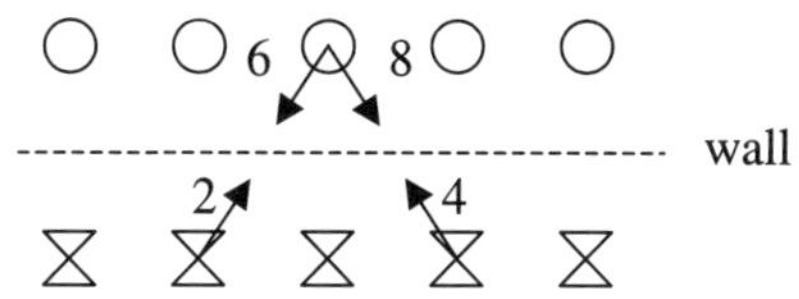

FIG. 6.7. Mid-site specular reflection.

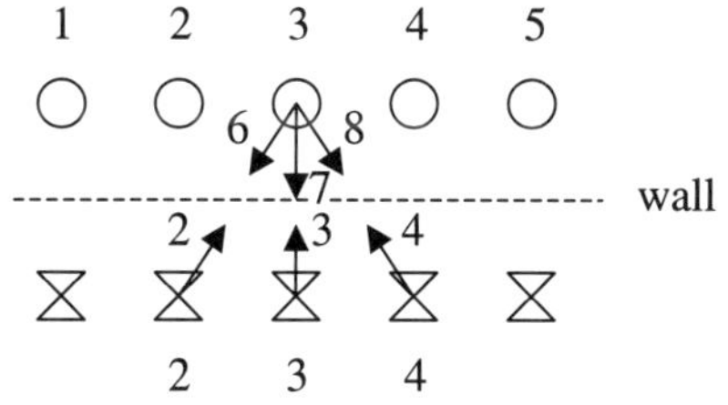

FIG. 6.8. Region of influence: The top site $n = 3$ interacts with the fluid sites 2, 3, 4 (left, mid, right) in the low-lying row. The wall (- - -) is in between.

A similar relation applies to the mid-grid version corresponding to the transformation

$$\begin{bmatrix} f_6(x,y) \\ f_7(x,y) \\ f_8(x,y) \end{bmatrix} = \begin{pmatrix} 0 & 0 & 1 \\ 0 & 1 & 0 \\ 1 & 0 & 0 \end{pmatrix} \begin{bmatrix} f_2(x-1,y-1) \\ f_3(x,y-1) \\ f_4(x+1,y-1) \end{bmatrix}. \tag{6.36}$$

Again, the bottom wall is handled with symmetrical arguments and the corners require special care.

6.2.4 *Frictional slip*

Often in applications involving liquid–gas flows, liquid droplets impinge on the solid wall and are only partially reflected back into the fluid, a certain fraction of molecules forming a film that moves along the wall subject to frictional drag.

This corresponds to no motion normal to the wall ($J_y = 0$) and damped tangential motion along it. The lattice analogue of this boundary condition is a blend of no-slip bounce-backs and free-slip reflections.

In terms of the boundary matrix B_{ij}, for the simple on-grid implementation (by now we know how to extend to the mid-grid implementation if we want to):

$$\begin{bmatrix} f_6(x,y) \\ f_7(x,y) \\ f_8(x,y) \end{bmatrix} = \begin{pmatrix} p & 0 & q \\ 0 & 1 & 0 \\ q & 0 & p \end{pmatrix} \begin{bmatrix} f_2(x,y) \\ f_3(x,y) \\ f_4(x,y) \end{bmatrix}, \tag{6.37}$$

where p and q, with $p+q=1$, are the proportion of bounce-back and specular reflections, respectively. The special choices ($p=1$, $q=0$) and ($p=0$, $q=1$) recover pure no-slip and free-slip, respectively. Note that by removing the condition $p+q=1$, absorbing/emitting boundaries can also be modeled.

If bounce-back is present in any proportion, $p>0$, frictional drag results at the wall that ultimately puts the fluid at rest. In fact, any $p>0$ does indeed produce a steady-state zero speed at the wall; however, the time it takes the fluid to attain this steady-state grows exponentially with $(1-p)$. This is easily seen by computing the momentum transfer to the wall

$$\delta J_x = (f_8 - f_2) - (f_6 - f_4) = -2p\,(f_2 - f_4)\,, \tag{6.38}$$

$$J_y = (f_2 + f_3 + f_4) - (f_6 + f_7 + f_8) = 0. \tag{6.39}$$

From this expression it is clear that any positive value of p results in a δJ_x opposite to J_x, a net frictional effect which produces exponential decay of the fluid momentum at the wall. Such an exponential decay has indeed be detected by Lavallée *et al.* [68] in early lattice gas and lattice Boltzmann simulations of straight channels. These authors also show that by the value of the coefficient p affects the effective location of the boundary where the fluid speed vanishes. This is not surprising, since in a truly microscopic picture the boundary is always diffuse and its effective location depends on the interaction potential between the wall atoms and the fluid molecules. This is a very active research topic in nano-engineering, microhydrodynamics and related disciplines dealing with systems with large surface-to-volume effects. The computational tool of choice is molecular dynamics [69] or Direct Simulation Monte Carlo for rarefied systems [100]. Both LGCA and LBE can provide a valuable complement to these *ab-initio* simulations, by giving access to larger scales once the expressions of p and q are given as a function of the wall parameters (temperature, rugosity, and so on).

6.2.5 *Sliding walls*

Another important and straightforward generalization of the free-slip boundary is the prescription of a given tangential wall speed $u(W) = u_w$, where W denotes the fluid speed at a generic wall W. A simple example is the shear flow driven by a sliding wall (planar Couette flow).

A useful method to handle sliding walls consists in using again adjustable reflection coefficients p, q and tuning them so as to make the fluid speed at the wall equal to the wall speed itself.

In full generality:

$$\begin{bmatrix} f_6(x,y) \\ f_7(x,y) \\ f_8(x,y) \end{bmatrix} = \begin{pmatrix} p & 0 & q \\ 0 & 1 & 0 \\ q & 0 & p \end{pmatrix} \begin{bmatrix} f_2(x,y) \\ f_3(x,y) \\ f_4(x,y) \end{bmatrix}. \tag{6.40}$$

The explicit computation of the normal current yields

$$J_y = f_2 + f_4 + f_3 - f_6 - f_7 - f_8 = (f_2 + f_4)(1 - p - q) = 0, \tag{6.41}$$

which delivers the normalization condition $p + q = 1$ for the case of a rigid wall.

Equating the tangential component to the wall current $J_w = \rho_w u_w$ yields

$$J_x = 2(1-p)(f_2 - f_4) = \rho_w u_w. \tag{6.42}$$

Contributions from f_1, f_5 are barred out since these populations do not contribute to the wall–fluid coupling.

The above equation delivers

$$q = \frac{\rho_w u_w}{2(f_2 - f_4)}, \tag{6.43}$$

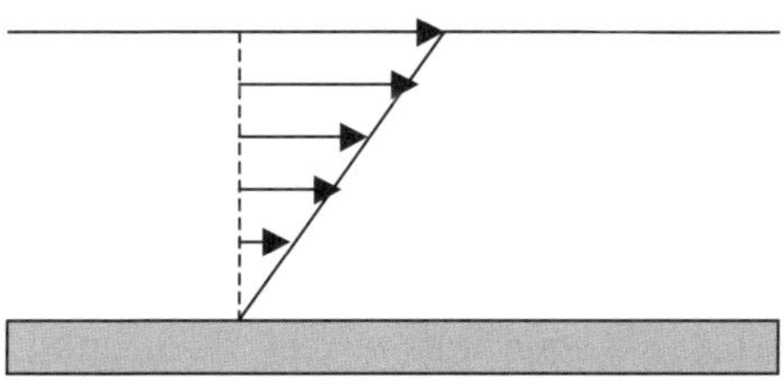

FIG. 6.9. Planar Couette flow. The upper wall moves tangentially to the flow and drags the underlying fluid.

where we have set $\rho_w = \rho/6$ since only four populations f_2, f_4, f_6, f_8 are involved. This expression is operational since the values of f_2 and f_4 are available from the actual simulation and can then be used to update the value of q 'on the fly'. A little further analysis helps us to gain a better physical insight. To a first-order approximation, we can replace f_2 and f_4 in eqn (6.43) with their equilibrium values. This yields

$$q \sim \frac{u_w}{2u}. \tag{6.44}$$

This equation shows that if the fluid is slower than the wall, $u < u_w$, then more particles are reflected ahead than bounced back, that is $q > 1/2$. This is plausible since the fluid needs to be accelerated in order to keep up with the moving boundary. The same equation also shows that if $u < u_w/2$, the coefficient q becomes larger than 1, a potentially dangerous situation since it corresponds to negative p. Therefore, in actual simulations, it may be advisable to adopt a soft, non-impulsive, start-up of the wall motion. At steady-state, $u = u_w$,[16] we obtain $p = q = 1/2$, namely an exact balance of bounce-back and specular reflections. This reminds us of experimental observations, dating back to Knudsen's work, according to which molecules impinging on a wall at any given incidence angle are reflected randomly in all directions.

6.2.6 *The Inamuro method*

Building upon the same physical picture Inamuro *et al.* [70] developed an elegant method to impose the Dirichlet boundary condition $u = u_w$.

The idea is to draw the reinjected populations from a local equilibrium with the same normal and different tangential speed at the wall: $v = v_w$, $u = u_w + u'$, where u', the so-called counterslip speed, is adjusted so as to achieve mass flux conservation on the wall: $\rho' u = \rho_w u_w$.

By treating the wall density ρ_w and the fictitious wall-equilibrium density ρ' as unknowns, one obtains the following equations:

$$\sum_i f_i = \rho_w, \tag{6.45}$$

$$\sum_i c_{ix} f_i = \rho_w u_w, \tag{6.46}$$

$$\sum_i c_{iy} f_i = \rho_w v_w. \tag{6.47}$$

In the above equations, the outgoing populations $f_{2,3,4}$ from the north wall are provided by the interior LBGK dynamics, whereas the incoming ones $f_{6,7,8}$ are drawn from the aforementioned equilibria.

This leaves us with three equations for the three unknowns ρ_w, ρ', u' which are readily solved to deliver the desired solution.

[16] We leave it to the reader to show that this relation holds true in the presence of dissipative effects.

Inamuro *et al.* tested this scheme for a 2D Poiseuille flow for a series of values of the relaxation parameter τ. Their results show that the no-slip condition is accurately reproduced for values of $0.5 < \tau < 20$ (see Fig. 6.10), whereas standard bounce-back would yield significant errors already beyond $\tau = 2$.

This is in line with physical expectations, since τ goes with the molecular mean free path. Besides its simplicity and physical soundness, the Inamuro method shines for its grid and dimensional independence. In fact, in D dimensions one can dispose of the $D + 1$ unknowns, wall density, equilibrium density and $D - 1$ tangential speed components by means of $D + 1$ constraints, namely D velocity components $u_a = u_{wa}$ and one from equating the wall density to the fluid one $\rho = \rho_w$.

Other strategies have been proposed [71, 72] which focus directly on the unknown populations rather than expressing them via adjustable local equilibria.

These strategies are more general but tend to be involved because they must cope with an inevitable mismatch between the number of unknowns and the number of macroscopic constraints. This mismatch is usually closed by heuristic rules involving a series of extra steps, including extrapolation and mass redistributions among the populations. As repeatedly stated, this extra freedom is a blame and a blessing at the same time. The reader interested in full details is directed to the original references.

6.2.7 *Moving walls*

The more general case of a wall moving along both tangential and *normal* directions implies a significant leap of complexity.

This is because the normal motion of the wall induces *compressed* fluid motion, whereby the fluid volume changes in time. A typical example is piston

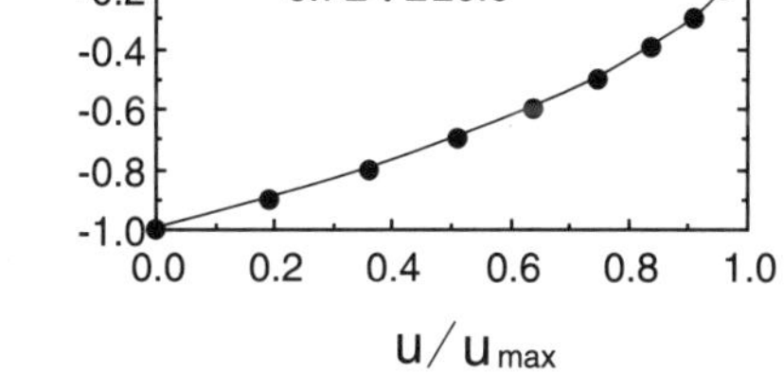

FIG. 6.10. Calculated velocity profile for Poiseuille flow as a function of the relaxation parameter τ. (After [70], courtesy of O. Inamuro.)

motion in a combustion engine.

This is a 'grand canonical' type of computation in which fluid sites must be erased or regenerated in the course of time depending on the wall position.

It implies some additional bookkeeping which may become very demanding if the wall motion proceeds along a generic direction not aligned with the grid coordinates. Such a problem is an order of magnitude higher in complexity and has been tackled by Ladd [73] for the simulation of particle suspensions (colloidals). It will be described shortly in Part III.

6.3 Open boundaries

Many situations of practical and theoretical interest involve *open* boundaries such as fluid inlets and outlets. For these open flows, it is common to assign a given velocity profile $u_{\text{in}}(y)$ at the inlet while at the outlet either a given pressure P_{out} or a no-flux condition normal to the wall, $\partial_n u = 0$, is imposed.

Let us consider the inlet boundary condition first.

A prescribed inlet flow is easily implemented by constantly refilling the buffer with the equilibrium population corresponding to the desired values of density and flow speed:

$$f_{\text{in}}(y) = f^{\text{e}}\left[\rho_{\text{in}}, u_{\text{in}}(y)\right]. \tag{6.48}$$

0	1	2	3	4	5	6	7	
n	n	n	n	n	n	n	n	north wall
i	f	f	f	f	f	f	o	
i	f	f	f	f	f	f	o	
i	f	f	f	f	f	f	o	
i	f	f	f	f	f	f	o	
s	s	s	s	s	s	s	s	south wall
0	1	2	3	4	5	6	7	

SKETCH 6.3. i: inlet site, o: outlet site.

The outlet is more delicate.

Naively, we might think that the zero-gradient condition can be imposed by simply copying the last but one column of the channel into the last one, (column 5 into column 6, in Sketch 6.3). Unfortunately, this does not work, at least, not unless the outlet is placed far enough downstream so as to allow the flow to settle down to a one-dimensional zero-gradient profile. If this condition is not met, trains of backward propagating disturbances are triggered, which may invade the whole domain and put the computation in jeopardy. A practical way of dealing with these disturbances is to apply the so-called '*porous plug*' boundary condition. The idea is that particles reaching the outlet do not necessarily leave the domain but instead they are reflected back with a probability r, which is adjusted so as to ensure mass conservation.

As usual, we replace the outgoing populations with their equilibrium values (which is tantamount to implicitly assuming no gradients at the outlet).

This yields:[17]

$$J_{\text{out}}^{\text{out}} = f_1 + f_2 + f_8 = \frac{\rho}{24}\left(6 + 12u\right), \tag{6.49}$$

$$J_{\text{in}}^{\text{out}} = f_5 + f_6 + f_4 = r\frac{\rho}{24}\left(6 - 12u\right). \tag{6.50}$$

By equating the total outgoing current with the inlet current, i.e., $J_x = J_{\text{out}}^{\text{out}} - J_{\text{in}}^{\text{out}} = \rho u_{\text{in}}$ and assuming zero transversal speed $v = 0$, we obtain r as function of u_{in} and the outlet speed u:

$$r = \frac{1 + 2u - 4u_{\text{in}}}{1 - 2u} = 1 + 4\left(\frac{u - u_{\text{in}}}{1 - 2u}\right), \tag{6.51}$$

where contributions to order $O(u^2)$ have been neglected for simplicity.

This relation shows that if the outlet flow matches exactly the inlet speed, no reflection is required ($r = 1$) because the flow is automatically settled to a no gradient configuration. Note also that if $u > u_{\text{in}}$, outgoing populations must be over-reflected ($r > 1$), in order to reabsorb the excess speed $u - u_{\text{in}}$. Assuming a one-dimensional local equilibrium at an open boundary is somewhat delicate, as it forces the flow to a configuration it might not be ready for if the outlet boundary is too close to the inlet.

One way (never tested by this author!) to mitigate the effects of this assumption could be to include residual velocity gradients in the expression of the reflection coefficient r. To this end, we just add the non-equilibrium components to the momentum balance at the outlet ($u_x \equiv \partial_x u$):

$$J_{\text{out}}^{\text{ne}} = f_1^{\text{ne}} + f_2^{\text{ne}} + f_8^{\text{ne}} = \frac{\rho}{24}\left(4u_x\right), \tag{6.52}$$

$$J_{\text{in}}^{\text{ne}} = f_5^{\text{ne}} + f_6^{\text{ne}} + f_4^{\text{ne}} = r\frac{\rho}{24}\left(4u_x\right). \tag{6.53}$$

By including these contributions in the expression of the total current J_x, it is easily shown that the expression of r gets an additional factor $2u_x/3$ in the denominator. This factor may correct for the presence of residual velocity gradients at the outlet.

6.4 Complex (misaligned) boundaries

All situations considered thus far dealt with simple boundaries aligned with the grid lines. The general case of boundaries cutting through the mesh is obviously more complicated.

[17] Beware: subscript 'in' in J refers to the incoming current at the outlet, which is why we have added the superscript 'out'.

6.4.1 *Staircased boundaries*

The quickest approach to a complex boundary is to 'staircase' the cutting boundary by replacing it with a zig-zagging contour lying entirely on the grid (a sort of 'lattice polymer'). This has the great merit of simplicity and truly honors the heralded assets of LGCA and LBE methods: easy handling of complex boundary conditions.

However, it also introduces an artificial rugosity of order $1/L$, L being a typical linear size of the obstacle in lattice units, which may deteriorate the accuracy of the near-wall flow computation.

Staircased boundaries are relevant to flows in porous media and for the sake of concreteness they shall be described in some more detail in the chapter devoted to this application.

6.4.2 *Extrapolation schemes*

An elegant and simple method to handle boundary conditions in a systematic rather than heuristic way is due to Chen and Martinez (CM) [78]. The method places the wall on boundary nodes and *lets them undergo the same collisional step as the fluid nodes*, with the important proviso that the equilibrium population is explicitly tuned on the desired wall speed u_w. In this picture, the wall is nothing but a portion of fluid with a prescribed equilibrium. To complete the streaming step, boundary nodes need to import information from solid nodes lying inside the solid wall. This information is obtained by a straightforward second-order interpolation:

$$f_{2,3,4}\,(\mathrm{N}) = 2f_{2,3,4}\,(\mathrm{S}) - f_{2,3,4}\,(\mathrm{F})\,. \tag{6.54}$$

s s s s s s
n n n n n n wall
f f f f f f

SKETCH 6.4. f: fluid, s: solid, n: north wall.

This scheme preserves second-order accuracy, does not depend on the specific lattice topology nor on its dimensionality and it is fairly easy to implement. The authors demonstrate its validity for the case of Poiseuille flow, Couette flow, lid-driven square cavity and others. They also point out potential generalizations to include von Neumann type boundary conditions.

6.4.3 *The picky stuff*

It is well known from kinetic theory that the presence of a wall produces a thin layer of the order of the mean free path (Knudsen layer) whose steady-state differs from the one attained in the bulk region (Kramers problem). On a discrete speed lattice the Kramers problem can be solved analytically, as shown by d'Humières and co-workers [74]. On a 6-state hexagonal lattice the main results are as follows:

- *Parallel/perpendicular orientation*
  For parallel and perpendicular boundaries, anisotropic Knudsen layers are completely suppressed by bounce-back or reflecting boundary conditions. However, the effective location of the wall is slightly displaced off-lattice with respect to the geometrical position.
- *General orientation*
  For a general orientation (irrational with respect to the lattice coordinate lines) there is an infinity of spurious modes that disappear only once the slope of the boundary becomes rational.

These spurious modes arise as a consequence of the spurious invariants of the lattice (see Part I) and must be watched carefully in applications dealing with flows whose dynamics are dominated by irregular, jagged boundaries.

For a clever treatment of boundaries with an arbitrary inclination to the lattice grid lines, see [75, 76].

6.4.4 *The surfel method*

A powerful method to deal with arbitrary boundaries is based on the notion of *surfels*, that is surface elements.[18] Surfels are elementary portions of the boundary cutting through a single volume cell, called *voxel*, and must viewed as *independent computational entities with a dynamics on their own.* They act as *flux scattering elements* receiving incoming fluxes from the flow and renstituting a corresponding set of fluxes honoring the basic conservation laws at the surface [77]. Let us clarify with an example. Consider a 2D cell of the D2Q9 lattice identified by the four vertices (A, B, C, D). The cell is cut through by a surfel (segment) joining points X and Y on sides AC and CD. The mass entering the surfel per unit time is given by (see Fig. 6.11):

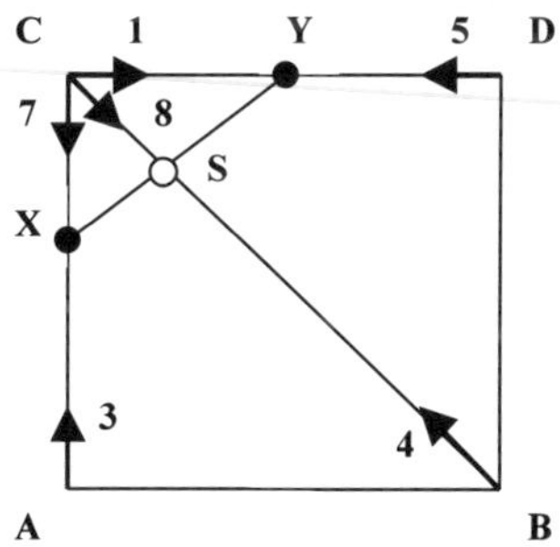

FIG. 6.11. A surfel $S = \overline{XY}$ joining west and north cell boundaries.

[18]TM EXA Corporation

$$M^{\text{in}} = a_7 f_7\,(C)\,\overline{AX} + a_8 f_8\,(C)\,\overline{SB} + a_1 f_1\,(C)\,\overline{YD}, \tag{6.55}$$

where $a_i = c_{ia}n_a$, n_a being the outward normal of the surfel. In order to ensure mass conservation (non-absorbing boundary) the surfel must renstitute an equal amount of mass M^{out}.

This is given by

$$M^{\text{out}} = a_3 f_3\,(A)\,\overline{XC} + a_4 f_4\,(B)\,\overline{SC} + a_5 f_5\,(D)\,\overline{YC}, \tag{6.56}$$

where the meaning of the notation is again apparent from Fig. 6.11.

By the same token, one writes down the momentum fluxes along x and y directions:

$$J_x^{\text{in}} = f_1\,(C)\,a_1\overline{YD} + f_8\,(C)\,a_8\overline{SB}, \tag{6.57}$$

$$J_x^{\text{out}} = f_4\,(B)\,a_4\overline{CS} + f_5\,(D)\,a_5\overline{YD}, \tag{6.58}$$

$$J_y^{\text{in}} = f_7\,(C)\,a_7\overline{AX} + f_8\,(C)\,a_8\overline{SB}, \tag{6.59}$$

$$J_y^{\text{out}} = f_3\,(A)\,a_3\overline{CX} + f_4\,(B)\,a_4\overline{CS}. \tag{6.60}$$

The three conditions

$$M^{\text{out}} = M^{\text{in}}, \quad J_x^{\text{out}} = J_x^{\text{in}}, \quad J_y^{\text{out}} = J_y^{\text{in}} \tag{6.61}$$

provide a 3×3 system for the 3 unknowns $f_3(A)$, $f_4(B)$, $f_5(D)$.

This procedure applies to generic shapes of the voxels but it requires a certain degree of sophistication to work in full generality. Since surfel dynamics is naturally expressed in terms of incoming–outcoming fluxes crossing the surface of each cell, the surfel method deploys its full power when formulated in the language of finite volumes (see in the sequel) [77]. Successful finite difference analogues are also reported in the literature [64, 65, 78].

6.5 Exactly incompressible LBE schemes

In many problems of engineering interest (especially in open flows) it is often convenient to specify the value of the pressure either at the outlet or at the domain inlet. Incompressible isothermal LBE schemes do not deal easily with pressure boundary conditions simply because density and pressure are tied up together by an ideal gas equation of state $P = \rho T$, where $T = c_s^2$ is a constant.

Since in the lattice c_s is $O(1)$, any pressure change is necessarily associated with a comparable change in density, which breaks the incompressible nature of the LBE flow. In conventional CFD this problem is usually side-stepped by allowing an artificially high sound speed $c_s \to \infty$, so that significant pressure drops ΔP can be sustained by small density changes $\Delta\rho = \Delta P/c_s^2$.

The conventional way to impose pressure drops in open flows consists of replacing the 'true' pressure gradient with a corresponding volume force F providing exactly the same momentum input to the flow. Such a surrogate procedure works fine as long as there is no need to know the detailed space–time distribution of the pressure field. If this information cannot be passed out, more sophisticated models are called for. Ideally, these models should be tailored on the requirement of strict incompressibility of the flow field: $\operatorname{div} \vec{u} = 0$.

Before mentioning how these strictly incompressible models work [79], it is worth mentioning a simple strategy [70] which apparently yields acceptable results within the simple class of quasi-incompressible (as opposed to strictly incompressible) LBE models.

The idea is simply to impose a density drop between inlet and outlet by augmenting the outgoing populations by a constant term:

$$f_{\text{in}} = f_{\text{out}} + C. \tag{6.62}$$

The constant C is computed by using the equilibrium values of the distribution function and imposing a prescribed value for the inlet density ρ_{in}. By doing so, the flow proves capable of sustaining minor density drops, yet sufficient to produce the pressure gradients needed by many incompressible applications.

As repeatedly stated, standard LBE methods are weakly compressible approximations of the incompressible Navier–Stokes equations. The continuity equation, rewritten in the form

$$\operatorname{div} \vec{u} = -\frac{\mathrm{D} \ln \rho}{\mathrm{D}t}, \tag{6.63}$$

shows that the LBE steady-states ($\partial_t = 0$) are affected by a compressibility error

$$E = |u_a \partial_a \ln \rho| \, . \tag{6.64}$$

This error is of order Ma^2, hence negligible for many practical purposes, but still annoying as it hinders the application of LBE to flows with significant pressure changes $\Delta P/P > Ma^2 \sim 0.1$. It would therefore be desirable to extend LBE so as to ensure exact incompressibility at steady-state.

A quick way to achieve this is to use the current density J_a to represent the flow speed u_a. Then, the LBE steady-state automatically delivers a solenoidal flow $\partial_a u_a = 0$ [13].

The trick is to modify standard LBE equilibria so as to fulfil the following equalities:

$$\sum_i f_i^{\mathrm{e}} = \rho, \quad \sum_i f_i^{\mathrm{e}} c_{ia} = u_a. \tag{6.65}$$

With specific reference to the D2Q9 model, the sought set of equilibria is readily found [72]:

$$f_0^{\mathrm{e}} = \frac{4}{9}\left(\rho - \frac{3}{2}u^2\right), \tag{6.66}$$

$$f_{i1}^{\mathrm{e}} = \frac{1}{9}\left(\rho + 3u_i + \frac{9}{2}u_i^2 - \frac{3}{2}u^2\right), \tag{6.67}$$

$$f_{i2}^{\mathrm{e}} = \frac{1}{36}\left(\rho + 3u_i + \frac{9}{2}u_i^2 - \frac{3}{2}u^2\right), \tag{6.68}$$

where $u_i = c_{ia}u_a$ and the subscripts 1, 2 denote the nearest-neighbors and diagonal links, respectively. The crucial ingredient is that in order to match the second eqn (6.65), at variance with standard LBGK, *the density ρ is no longer prefactoring the polynomials in u_i*. Zou and He [72] used the above *D2Q9i* (*i* stands for incompressible) for various two-dimensional flows and demonstrate a dramatic drop of the compressibility error, from about 10^{-4} to 10^{-9}.

Coming back to pressure boundary conditions, the good news is that one is free to accept a space-dependent density, hence pressure, without hindering incompressibility simply because the compressibility errors has been 'swallowed' by the new definition of u_a. The flip side of the medal is that the trick only works at steady-state, since there is no way to reabsorb the time derivative of the fluid density. In this respect, these exactly incompressible LBGK models are best thought of as *false-transient, iterative schemes* similar to those widely in use for steady-state CFD computations [80]. Further development is surely needed to make these schemes competitive against well-rugged steady-state CFD solvers, such as multigrid methods. LBE is the natural-born weakly compressible dynamic method, and turning it into a competitive tool for steady-state incompressible flows is no mean feat. However, given the practical importance of this subject for engineering fluid dynamics, this effort is certainly a worthy one.

6.6 Exercises

1. A particle hits a solid wall at a given incident angle θ. Compute the particle-to-wall momentum and energy transfer as a function of the reinjection angle. How would you represent this on a lattice?
2. Write down the explicit Inamuro solution for sliding walls and analyze the solution as a function of the wall speed u_w. Discuss the admissible range of wall speeds.
3. Compute the explicit solution of the 3×3 system (6.45) and discuss its positivity domain.
4. Prove the expression (6.63).
5. Compute the explicit solution of the 3×3 system (6.61) and discuss how well it is posed.

7

FLOWS AT MODERATE REYNOLDS NUMBER

In this chapter we present the application of LBE to the problem of transitional flows at moderate Reynolds numbers. This is an important area of theoretical and applied fluid mechanics, one that relates mainly with the onset of nonlinear instabilities and their effects on the transport properties of the unsteady flow configuration. The regime of Reynolds numbers at which these instabilities take place is usually not too high, of the order of thousands, and consequently basically within reach of present day computer capabilities. Nonetheless, following the entire evolution of these highly dynamic transitional flows requires very long integrations with short time-steps which demand conspicuous computational power. Which is why, again, efficient numerical methods are in great demand. As we shall see, LBE has good credentials to meet this demand.

7.1 Moderate Reynolds flows in simple geometry

Let us consider a rectangular channel of length L and height H. The fluid enters the channel from the left side (inlet) and flows steadily down the channel under the effect of a constant pressure drop $\Delta P = P_{\rm i} - P_{\rm o}$ between the inlet and outlet section. At the top and bottom boundaries the flow speed vanishes (no-slip boundary condition) as a result of the drag exerted by the rigid walls. This is how dissipation balances the momentum input from the pressure gradient. If the Reynolds number is small enough, typically below 2000 in a real pipe, occasional fluctuations are unable to break the one-dimensional symmetry of the flow so that the nonlinear terms of the Navier–Stokes equations, $(\partial_b u_a u_b)$, vanish identically and have no effect on the evolution of the flow. Under these conditions,

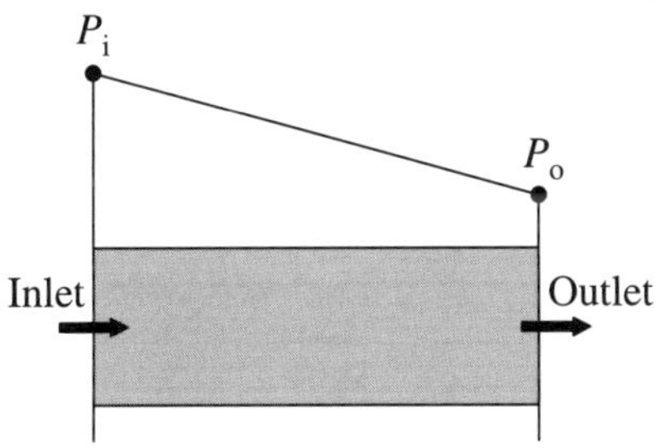

FIG. 7.1. Sketch of channel duct. The fluid enters from the left (inlet) and proceeds rightwards to the outlet under the effect of the pressure gradient.

the steady-state Navier–Stokes equations reduce to a simple momentum balance between dissipation and pressure gradients (Stokes equation):

$$\nu \partial_{yy} u = -\frac{\partial_x P}{\rho}. \tag{7.1}$$

Note that the one-directional flow configuration $u = u(y)$, $v = 0$, automatically fulfils the incompressibility condition

$$\partial_x u + \partial_y v = 0. \tag{7.2}$$

Within LBE, this divergence-free condition is fulfilled only approximately, typically to order Mach squared, because, as pointed out in Part I, *LBE is a weakly compressible approximation of the NSE.* We shall comment on this point in more detail later.

On the assumption that the pressure gradient is constant throughout the flow, say $G = (P_i - P_o)/\rho L$ = constant, the linear Stokes equation is easily solved to yield:

$$u(y) = U_0 \left(1 - \eta^2\right), \tag{7.3}$$

where

$$U_0 = \frac{GH^2}{8\nu}, \tag{7.4}$$

$$\eta = \frac{2y}{H}, \quad -\frac{H}{2} < y < \frac{H}{2}. \tag{7.5}$$

The corresponding stress tensor reduces to a single component

$$\tau_{xy} = \frac{\rho \nu U_0 \eta}{H}, \tag{7.6}$$

which attains its extrema at the wall ($\eta = \pm 1$) and vanishes in the centerline of the flow, $\eta = 0$.

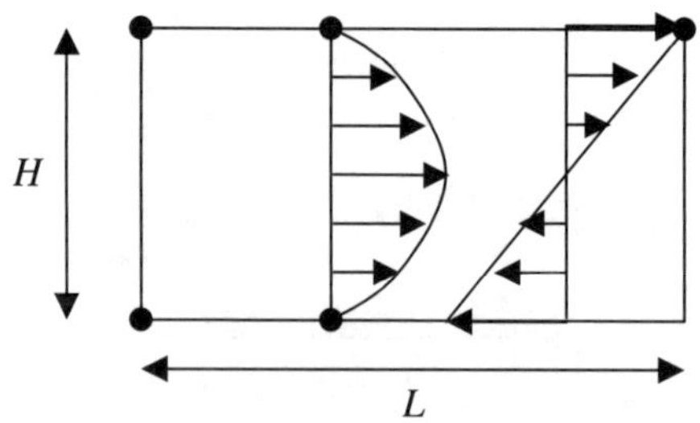

FIG. 7.2. Sketch of the Poiseuille flow and the corresponding stress tensor.

This solution is stable to transverse perturbations for Reynolds numbers up to about 2000. Beyond this value, dissipation is no longer able to damp spontaneous fluctuations and instabilities develop which drive the transition to turbulence.[19] The Poiseuille flow test is often used to calibrate the flow viscosity. In practice, one selects a given viscosity via the theoretical expression given in previous chapters, $\nu = c_s^2(1/\omega - 1/2)$, and verifies that the maximal speed attained by the flow agrees with the expression (7.4). In a way, the LBE code serves as a 'numerical viscosimeter' [81].

7.2 LBE implementation

Let us now present in some more detail how to set up a LBE simulation in practice. For the sake of concreteness let us refer to the 2D version of the FCHC LBE, namely a nine speed model (rest particles included) with elementary mass 4 for rest particles and direct links, and mass 1 for diagonal links.

The first problem we face in the LBE implementation is the representation of the pressure drop G. This bears on a rather delicate theoretical issue that was left silent in the theoretical presentation. As we have learned in Part I, the LBE model applies to quasi-incompressible flows because of the low-Mach expansion that lies behind the whole LBE theory presented so far.

Another thing we have learned is that since there is no potential energy between the LBE 'molecules', the LBE fluid obeys an ideal gas equation of state $P = \rho T$. Since the flow is isothermal (that is, $T = c_s^2 = \text{constant}$), it is clear that a regime of no density variation, as required by incompressibility, implies no pressure variations as well. So, the conundrum is: *where does the pressure gradient come from?*

The answer is that the pressure gradient $\partial_a P$ in LBE is entirely due to pressure fluctuations acting on top of a uniform background P_0, namely

$$\partial_a P = \partial_a P_0 + \partial_a p' \sim \partial_a p', \tag{7.7}$$

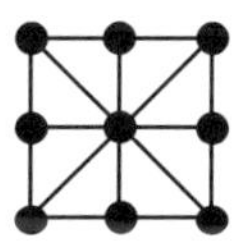

FIG. 7.3. The 9-state 2D FCHC lattice.

[19]Practical hint: In numerical practice, it is often hard to trigger these instabilities even well beyond the critical threshold. A useful remedy is to generate artificial perturbations of the initial conditions, such as appropriate combinations of sines and cosines. In planning these perturbations, one should not forget compliance with the divergence-freedom condition.

where the fluctuations $p'/P_0 = \rho'/\rho_0$ are of the order of the Mach number squared. In terms of kinetic theory, pressure gradients are entirely in charge of the non-equilibrium diagonal component of the momentum flux tensor P_{ab}^{ne}. As a result, it is difficult to sustain a significant and systematic pressure gradient within the LBE flow. *This is the price to pay for being dispensed from solving the pressure field for each instantaneous flow configuration* as in standard Computational Fluid Dynamics (CFD). The non-equilibrium kinetic fluctuations may break divergence-freedom, but, owing to the quadratic dependence on the flow speed, this violation can normally be tolerated provided the Mach number is sufficiently small, say below 0.1.

The lack of exact incompressibility is not unique to LBE, but a common feature of all weakly compressible approximations of the Navier–Stokes equations. What is unique to LBE, and makes life a bit more difficult here, is that the LBE sound speed is not so easily disposable as in conventional CFD, but it is frozen to values $c_s \sim O(1)$ in lattice units.

As a result, at variance with conventional computational fluid dynamics, it is not possible to send the sound speed virtually to infinity so as to support significant pressure drops with only minor attendant density changes $\delta\rho = \delta P/c_s^2$ (*artificial compressibility* methods [82]).

It is worth recalling that the above relation is nothing but the definition of the sound speed, namely the pressure change per unit change in density. As a result, very small violations of the incompressibility conditions can be achieved even in the presence of significant pressure changes. According to the continuity equation, rewritten in the form

$$\partial_a u_a = -u_a \partial_a \ln \rho, \tag{7.8}$$

we see that departures from divergence-freedom (solenoidality) are correspondingly small. The standard way out of this difficulty is to *mimic* the pressure gradient by means of an *equivalent volume force*, say F_a, designed in such a way as to produce the same momentum input to the flow as the true gradient.

Formally, this is tantamount to invoking an external force term F_a in the Boltzmann equation. The effect of this force on the generic macroscopic quantity $\Phi(x,t) \equiv \int f(x,v,t)\phi(v)\, dv$ is given by

$$\frac{\delta\Phi(x,t)}{\delta t} = \langle F_a \frac{\partial f}{\partial v_a}, \phi(v)\rangle = \langle -f, \frac{\partial}{\partial v_a}(F_a \phi)\rangle, \tag{7.9}$$

where $\phi(v)$ is any suitable function of the velocity coordinates and $\langle \cdot,\cdot \rangle$ denotes scalar product in velocity space.

It is therefore possible to tailor the external force in such a way as to impose the required conditions on the macroscopic variation $\delta\Phi$.

In a Poiseuille flow this is easy: all we need is to bias the collision rule so as to enhance the populations moving along the pressure gradient and place a corresponding penalty on counterstreamers. Let Δf_i be the bias per population at a given site, the modification reads as follows:

$$f_i' = f_i + \Delta f_i, \tag{7.10}$$

where it is understood that f_i refers to the state *after* streaming and collision.

For convenience, we define a set of allied variables $g_i \equiv \Delta f_i/\Delta t$, the population change rates.

With reference to a two-dimensional projection of the FCHC hypercube, it is readily checked that mass and momentum conservation are fulfilled under the following conditions (counterclockwise numbering of the discrete speeds is assumed, starting with rightward motion $\vec{c}_1 = (1,0)$):

$$g_1 = 4g_2 = 4g_8 = \tfrac{1}{12} g\rho, \tag{7.11}$$

$$g_5 = 4g_4 = 4g_6 = -\tfrac{1}{12} g\rho, \tag{7.12}$$

$$g_0 = \quad g_3 = \quad g_7 = 0, \tag{7.13}$$

where the (frequency) amplitude g is fixed by the condition of imparting the desired momentum input to the fluid.

The mass input (per unit time and volume) due to the source term is $\dot{\rho} \equiv \sum_i g_i$ and it sums to zero as it should. The same does the transversal momentum input $\dot{J}_y \equiv \sum_i g_i c_{iy} = 0$.

The longitudinal momentum input is given by $\dot{J}_x^+ \equiv \sum_i g_i c_{ix} = \rho g c$ where $c = 1$ is explicitly retained to help dimensional transparence.

The actual value of g is singled out by equating the momentum input per unit volume $gc\rho/2$ to the corresponding viscous loss $\dot{J}_x^- = \rho\nu\partial_{yy}u$.

Upon using the expression (7.3) for the velocity field, we obtain

$$\dot{j}_x^- = \frac{K\rho\nu U_0}{H^2}, \tag{7.14}$$

where $K = 1/8$ for the specific geometry in point.

The final solution reads:

$$g = K\frac{U_0}{c}\frac{\nu}{H^2}. \tag{7.15}$$

By introducing the channel traversal time $t_T = H/U$, the above relation can be recast in the more intuitive form

$$gt_T = K\frac{\nu}{cH}. \tag{7.16}$$

This solves the problem of injecting the desired amount of momentum input within the flow.

On practical grounds, the change in the computer program is really trivial, just a few extra lines to add the source term g_i to each population after (or within) the collision routine.

A few remarks are in order.

First, it should be appreciated that the procedure outlined here is fairly general and easily extends to more complex situations described by space, time and also state-dependent forces of the form $F_a(x, t; M(x, t), \nabla M(x, t))$ where M denotes any generic macroscopic quantity. *This feature proves exceedingly useful to enrich the basic LBE schemes with additional physics and it has been used for a variety of generalized fluid dynamic applications such as: magnetohydrodynamics, thermal convection, granular flows, suspensions, flows with chemical reactions, flows with phase transitions, to name but a few.* A selected collection of these applications will be presented in the 'beyond' part of this book.

The procedure is general, but one must nonetheless make sure that the source term does not spoil stability. With specific reference to the case in point, one must guarantee that g is sufficiently small to prevent counterstreaming populations from going negative (or co-streaming ones to exceed the unit value).

Let us analyze this issue in a more detail. The stability condition on the external forcing terms amounts to requiring that the absolute change of the current density in a single time-step (made unit here and throughout) be a small fraction of the current itself. In equations:

$$\frac{\dot{J}}{J} \ll 1 \tag{7.17}$$

or, in view of the expression (7.15), $K\nu U/H^2 \ll 1$.

It is easily checked that for all parameter values of interest this is safely fulfilled since: $\nu < O(0.1)$, $U \sim O(0.1)$ and $H \sim O(100)$, so that the left-hand side is of order 10^{-6} or less. A similar analysis applies to the populations: $\delta f/f \sim (\rho\nu UK/H^2)/\rho \sim KU\nu/H^2$ which is again much smaller than 1 under all conditions of interest. Note that the group ν/H^2 is the inverse momentum diffusion time τ_d, and consequently the inequality (7.17) is nothing but the familiar stability condition $\Delta t < \tau_d$ typical of explicit schemes.

7.3 Boundary conditions

Coming back to the Poiseuille flow simulation, we still have to impose boundary conditions. To make life easy, we select periodic conditions at the inlet/outlet and no-slip on the top and bottom rigid walls. Periodic boundary conditions are implemented by simply reinjecting at the inlet the populations leaving the outlet. With reference to the Sketch 7.1,

$$f_{1,2,8}\,(\mathrm{I}) = f_{1,2,8}\,(\mathrm{W})\,, \tag{7.18}$$

$$f_{4,5,6}\,(\mathrm{O}) = f_{4,5,6}\,(\mathrm{E})\,, \tag{7.19}$$

where I and O denote the Inlet and Outlet columns and W and E the leftmost and rightmost (West and East) fluid columns, respectively. No-slip boundary conditions are implemented by bouncing-back the populations leaving the walls:

6	n	n	n	n	n	n	n	n	n	n	n	n
5	i	f	f	f	f	f	f	f	f	f	f	o
4	i	f	f	f	f	f	f	f	f	f	f	o
3	i	f	f	f	f	f	f	f	f	f	f	o
2	i	f	f	f	f	f	f	f	f	f	f	o
1	i	f	f	f	f	f	f	f	f	f	f	o
0	s	s	s	s	s	s	s	s	s	s	s	o
	0	1	2	3	4	5	6	7	8	9	10	11

SKETCH 7.1. Map of a channel flow with 10×5 fluid sites, each with its identity card. f: fluid site, i: inlet buffer, o: outlet buffer, n: north buffer, s: south buffer.

$$f_{6,7,8}\,(\mathrm{N}) = f_{2,3,4}\,(\mathrm{N}-1)\,, \tag{7.20}$$

$$f_{2,3,4}\,(\mathrm{S}) = f_{6,7,8}\,(\mathrm{S}+1)\,, \tag{7.21}$$

where N and S denote the North and South rows and N − 1 and S + 1 the immediate lower-lying and upper-lying fluid rows, respectively.[20] This naive implementation is less than perfect in terms of accuracy but good enough for the present purposes.

A simple LBE Poiseuille flow is represented in the Fig. 7.4 which refers to the following parameters: $L = 128$, $H = 32$, $\nu = 0.1$, $U_0 = 0.1$. This has been produced with the sample code given in Part I, which the reader is kindly recommended to try on his own.

7.4 Flows past obstacles

The Poiseuille flow is an easy test because only the linear components of the Navier–Stokes equations take part to the dynamics of the system, the nonlinear ones being consistently left dormant by the symmetries of the problem.

This symmetry spontaneously breaks down for Reynolds numbers above 2000, marking the transition to chaos (loss of temporal coherence) and subsequently full turbulence (loss of temporal and spatial coherence). The loss of symmetry, and the resulting instabilities can be triggered at much lower Reynolds in the presence of solid obstacles. This type of turbulence is often sought after in practical engineering devices aimed at heat and mass transfer, such as heat exchangers or catalytic converters, simply because turbulence means enhanced transport,

[20]Practical remark: it is good practice to assign each node of the lattice its own specific identity card, say Fluid, Boundary, Inlet, Outlet and so on (see Sketch 7.1). This notation, if redundant for simple geometries like the one discussed here, becomes instrumental to an efficient management of the data structures appropriate for more complex geometries.

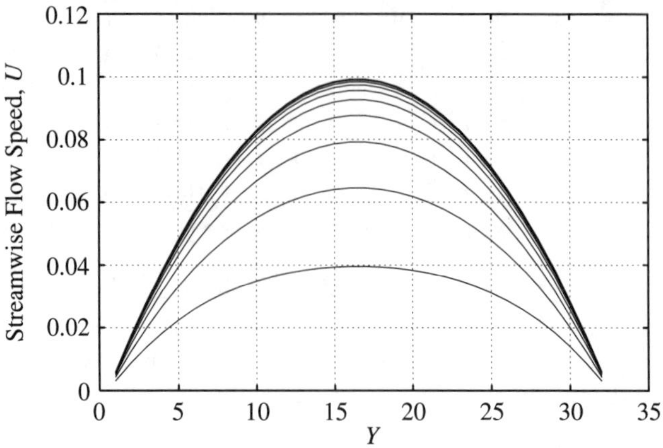

FIG. 7.4. Time development of a LBE-simulated Poiseuille flow.

hence more efficient heat and mass transfer. As an example of such a situation, we shall discuss the flow in a plane channel containing a periodic array of identical obstacles along the streamwise direction.

As anticipated above, the idea is to promote flow instabilities at low Reynolds numbers, about one order of magnitude below the critical Reynolds of a free channel flow. This is achieved by exciting large scale instabilities known as Tollmien–Schlichting waves triggered by the shear layer instabilities past the thin plate [83].

This flow is obviously much richer than the laminar Poiseuille flow since all nonlinear terms are now in action. Their effect is well visible in terms of coherent structures (vortices) ejected in the wake of the obstacle. From the point of view of the LBE implementation, all proceeds exactly as in the case of the Poiseuille flow, with the minor addition of an internal boundary, the solid obstacle, thin plate. This is easily accommodated in the LB code by adopting the same technique used for the north–south walls of the channel.

This is a nice feature of LBE: *the fully nonlinear regime does not require any extra coding.* For the sake of concreteness we shall again refer to the 2D version of the FCHC lattice. Assuming for simplicity the thin plate consisting of a vertical shell spanning h sites in height ($y_s < y < y_n$) and only one lattice unit in width ($x_c < x < x_c + 1$), the non-slip internal boundary condition reads as follows:

- West-side story:

$$f_5(x_c, y) = f_1(x_c, y), \qquad y = y_s, \dots, y_n, \tag{7.22}$$

$$f_6(x_c, y) = f_2(x_c, y), \qquad y = y_s, \dots, y_n, \tag{7.23}$$

$$f_4(x_c, y) = f_8(x_c, y), \qquad y = y_s, \dots, y_n. \tag{7.24}$$

- East-side story:

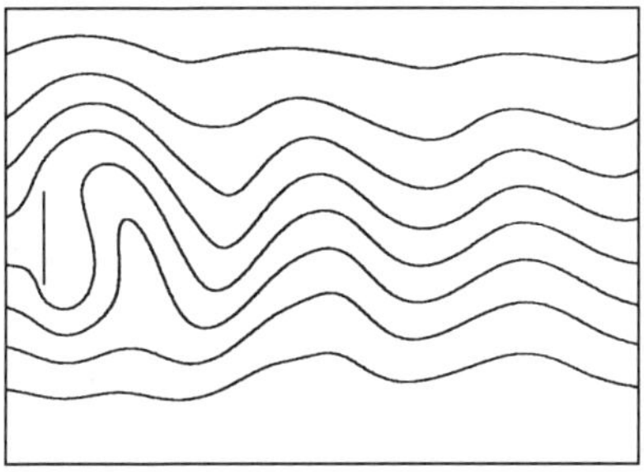

FIG. 7.5. Streamlines of a moderate Reynolds flow past a thin plate, (after [41]).

$$f_1(x_c+1,y) = f_5(x_c+1,y), \qquad y = y_s,\ldots,y_n, \tag{7.25}$$

$$f_8(x_c+1,y) = f_4(x_c+1,y), \qquad y = y_s,\ldots,y_n, \tag{7.26}$$

$$f_2(x_c+1,y) = f_6(x_c+1,y), \qquad y = y_s,\ldots,y_n. \tag{7.27}$$

Corners require special care, as witnessed by the different bounds of the y intervals. Since we consider a periodic array of obstacles, the inlet and outlet boundary conditions are cyclic and can be implemented exactly as in the case of the Poiseuille flow.

A specific example of LB simulations for this type of flow was first presented in [84] for a series of plates 1:10 of the channel height H, centered and laid down a distance $3.3\,H$ apart along the streamwise direction.

The stability of the flow was inspected by placing a series of probes at $x_n = x_c + L/n$, $n = 8, 4, 2$ and $y = y_c$, and monitoring the transversal component of the flow speed as a function of time $v = v(t)$. Under stable conditions this signal

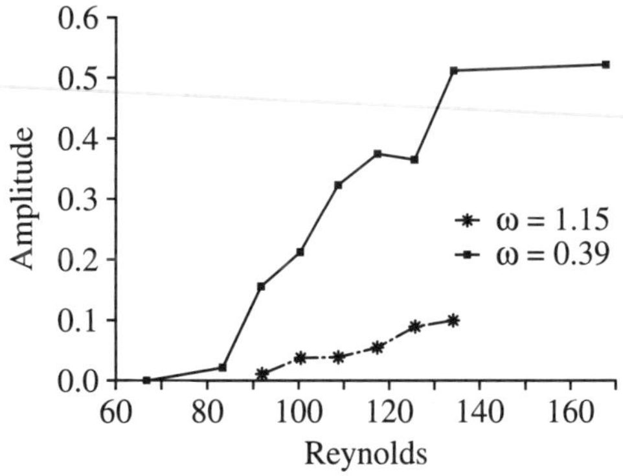

FIG. 7.6. Amplitude of the bifurcated modes of a moderate Reynolds flow past a thin plate (after [84]), as a function of the Reynolds number.

is flat, only to take an oscillatory waveform $v = A\sin(\omega t)$ for Re above a critical value Re_c.

As the Reynolds number is furthered a second harmonic develops, then a third one and so on until the flow becomes turbulent. The typical dependence of the harmonic amplitudes on the square root of the excess Reynolds number $Re - Re_c$ predicted by the Landau–Ginzburg theory is qualitatively visible in Fig. 7.6.

Extensive work on time-dependent LBE flows past obstacles can be found in L. S. Luo's thesis work [85].

7.5 More on the pressure field: Poisson-freedom

Before closing, some further remarks on the pressure gradient implementation is in order. Earlier on in this chapter, we introduced the notion of 'functional equivalence' between the pressure gradient and a volume force designed in such a way as to produce the same momentum input. For the Poiseuille flow, such an equivalence makes perfectly sense since we already know the solution, and we can gauge the force accordingly.

In general, this equivalence is ill-posed: how can we possibly compute a force equivalent to an unknown field?

Still, operationally the procedure makes sense. The reason is that the pressure field in an incompressible flow *does not obey a prognostic equation* [86] but rather serves as a dynamic constraint enforcing instantaneously the divergence-freedom of the flow. This constraint acts instantaneously precisely because sound waves propagate at a virtually infinite speed in an incompressible flow.

In mathematical terms this is readily checked by taking the divergence of the Navier–Stokes equation and taking into account the solenoidal condition (7.2). This yields

$$\Delta P = -\nabla \cdot (\vec{u} \cdot \nabla)\,\vec{u}, \tag{7.28}$$

where ∇ is the spatial gradient.

This shows that once the velocity field is known, the pressure field follows instantaneously from the solution of the above elliptic Poisson equation. In terms of our equivalence procedure, this means that we are entitled to replace the term ∇P with an equivalent volume force, *as long as this force does not break the incompressibility condition.*

This is a necessary but not sufficient condition, though. To be quantitatively consistent, the volume force should match exactly the gradient of the pressure field everywhere within the flow. To this end, one should solve the Poisson problem at each time-step, something we would rather steer clear from, since this is a very costly operation taking a major fraction of the whole CPU time in CFD codes.

The operational short cut is to give up the idea of matching ∇P pointwise, achieving nonetheless the same Reynolds number. To that purpose, all we need is to adjust the volume force so as to attain the desired Reynolds based on some

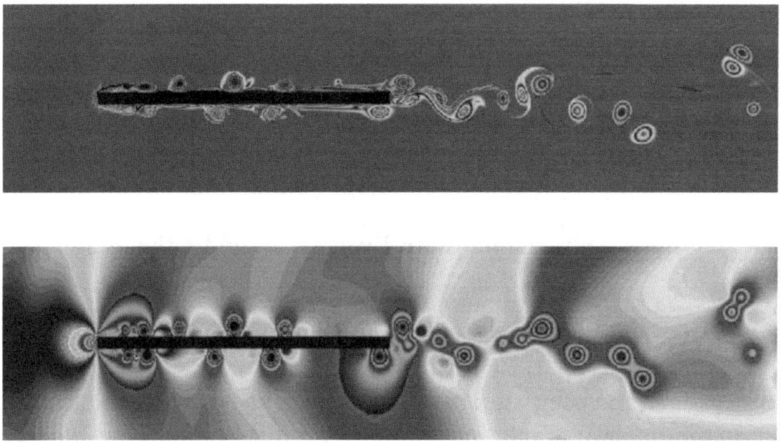

FIG. 7.7. Vorticity and pressure patterns of a von Karman street past a plane slab. (Courtesy of L. S. Luo.)

average flow speed. This average flow speed cannot be known beforehand, but it can easily be adjusted in a few cut-and-try experiments. This is not very elegant, but certainly more efficient than solving the Poisson problem at each time-step. Just to clarify things, let us refer to the transitional flow past a plate discussed previously.

We know beforehand that by calibrating the volume force on a Poiseuille flow, according to eqn (7.15), we would obtain an effective Reynolds number smaller than desired simply because the internal plate contributes a lot of additional dissipation. Let ν_{eff} be the effective viscosity of the flow accounting for the additional dissipation and call $E = \nu_{\text{eff}}/\nu$ the enhancement factor over the molecular value. We do not know this value exactly, but very often we have some educated guess for it, suppose a factor $E = 2$. What we can do, then, is to enhance the volume force by a factor E and achieve a Reynolds in the vicinity of the desired one. This permits computing integral figures such as drag and lift coefficients of solid bodies of assorted shapes. A practical way of doing this is to define a a control volume, say an enboxing rectangle, and compute the momentum change per unit volume across the box as a surface integral of the momentum flux tensor P_{ab}:

$$F_a = \int_{S_{\text{ext}}} P_{ab}\, \mathrm{d}S_b + \int_{S_{\text{int}}} P_{ab}\, \mathrm{d}S_b, \tag{7.29}$$

where $P_{ab} = \rho u_a u_b + P\delta_{ab} + P'_{ab}$ is the Navier–Stokes momentum flux tensor and 'ext/int' label the external boundaries of the control volume (the perimeter of the box in our example) and the internal ones (the surface of the body, a straight segment in our case).

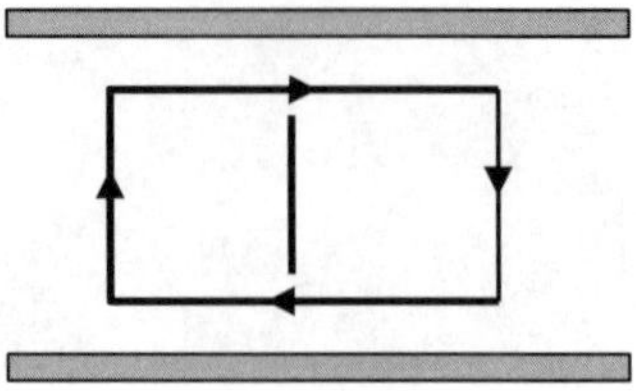

FIG. 7.8. Sketch of a control volume around a thin plate.

A nice feature of the LB method is that *the momentum flux tensor is available locally* from its very definition: $P_{ab} = \sum_i f_i c_{ia} c_{ib}$.

Consequently, the computation of the surface integrals does not involve any explicit space derivative, since these are implicitly contained in the non-equilibrium component of the momentum flux tensor. The convective term $\rho u_a u_b$ contributes nothing to the external surface integral if the control volume is large enough to credit the assumption that the flow is one-directional along all four sides of the box. On the top and bottom surfaces the integral is identically zero because the vertical component $v = 0$. At the inlet/outlet surfaces the integrals are nonzero but they cancel each other on account of the continuity equation $u_{\text{in}} = u_{\text{out}}$. The integral on the body surface is also identically zero by virtue of the no-slip boundary condition $u = v = 0$.

Once the equivalent pressure drop $\Delta P = F_x/H$ across the control volume is computed, one obtains the desired value of the drag coefficient $C_D = 2\Delta P/(\rho U^2)$.

The strength of this procedure is that all 'universal' quantities, the ones controlled basically by the Reynolds number alone, can be computed even if the external force does not solve the Poisson constraint pointwise.

Coming back to general issues, it is clear that the volume–force procedure is ill-suited if we are to be keen on a precise value of the Reynolds number, but works quite well if the scope is rather a parametric scan over a range of Reynolds numbers. Since the latter scenario is at least as relevant as the former one, the volume–force approach appears perfectly justified in spite of its markedly pragmatic character.

This leaves a loose end for those engineering studies where the internal distribution of the pressure field is needed in detail. Even more so, if one considers that very often, engineering practice requires the specification of pressure boundary conditions.

The weakly compressible LBE is ill-positioned to meet these needs, and must be turned either into an *exactly* incompressible method or into a thermodynamically consistent compressible method.

The latter goal represents a significant challenge and will be discussed in the final part of this book. As for the former, a few proposals have appeared in the literature [87], but in this author's opinion, this subject is still highly in flux and the discussion of these methods would defy our commitment to simplicity.

Consequently, the interested reader is directed to the original papers [87].

In conclusion, although a lot more comparative work remains to be done, it appears as though LBE is a simple and promising tool for the efficient computation of isothermal, incompressible, transitional flows. Much of this efficiency stems from doing away with the Poisson problem for the pressure field. This ensures high numerical efficiency, which is mostly treasured for long-time integrations of dynamically unstable flows. The flip side is the inability to reconstruct the internal structure of the pressure field if the incompressibility condition has to be forced strictly, and not just up to fluctuations of the order of Ma^2. This places a limit on the use of LBE for the assessment of non-universal features of the fluid flow, but makes it an excellent candidate to capture the universal ones.

7.6 Exercises

1. Set up the thin plate calculation using the sample code provided in Part I and run it for various Reynolds numbers between 10–100. Monitor the velocity field in a series of stations along the centerline $y = H/2$. At what value of Re do you start seeing a nonzero vertical speed v?
2. For the above simulations, monitor the local density fluctuations. What values do you expect to see: 10^{-1}, 10^{-3}, 10^{-6}?
3. Run the case of a point-like obstacle and compare the patterns of the velocity and vorticity fields of the thin plate. Which one gets farther? Why?

8

LBE FLOWS IN DISORDERED MEDIA

The study of transport phenomena in disordered media is a subject of wide interdisciplinary concern, with many applications in fluid mechanics, condensed matter, and life and environmental sciences as well. Flows through grossly irregular (porous) media is a specific fluid mechanical application of great practical value in applied science and engineering. It is arguably also one of the applications of choice of the LBE (and LGCA) methods. The dual field–particle character of LBE shines brightly here: the particle-like nature of LBE (populations move along specific particle trajectories) permits a transparent treatment of grossly irregular geometries in terms of elementary mechanical events, such as mirror and bounce-back reflections. On the other hand, the field-like nature of LBE (populations carry smooth, hydrodynamic information) permits achievement of fluid dynamic behavior also in tiny interstitial regions of the flow which are hardly accessible to the macroscopic approach. These assets were quickly recognized by researchers in the field, and still make of LBE (and LGCA) an excellent numerical tool for microhydrodynamic flows in porous media.

8.1 Flows through porous media

Flow through porous media is a typical *multiscale* phenomenon, encompassing several orders of magnitude in scale and involving three and perhaps even four basic levels of description:

- *Microscopic* (molecular level);
- *Mesoscopic* (single pore level);
- *Macroscopic* (many pores, sample size level);
- *Megascopic* (field size level).

An all-embracing approach is clearly out of the question, and consequently specific techniques have been developed for each of these levels.

At the mega/macroscopic level the internal structure of the porous media is ignored and all microscopic knowledge is lumped into coarse-grained quantities, such as the volume of solid/fluid, via so-called *homogenization* procedures. A typical output of this analysis is the aptness of the porous media to let fluid flow across it, as a function of the aforementioned average quantities. The megascopic level is usually described by a heterogeneous assembly of macroscopic units, with locally varying transport properties.

The chief relation at the macroscopic level is Darcy's law:

$$Q = -\frac{K}{\mu}\frac{\Delta P}{L}, \tag{8.1}$$

where $Q = UA$ is the flow rate across the surface of area A, driven by a pressure drop ΔP across the length L. Here μ is the dynamic viscosity of the fluid and K, the sought macroscopic transport property, is the so-called *permeability* of the porous media. In general K is a tensorial function of the spatial location $\vec{x}$, $K_{ab}(\vec{x})$ (heterogeneous media), but we shall restrict our analysis to the simple homogeneous isotropic case.

The mesoscopic methods come mainly from statistical mechanics, say network and percolation theory [88]. At this scale, the porous media is represented as a network of micropores traversed by the fluid flow. The permeability of each single micropore/micropipe takes the form:

$$k(h) = Ch^2, \tag{8.2}$$

where h is the pore size (width of the micropipe) and C is a numerical constant depending on the specific geometry of the microchannel. For instance, as shown in the previous chapter, the Poiseuille flow between parallel plates yields $C = 1/8$. Assuming the pore network obeys a probabilistic distribution $P(h)$, the effective permeability is obtained as a weighted average over this distribution

$$K = \int_{h_{\min}}^{h_{\max}} P(h)\, k(h)\, \mathrm{d}h, \tag{8.3}$$

where $h_{\min}$ and $h_{\max}$ are the smallest and largest size of individual pores, respectively.

Of course, at this level, an educated guess for the probability distribution $P(h)$ is needed. Typically, this is taken in the form of a power-law, reflecting an assumption of fractality of the underlying porous media. This approach is elegant and well placed to borrow most of the powerful tools of modern statistical mechanics, such as percolation theory, renormalization group methods and others [89]. Nevertheless, none of these powerful methods can account for the full details of what happens inside the micropipes. In line with percolation theory, the typical network model would place solid/void sites on a lattice (not to be confused with lattice gas cellular automata, or lattice Boltzmann!) and give a probability p that a given pipe be traversed by the fluid. The factor p responds to morphological constraints, but is totally unaware of hydrodynamic details inside the pipe. What happens inside the channel is taken for granted, in the sense that

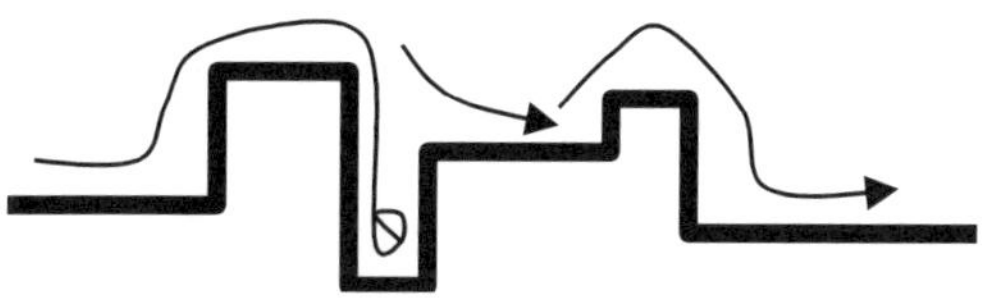

FIG. 8.1. Sketch of geometrical microdefects on a solid wall.

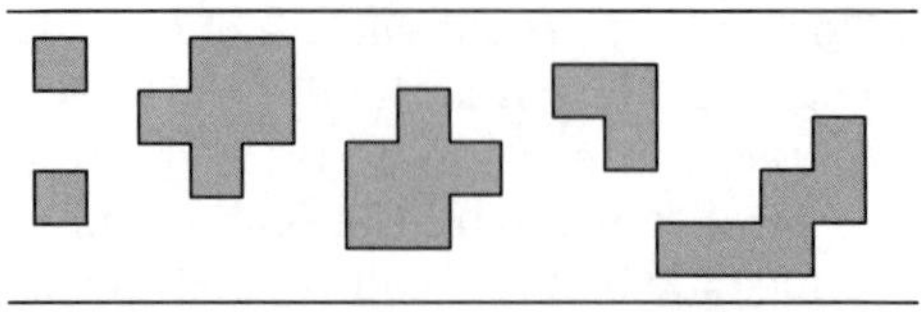

FIG. 8.2. Sketch of a disordered media.

a fluid link is associated with a microscopic permeability which is to be taken as an input parameter of the model. On the other hand, if (as is often the case) the network topology is complex and tortuous, the microscopic flow inside the channel may depart significantly from simple Poiseuille flow and so might the micropermeability $k(h)$.

In other words, eqn (8.3) is made up of two parts, the distribution function $P(h)$, which carries the *morphological* information, and the micropermeability $k(h)$ which carries the *hydrodynamic* one.

The network approach focuses on the former and takes the latter for granted. This is certainly wise, but not foolproof. In fact, as the network complexity increases, the two parts, morphological and hydrodynamical, can no longer be treated independently. Microflows inside tortuous meanders may give rise to additional losses of momentum (*microturbulence*) which might sum up coherently to produce appreciable variations of the coefficient in front of Darcy's law, and sometimes, even yield nonlinear deviations from Darcy's law altogether.

This is precisely the framework in which LGCA and LBE may prove valuable sources of *genuinely microhydrodynamic information*. By simulating the Navier–Stokes equations in actual microgeometries, they can tell us whether microhydrodynamic details are irrelevant to Darcy's law or if they produce significant deviations instead.

8.2 LBE flows through porous media

Let us now discuss how to set up a LBE simulation of porous media. We shall consider a square sample of porous material (say a rock) and put across it a low Reynolds two-dimensional fluid flow driven by an external pressure gradient $G = \Delta P/L$,[21] L being the linear size of the sample.

For the sake of simplicity the internal solid is represented by a collection of squares. This is of course only a sketch of a real porous media, but one that fits our present purposes well. In general, full detailed knowledge of the porous structure can be encoded into a porosity function $p(\vec{x})$, a Boolean array flagging the fluid/solid status of the lattice site $\vec{x}$:

$$p(\vec{x}) = 1 \quad \text{(fluid)}, \tag{8.4}$$

$$p(\vec{x}) = 0 \quad \text{(solid)}. \tag{8.5}$$

[21] We assume an incompressible fluid whose density can be safely taken as $\rho = 1$.

In the jargon of computer graphics this is known as a 'bitmap'; that is, a map of bits (binary digits). From knowledge of this bitmap, one can derive all statistical indicators, such as for instance the spatial autocorrelation

$$A(\vec{r}) = \sum_{\vec{x}} p(\vec{x} + \vec{r})\, p(\vec{x})\,. \tag{8.6}$$

Of particular interest for practical applications is the *void fraction* ϕ, defined as the void (fluid) versus total volume of the sample:

$$\phi = \frac{V_f}{V_s + V_f} \tag{8.7}$$

and the *solid fraction*, its complement to one: $\phi_s \equiv 1 - \phi$. Void fraction and *porosity* will be used interchangeably in the sequel.

It is easily checked that $\phi = A(0)$.

In nature these values change greatly from material to material. A few actual figures, from [90], are reported in Table 8.1.

The porosity is only a zeroth-order indicator of the porous geometry. Other useful quantities are the minimum, maximum and average grain size, where by grain we mean a cluster of solid sites (squares in our simple example) [89]. Another important property of the porous structure is the *tortuosity*, roughly speaking, a measure of the average curvature and torsion of the micropipes. To make life easy, here we shall consider a *monodisperse* media where all grains (squares) have the same size h, with their centers laid down randomly across the sample (random media).

The chief theoretical (and practical) question is to predict how much fluid we can push across a sample of section A by applying a given pressure drop ΔP across the longitudinal size L of the sample. As we said, this is measured by the *permeability* K of the rock sample defined as ratio between the flow rate $Q = UA$ and the driving pressure gradient $G = \Delta P / L$.

Generally, the permeability K is sought in the form of a functional (constitutive) relation with the *porosity* ϕ:

$$K = K(\phi)\,. \tag{8.8}$$

TABLE 8.1. Typical porosity of some representative real materials.

Material	Porosity (%)
Sandstone	10–20
Clay	45–55
Gravel	30–40
Soils	50–60
Sand	30–40

This is of course only a simplification that does not take into account other important morphological factors, but it is nonetheless very useful for many practical purposes. The correlation $K = K(\phi)$ is either measured experimentally or, alternatively, it is derived analytically by means of statistical models based on network representations of the porous geometry [88].

The range of validity of these models is typically associated to very low or very high solid fractions ϕ_s, intermediate, non-asymptotic regions being hardly accessible to analytical treatment. This leaves wide scope for the application of direct numerical simulation.

8.3 Setting up the LBE simulation

As for all simulations, one has first to decide about initial and boundary conditions. The latter typically correspond to a uniform flow $u(x, y) = U$, $v(x, y) = 0$. The initial populations can then be taken in the form of local equilibria corresponding to the above velocity field.

Next come the boundary conditions.

External boundaries can be dealt with as for solid bodies, the main practical issue being the LBE implementation of internal solid boundary conditions, the porous media. An effective strategy consists of tagging all boundary solid sites with a special flag and then associating to each of these nodes a list with the discrete speeds involved in the boundary procedure. Once this topological information, which is built once and for all at the outset, is available, the boundary conditions are implemented via a straightforward sweep over the boundary nodes. A possible pseudo-code illustrating the idea is given below:[22]

```
=======================================================================
For b = 1,NB{              /* sweep thru all boundary nodes           */
  xs   = ISOLID(b)
  ys   = JSOLID(b)

  For k=1,NF(b){           /* scan thru all active links connecting the
                              given boundary node with a fluid site   */
    xf=IFLUID(b,k)
    yf=JFLUID(b,k)
    is=speed(b,k)          /* set of discrete speeds emanating from
                              the solid boundary site b               */
    if=Mirror(is)          /* set of mirror conjugate discrete
                              fluid speeds                            */
    F(is,xs,ys)=F(if,xf,yf)    /* bounce-back */
  }
}
=======================================================================
```

[22] *Cave canem*! This pseudo-code is only meant to convey the general ideas in the simplest possible way. As such, it has not been tested as a real piece of code, a task which is left to the brave reader.

The data structure is as follows:

- `NB`: number of boundary solid sites.
- `NF(b)`: number of active links at site b (discrete speeds) connecting the boundary node b to fluid sites.
- `ISOLID(b)`, `JSOLID(b)`: indirect address of the x, y Cartesian coordinates of the solid boundary site.
- `IFLUID(b,k)`, `JFLUID(b,k)`: indirect address of the x, y Cartesian coordinates of the k-th fluid site interacting with the boundary site along the link identified by the mirror conjugate discrete-speed indices if, is.
- `is=speed(b,k)`: actual index of the discrete speed emanating from solid site site b along the k-th entry of the set of active links.
- `if`: actual index of the discrete speed emanating from the fluid site interacting with b along the link (is, if).
- `F(i,x,y)`: i-th population at Cartesian site x, y.

For the sake of concreteness let us illustrate the idea with an example:

```
7   f  f  f  f  f  f  f
6   f  f  f  b  b  b  f
5   f  f  f  b  s  b  f
4   f  f  f  b  s  b  f
3   f  f  f  B  b  b  f
2   f  f  f  f  f  f  f
1   f  f  f  f  f  f  f

    1  2  3  4  5  6  7
```

SKETCH 8.1. Numbering of the fluid (f), solid boundary (b) and dead-solid (s) sites.

At the south-west corner of the solid block ($xs = 4$, $ys = 3$), denoted by B, we have (as usual, counterclockwise numbering starting from east is assumed):

```
xs=4, ys=4, b=1:

               xf  |  yf
               ----|----
nw:  k=1,      xs-1 , ys+1
w :  k=2,      xs-1 , ys        f
sw:  k=3,      xs-1 , ys-1      f B
s :  k=4,      xs   , ys-1      f f f
se : k=5,      xs+1 , ys-1
```

This means that only the north-west, west, south-west, south and south-east fluid neighbors are involved in the population exchanges needed to set the desired boundary condition, typically $u = v = 0$ at the solid site.

An elementary way to impose this boundary condition is simply to bounce-back the populations along the direction they were coming from (standard counterclockwise speed numbering is assumed, starting from eastward propagation).

```
F(is,xs,ys) = F(if,xf,yf)  for:

            is | if
          ----|----
nw:         4    8
w :         5    1
sw:         6    2
s :         7    3
se :        8    4
```

By scanning across the whole list of nodes we can store the set of discrete directions pointing to fluid sites and indices of the corresponding fluid sites. This is a bit of a cumbersome pre-processing operation, but fortunately one that needs be performed only once.

All we need to do at run time is to read-off this information from the list and process only *active* sites, namely fluid sites and solid boundary sites. This data structure is quite general and permits the addressing of fairly irregular and disordered geometries.

Apart from the issue of indirect addressing (which could be dispensed with by using more flexible data structures in C language), it is also fairly efficient.[23] Care needs to be exercised to guard against topological defects, bottlenecks or dead ends, such as the one depicted in Sketch 8.2.

It is clear that this geometry does not allow *any* net flow across the sample simply because there is no fluid path (a path composed by fluid sites only) connecting the inlet and outlet sections of the sample. One such path materializes just by removing a single solid site, the one labeled S in the above picture.

Even then, one has to be careful because the communication channel between fluid regions is only one lattice site wide, which makes the whole issue of a hydrodynamic representation highly questionable, unless a very high R^* (small free path) is adopted. Differently restated, *small pores may not be amenable to*

[23]This data structure is a bit spendthrift in computer storage. Reserving a memory location also for dead-solid sites (internal solid sites) is a waste of memory, especially at high solid fractions. To cope with this waste, one could think of including only active sites, namely fluid sites and solid boundaries. This stores only just what is needed. The resulting data structure is more cumbersome, however, since now *all* sites, even the internal fluids, need topological data (the list of interacting neighbors). Thus, part of the memory savings must be surrendered anyway.

```
s s s s f f f s s s
s s s s f f f s s s
s s s s s f s s s s
f f f f f f S f f f
s s s s s s s s s s
```

SKETCH 8.2. A pathological porous media with no fluid channel connecting the inlet to the outlet sections.

hydrodynamic treatment due to high local Knudsen numbers and the attendant free-slip effects (rarefied gas regime).

It is therefore important to identify the smallest channel width supporting fluid-like behavior. This width marks the gap in scale between the microscopic and mesoscopic levels, which is why the shorter the better. Detailed computational studies show that 4 lattice sites is basically the minimal channel width supporting Poiseuille-like behavior within a single channel. This is a good figure because it shows that the LBE fluid approach can be taken down almost to the level of single lattice spacing.

Let us illustrate the point by means of concrete numbers.

Consider water flowing in a cubic rock sample one centimeter in linear size ($L = 1\,\mathrm{cm}$) and assume a monodisperse media with $h = 10$ micron pore size traversed by an average flow speed $U = 0.1\,\mathrm{cm/s}$. Taking for simplicity $\nu = 0.01\,\mathrm{cm^2/s}$ as a kinematic viscosity of the water, yields a global Reynolds $Re_L = 0.1/0.01 = 10$, and a local Reynolds $Re_h = 0.01$. Assuming an average molecular speed $v = 10^4\,\mathrm{cm/s}$, the resulting mean free path in water at room temperature is about $l \sim \nu/v = 10^{-6}\,\mathrm{cm} = 0.01\,\mu\mathrm{m}$, which means a local Knudsen number $Kn_h = 0.01/10 = 0.001$. No question that in the physical world hydrodynamics applies to this microchannel flow.

What about the lattice flow?

Assuming we give just one lattice spacing to the smallest pore (in fact we need at least four, but let us look at the order of magnitude alone for the sake of simplicity), we need $N = L/h = 1000$ lattice sites per side. This is approximately the best one can do with the present computer technology in three dimensions.

The resulting lattice pitch is $\mathrm{d}x = h = 0.001\,\mathrm{cm}$, namely 10 microns. Assuming for simplicity the sound speed $c_s \sim 10^4\,\mathrm{cm/s}$ the corresponding lattice time-step is given by $\mathrm{d}t = \mathrm{d}x/c_s = 0.001/10^4 \sim 10^{-7}\,\mathrm{s} = 0.1$ microseconds.

Note that these quantities set the lattice units: $\mathrm{d}t_{\mathrm{lattice}} = \mathrm{d}x_{\mathrm{lattice}} = 1$.

The lattice speed going with this time-step is given by $u_{\mathrm{lattice}} \equiv u/c_s \sim 10^{-5}$ (in lattice units) and the lattice viscosity $\nu_{\mathrm{lattice}} = \nu\,\mathrm{d}t/\mathrm{d}x^2 = 0.01 \times 10^{-7}/10^{-6} = 0.001$.

Summarizing, in order to keep full consistency with the physical units the lattice simulation should proceed with the following parameters (ℓ.u. means lattice units):

- $N = 1000$ grid points/linear size;
- $\mathrm{d}x = 1\,\ell.\mathrm{u.} = 10\,\mathrm{micron}$;
- $\mathrm{d}t = 1\,\ell.\mathrm{u.} = 0.1\,\mathrm{microseconds}$;
- $u = 10^{-5}\,\ell.\mathrm{u.} = 0.1\,\mathrm{cm/s}$;
- $\nu = 10^{-3}\,\ell.\mathrm{u.} = 0.01\,\mathrm{cm^2/s}$.

All is well, except that in order to simulate, say, 1 physical second we need 10 million time-steps! In order to guarantee correct hydrodynamics (local Knudsen 0.001) LBE is forced to tick really fast, with a consequent heavy loss of efficiency.

How to get around such an awkward state of affairs?

First, let us come to grips with the reasons of such an inefficiency. This traces back to a lattice sound speed that cannot be made higher than $O(1)$. *The result is a very small time-step because the numerical scheme is bound to track sound waves.* This is again the issue of artificial compressibility discussed in the previous chapter. In a truly incompressible scheme sound waves propagate at infinite speed and the numerics can be instructed not to even try to track them, just let them go! The price is that the pressure field has to be computed via a costly elliptic Poisson problem. The reward is large time-steps.

Since we would like to do away with the Poisson problem, a viable strategy is to **make the lattice Mach number much higher than the real one** by raising the lattice flow speed u and the lattice viscosity ν both at the same time so as to *keep the same Reynolds number.*

Given the relation

$$Re = \frac{Ma}{Kn}, \tag{8.9}$$

it is clear that raising the Mach number at a given Reynolds number means increasing the Knudsen number as well. This is the stage where hydrodynamics is endangered. In our specific example, since the Knudsen number is about 0.001 we may hope to be able to raise it by, say, one or two orders of magnitude and still observe hydrodynamic behavior at $Kn = 0.01 \sim 0.1$. On a ski-oriented analogy, we could say that real molecules perform a special slalom, whereas lattice molecules go for a giant one!

The strategy is illustrated in Fig. 8.3.

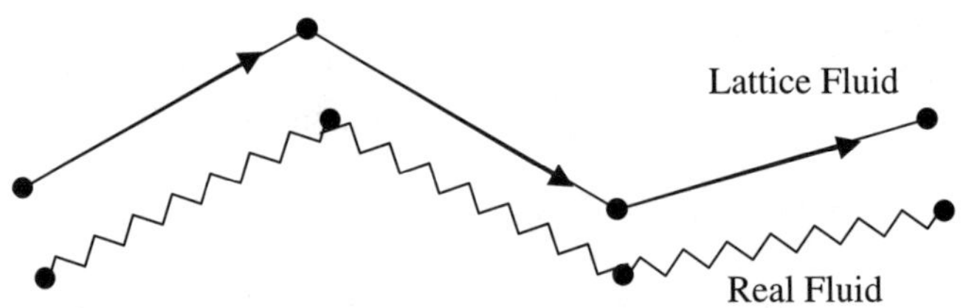

FIG. 8.3. Tight slalom: real fluid. Coarse slalom: lattice fluid.

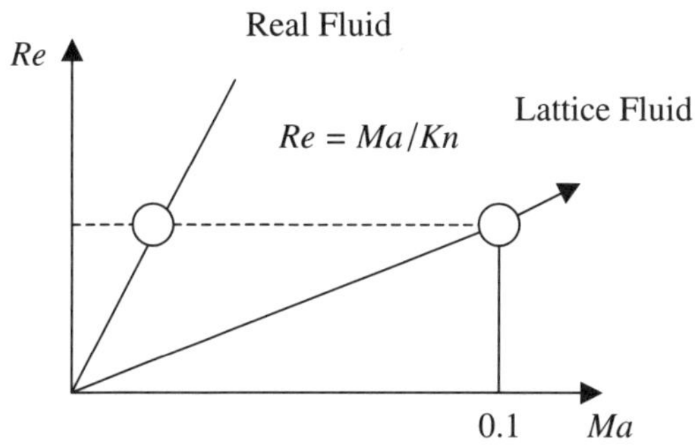

FIG. 8.4. Reynolds and Knudsen numbers as a function of the Mach number. Real and lattice fluids work at the same Reynolds but the lattice Knudsen is much higher.

What we are taking here is a bold step, namely to assume that hydrodynamics sets in *just a few* mean free paths. This assumption, which is supported by several molecular dynamics simulations of simple fluids, is *key* to the whole lattice approach. *Should hydrodynamics unforgivingly ask for hundreds of mean free paths to set in, the lattice approach would be doomed to failure because of unacceptable inefficiency.*

Early numerical experiments with the FCHC–LBE scheme were performed to assess up to what Knudsen number lattice hydrodynamics is tenable [91]. These tests consisted in running a Poiseuille flow in a rectangular channel of decreasing width and monitor at what width h the simulation starts to depart from the hydrodynamic relation

$$Q \propto \frac{Gh^2}{\nu}. \tag{8.10}$$

These experiments indicate visible departures from hydrodynamics for Knudsen numbers well above 0.01.

This places obvious constraints on the granularity of the porous media. In order to observe hydrodynamic behavior, the smallest pore size should cover at least 4–5 lattice units: with $R^* = 10$, $l_\mu = 0.1\,\mathrm{d}x$, $h \sim 5\,\mathrm{d}x$, whence $Kn_h = 1/50$. To have a granularity $h/L \sim 0.01$, the computational box should consequently cater for 500 lattice sites, which is close to the limits of the largest present day supercomputers.

8.4 Deposition algorithm

An important ingredient of LBE simulations of porous flow is the preparation of the porous geometry. For simple geometries such as the one described here this task is simple:

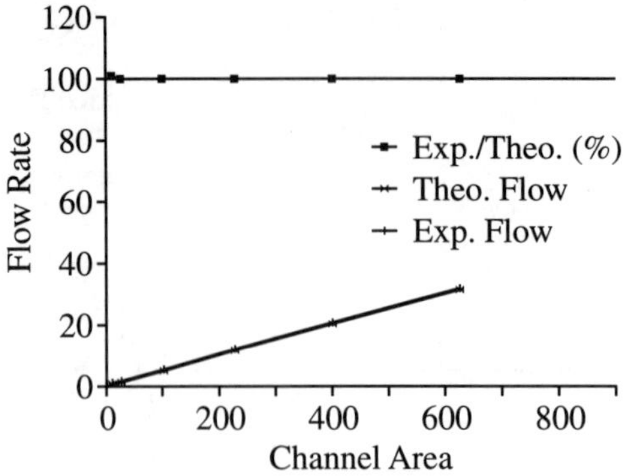

FIG. 8.5. Theoretical and numerical flow rates as a function of the channel area for $R^* = 7.57$, $u = 0.05$ (after [84]).

1. First, define a coarsening length l and a corresponding coarse-grained lattice made up of $(N/l)^3$ boxes of side l, with $l \gg h$, h being the size of solid feature (a cubelet, either solid or fluid).
2. At a random location $\vec{x}_l$ within each box B_l place the center of a solid cubelet.
3. For each box B_l, flip a random number r between 0 and 1. If $r < q$, place the solid cubelet, otherwise move to the next box. Here $0 < q < 1$ is a user-specified parameter which controls the desired solid fraction.

The following procedure grows a monodisperse porous media with an average solid volume

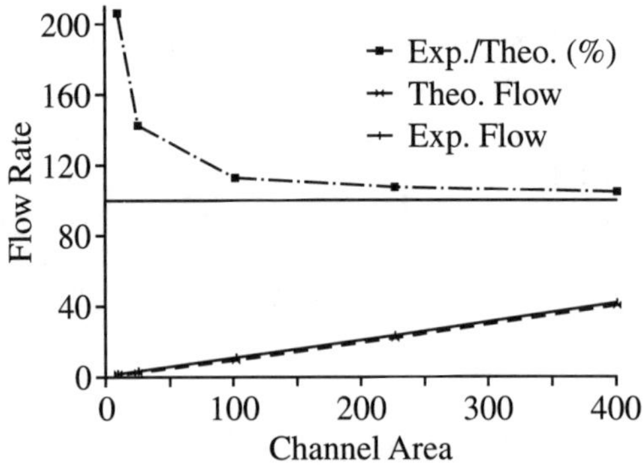

FIG. 8.6. Theoretical and numerical flow rates as a function of the channel area. Significant departures are visible for narrow channels (after [84]).

$$V_s \sim \left(\frac{L}{l}\right)^3 \times h^3 \times q \tag{8.11}$$

corresponding to an average solid fraction

$$\phi_s = \frac{V_s}{L^3} = \left(\frac{h}{l}\right)^3 \times q. \tag{8.12}$$

This relation fixes the free parameter q in terms of the desired solid fraction. This is just a 'vanilla' version of a deposition algorithm. Much more sophisticated procedures designed in such a way as to conform to realistic morphologies can be found in the specialized literature [89, 92, 93].

8.5 Numerical simulations

The first LBE simulation of porous media was performed on a low resolution 32^3 cubic lattice [94].

These simulations permitted assessment of the validity of Darcy's law and also provided a reasonable estimate of the permeability as a function of the porosity. These early investigations were subsequently refined by Cancelliere *et al.* [95] who doubled the resolution by a factor of 2 along each direction (64^3). The enhanced resolution allowed a better representation of the microgeometry, namely a random collection of overlapping spheres (see Fig. 8.7).

The spheres were represented in 'Legoland' format; that is, as a sequence of cubelets 4 lattice units in size. Since the centers of the spheres were placed at random on a 64/4 coarse grid, some degree of overlapping between the spheres was obtained for all but the most diluted porous media. This was desirable

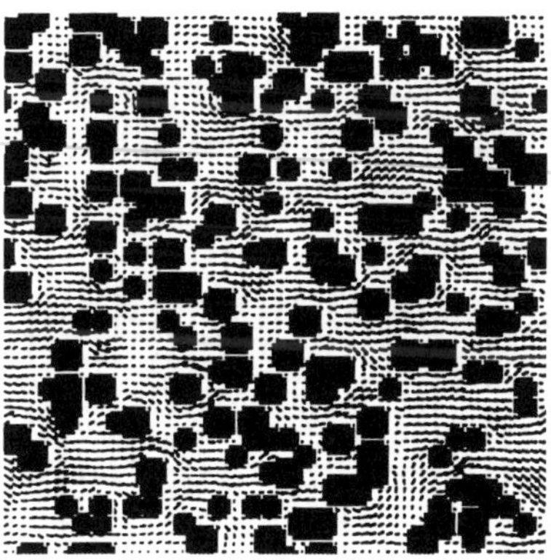

FIG. 8.7. Cross-section of a three-dimensional LBE flow in a porous media (after S. Succi, Automi Cellulari, *Collana Informatica Domani* **710**(5), 148, 1991).

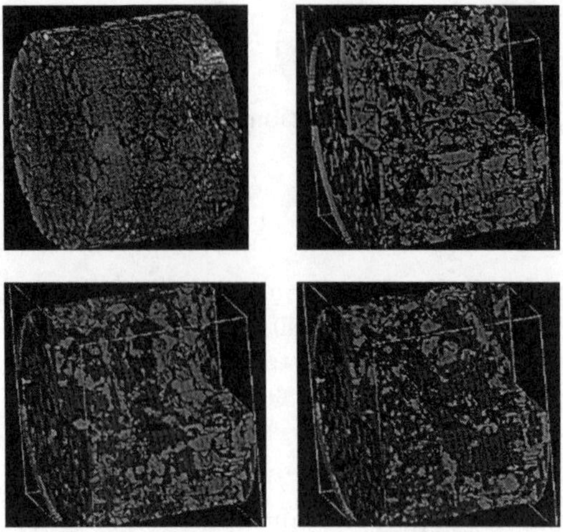

FIG. 8.8. A sandstone sample used in multiphase LBE simulations through porous media (after [50]).

since the prime focus of this study was porous media at high solid fraction, typically beyond 0.8. In spite of the rough representation of the spherical surface, the simulations provided a fairly good match with analytical results at both low and high solid fractions. They also showed that the intermediate porosity regime, unaccessible to analytical computations, connects rather smoothly the low and high solid fraction asymptotic regions. A number of authors, primarily D. Rothman and co-workers at MIT, and G. Doolen, S. Chen and co-workers at Los Alamos, extended these simulations to the case of complex multiphase flows [92, 96], still today one of the applications of choice for both LGCA and LBE simulations.

Generally speaking, there is a wide consensus that LGCA and LBE are among the best methods for simulating microhydrodynamic flows in porous media [97, 98].

8.6 Synthetic matter and multiscale modeling

Besides the issue of numerical resolution, the simple numerical example presented in this chapter highlights a general feature of LBE simulations, namely the fact that the *focus is invariably on dimensionless numbers, such as Reynolds, Mach, Knudsen, rather than on specific values of the physical parameters such as sound speed, viscosity and others.* Thus, with LBE one should always be prepared to work with *synthetic matter* rather than real materials. The assumption is that real and synthetic matter belong, loosely speaking, to the same class of universality.

Real materials are left for 'hi-fi' microscopic methods such as molecular dynamics and, possibly, direct Monte Carlo [99]. In fact, LGCA have sometimes been kind of dismissed as 'gas analogues', with little impact on real life situations begging for physical realism [100].

In this author's view, this statement deserves a little amendment.

LGCA and LBE are indeed 'only' gas analogues, in that they certainly can *not* be used to compute the actual viscosity of air, water or whatever.

But, it is precisely the flip side of this medal that buys them the capability to describe meso and macroscopic scales that are off limits to truly microscopic methods. Some physics is surely lost along the way, but hopefully not that associated with universal features governed by dimensionless numbers rather than specific physical properties. This is why LBE and LGCA qualify as excellent candidates for *multiscale* problems involving flows in disordered media.

8.7 Exercises

1. What is the physical meaning of the autocorrelation function $A(r)$?
   Hint: the product $p(x)p(y)$ is always zero unless both x and y fall on fluid sites.
2. What value of R^* is needed to achieve hydrodynamic behavior (Darcy's law) on a $h = 4$ sites wide micropore?
3. Based on pseudo-code illustrated in this chapter code up your own porous media simulation into the Fortran code provided in Chapter 5. Run and enjoy the simulation!

9

TURBULENT FLOWS

Knocking on Heaven's doors ...
Bob Dylan

In the previous chapter we discussed reasons why LBE represents an interesting alternative to traditional computational fluid dynamic methods for the numerical simulation of universal features of fluid flows. In this chapter we present an application of LBE methods to one of these problems 'par-excellence', incompressible turbulence.

9.1 Fluid turbulence

Turbulence is that peculiar state of matter (gases and liquids) in which the simultaneous presence of many active scales of motion makes the mid/long-term behavior of the system very hard to predict. Weather forecasting is probably the most telling example in point.

The importance of turbulence, from both theoretical and practical points of view, cannot be exaggerated. Besides the purely intellectual challenge associated with the predictability of complex nonlinear system dynamics, the practical relevance of turbulence is even more compelling if we think of the pervasive presence of fluids across most human and natural activities: cosmological and geophysical flows, weather forecasting, gas flows in car engines, blood flow in our body, electronic flows in solid-state devices, to cite but a few. In the last few decades numerical simulation has played an increasing role in furthering the frontier of this over-resilient problem. To appraise the potential and limitations of the numerical approach to fluid turbulence, it is instructive to revisit some basic facts about the physics of turbulent flows.

The degree of turbulence of a given flow is commonly expressed in terms of a single dimensionless parameter, the Reynolds number, defined as

$$Re = \frac{UL}{\nu}, \tag{9.1}$$

where U is a typical macroscopic flow speed, L the corresponding spatial scale and ν is the kinematic molecular viscosity (length squared over time) of the fluid. The Reynolds number measures the relative strength of advective over dissipative phenomena in a fluid flow: $Re \sim u\nabla u/(\nu\Delta u)$. Given the fact that many fluids have a viscosity around $0.01\,\text{cm}^2/\text{s}$ and many flows of practical interest work at speeds around and above $U = 1\,\text{m/s}$ within devices sized around and above $L = 1\,\text{m}$, it is readily seen that $Re = 10^6$ is commonplace in real life applications.

To date, turbulence energetics is best understood in terms of an *energy cascade* from large scales ($l \sim L$) where energy is fed in, down to small scales where dissipation takes place (see Fig. 9.1).

This cascade is driven by the nonlinear mode–mode coupling in momentum space associated with the advective term of the Navier–Stokes equations, $u\nabla u$. The smallest scale l_k reached by the energy cascade is known as the Kolmogorov length, and marks the the point where dissipation takes over advection and organized fluid motion dissolves into chaotic particle motion. According to the celebrated Kolmogorov (1941) scaling theory [101], the Kolmogorov length is estimated as

$$l_k \sim \frac{L}{Re^{3/4}}. \tag{9.2}$$

The estimate (9.2) follows straight from Kolmogorov's assumption of a *constant energy flux across all scales of motion.*

The qualitative picture emerging from this analysis is fairly captivating: a turbulent flow in a box of size, say L, at a given Reynolds number Re, consists of a collection of $N_k \sim (L/l_k)^3$ Kolmogorov eddies, each consisting of $N_m \sim (l_k/l_\mu)^3$ molecular 'trajectories'. By identifying each Kolmogorov eddy with an independent degree of freedom (quantum of turbulence), we conclude that the number of degrees of freedom involved by a turbulent flow at Reynolds number Re is given by:

$$N_{\text{dof}} \sim Re^{9/4}. \tag{9.3}$$

According to this estimate, even a modest flow with $Re = 10^4$ (less than a kitchen water faucet) features one billion degrees of freedom, enough to saturate the most powerful present day computers! This sets the current bar of Direct Numerical Simulation (DNS) of turbulent flows, manifestly one falling short of several decades to meet the needs raised by most real life applications.

The message comes down quite plainly: *computers alone won't do!*

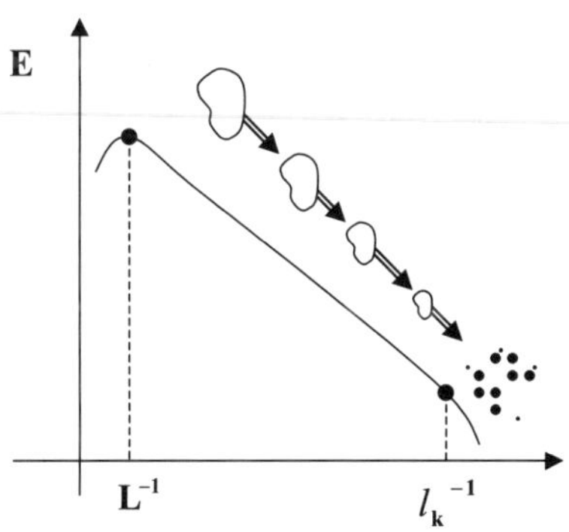

FIG. 9.1. Energy cascade in a turbulent flow: the dam analogy with energy flowing downhill from large to small scales.

Of course, this does not mean that computer simulation is useless. Quite the contrary, it plays a pivotal role as a complement and sometimes even an alternative to experimental studies.[24]

9.1.1 *Two-dimensional turbulence*

The earliest 'turbulent' LBE simulations were performed for two-dimensional flows [51]. Like most nonlinear systems, turbulent flows are fairly sensitive to the dimensionality of the space they live in. Indeed, two-dimensional turbulence differs considerably from three-dimensional turbulence. This is due the presence of an infinite number of invariants which can only exist in 'Flatland'. They read as follows:

$$\Omega_{2p} = \int_V |\omega|^{2p} \, \mathrm{d}V, \quad p = 1, 2, \ldots, \tag{9.4}$$

where

$$\vec{\omega} = \nabla \times \vec{u} \tag{9.5}$$

is the *vorticity* of the fluid occupying a region of volume V. By taking the curl of the Navier–Stokes equations, one obtains

$$D_t \omega_a = \nu \Delta \omega_a + \omega_b \partial_b u_a. \tag{9.6}$$

It is easily seen that the second term on the right-hand side, a source/sink of vorticity, identically vanishes in 2D. As a result, in the inviscid limit $\nu \to 0$, vorticity is a conserved quantity (topological invariant), and so are all its powers. The fact that vorticity, or better said, *enstrophy* Ω_2, is conserved has a profound impact on the scaling laws of 2D turbulence. It can be shown that the enstrophy cascade leads to a fast decaying (k^{-3}) energy spectrum, as opposed to the much slower $k^{-5/3}$ fall off of 3D turbulence. This regularity derives from the existence of long lived metastable states (vortices) that manage to escape dissipation for quite long times. The dynamics of these long-lived vortices has been studied in depth by various groups and reveals a number of fascinating aspects whose discussion goes beyond the scope of this book. For a recent account on modern turbulence theories see [102].

9.1.2 *Turbulence and kinetic scales*

Since the Kolmogorov length goes to zero in the limit of an infinite Reynolds number, one might wonder whether fully turbulent incompressible flows run against a lack of separation between hydrodynamic and kinetic scales similar to those encountered in the propagation of shock fronts in gas dynamics.

[24] As pointed out by P. Moin and K. Mahesh, 'DNS need not obtain real life Reynolds numbers to be useful in the study of real life applications.' (P. Moin and K. Mahesh, Direct numerical simulation: a tool in turbulence research, *Ann. Rev. Fluid Mech.* **30**, 539, 1998.)

It is easy to show that, at least within the Kolmogorov theory, just the opposite is true: the higher the Reynolds number, the better is the separation between kinetic and hydrodynamic scales. This follows from the simple relation:

$$Re = \frac{Ma}{Kn}. \tag{9.7}$$

By accepting that the particle mean free path is simply given by $l_\mu = \nu/c_s = L\,Ma/Re$, we immediately conclude that the hydrodynamic to kinetic scale separation l_k/l_μ scales like $Re^{1/4}/Ma$, hence goes to infinity at infinite Reynolds number at any given Mach number. This matches the intuitive notion that kinetic relaxation of molecular trajectories around a given fluid eddy proceeds faster for the smallest eddies than for the largest ones.

This means that *incompressible fluid turbulence does not require a detailed description of the kinetic relaxation mechanisms* and provides further motivation for the use of LBE and LBGK methods for the numerical simulation of incompressible turbulence.

9.2 LBE simulations of two-dimensional turbulence

Simulating a homogeneous incompressible turbulent flow with LBE does not require any additional techniques besides those with which we have already been familiarized in the previous chapters.

Since we are looking at a tiny internal squarelet of flow, the effects of real boundaries must be minimized and simple periodic boundary conditions can be adopted.

The second ingredient is initial conditions.

A customary choice is to start with a turbulent initial flow whose energy spectrum decays with some power law, say $E(k) \sim k^{-m}$, while the phase is made random.

On a regular grid: $x_j = j\Delta x$, $y_l = l\Delta y$, and a reciprocal grid $k_{x,m} = m\Delta k_x$ $k_{y,n} = n\Delta k_y$, with $\Delta x\Delta k_x = \Delta y\Delta k_y = 2\pi$, the following procedure applies:

1. *Define an initial velocity field* $u_a(x_j, y_l, 0) = \mathrm{IFT}[u_a(m,n)]$.
   Here IFT denotes the inverse Fourier transform:

$$u_a(j,l) \equiv u_a(x_j, y_l) = \sum_{m,n} \mathrm{e}^{\mathrm{i}(k_{x,m}x_j + k_{y,n}y_l)} u_{mn}.$$

2. *Define the initial populations.*
   The initial populations are obtained as equilibria associated with $u_a(x,y,$ $t=0)$:

$$f_i(x,y,0) = f_i^{\mathrm{e}}\left[u_a(x,y,t=0)\right].$$

Next, one decides between freely decaying or forced turbulence.

Free decay does not require any additional ingredient. In the case of forced turbulence, one can still start from the initial configuration described above, but turbulence must be sustained by a (large-scale) forcing of the form

$$F_x\,(y;k_f) = F_0 \sin\,(k_f y)\,, \tag{9.8}$$

$$F_y = 0. \tag{9.9}$$

This forcing is a Dirac delta in k space and corresponds to monochromatic energy input at scale $l_f = 2\pi k_f/L$. Eventually a broad-band excitation can be used over a whole range of wavenumbers.

The nominal Reynolds number is estimated as we did for the Poiseuille flow; that is, by ignoring nonlinear terms and solving the Stokes equation.

The sinusoidal forcing yields a sinusoidal velocity profile of amplitude

$$U_0 = \frac{F_0}{\nu k_f^2}. \tag{9.10}$$

Due to nonlinear effects, the flow develops its own bootstrap viscosity so that the observed velocity amplitude is typically only a fraction of U_0 in the above equation. This empirical fraction is then used to upscale the forcing term accordingly. Also, it should be realized that in homogeneous isotropic turbulence the role of U_0 is played by the rms speed $u_{\mathrm{rms}} = \sqrt{\langle u^2 \rangle}$.

A correct understanding of the LBE units is key to the interpretation of the simulation results. Consider a reference 1024^2 simulation, and compare this with a pseudospectral method, the most popular technique for this type of flows (see Appendix B) [103].

In the LBE code, space is normalized in units of the lattice pitch, $L_{\mathrm{LBE}} = 1024$, speed $u_{\mathrm{LBE}} = u_{\mathrm{rms}} \equiv \sqrt{\langle u^2 \rangle}$, whereas for the pseudospectral codes the common normalization is $L_{\mathrm{PS}} = 2\pi$ (the size of the periodic box) and $u_{\mathrm{PS}} = 1$, so that the ratio $t_{\mathrm{PS}} = L_{\mathrm{PS}}/2\pi u_{\mathrm{PS}}$, defining the typical large eddy lifetime, is made one in pseudospectral units. This means that the LBE and pseudospectral clocks are related as follows: $t_{\mathrm{LBE}}/t_{\mathrm{PS}} = (1024/2\pi)(1/u_{\mathrm{rms}})$. Since $u_{\mathrm{rms}} \sim 0.2$ (we cannot exceed the lattice sound speed), we obtain $t_{\mathrm{LBE}}/t_{\mathrm{PS}} \sim 4000$.

In other words, it takes about four thousand LBE time-steps to cover a large eddy turnover time $L_{\mathrm{PS}}/2\pi u_{\mathrm{PS}}$, i.e., the typical lifetime of a large-scale, energy containing eddy.[25]

Thus, a typical simulation spanning, say, ten spectral units requires a few tens of thousands of LBE time-steps. Since a typical LBE code involves $O(100)$ floating-point operations/site/step, a single time-step sweeping out a 1024^2 lattice takes approximately 1 second of a mid-range workstation (say 100 Mflops/s). As a result, a complete LBE simulation takes about 40 000 seconds, namely about

[25] Practical warning: Since the LB clock is configuration-dependent via the rms speed, it cannot help but meandering in time by some few per cent. This is something to bear in mind when making detailed comparisons between the two methods: the two clocks drift slightly apart!

half a day. The earliest LBE simulations of homogeneous isotropic incompressible turbulence were performed on a 2D decay experiment [51]. The main goal was to assess the fidelity of the (then) newborn method for moderately high Reynolds flows and appraise its computational efficiency against state of the art pseudospectral methods. The specific experiment consisted of a 512^2 simulation with $\lambda = -1.849$ (using LBE with enhanced collision) corresponding to $\nu = 0.05$, and $Re \sim 1000$.

A typical LBE vorticity field is shown in Fig. 9.2.

This figure highlights the presence of various vortices at all scales below the forcing length $L/4$. The global indicators of turbulence dynamics, the total energy $E(t)$ and the total enstrophy $\Omega(t)$ provide a good agreement with pseudospectral calculations (see Fig. 9.3).

Despite this nice global agreement, the vorticity map betrays the absence of truly small scale structures as compared with the corresponding pseudospectral calculations. This sensation is made quantitative by inspecting the energy spectrum as shown in Fig. 9.4.

From this spectrum one notes that the inertial k^{-3} range is satisfactorily represented. However, the high k region presents a a tiny but persistent energy tail at $k > 128$. Such a phenomenon is also visible in pseudospectral calculations, as witnessed by the energy spectra of LBE and pseudospectral simulations at $L = 128^2$ resolution.

A *tentative* explanation is as follows. High resolution pseudospectral simulations do not generally use plain Laplacian dissipation but so-called *hyperviscous* dissipators of the form $\nu_p \Delta^p$, with p some integer greater than one. The task of these hyperLaplacians is to sharpen the dissipative boundary, namely keep dissipation at a minimum below the dissipative Kolmogorov wavenumber $k_d = 2\pi/l_k$ and raise it sharply above it. In k space: $\nu(k/k_d)^{2p}$.

The net result is that large structures are less affected by dissipation whereas

FIG. 9.2. Vorticity map of a 512^2 LBE turbulent flow simulation (after [51]).

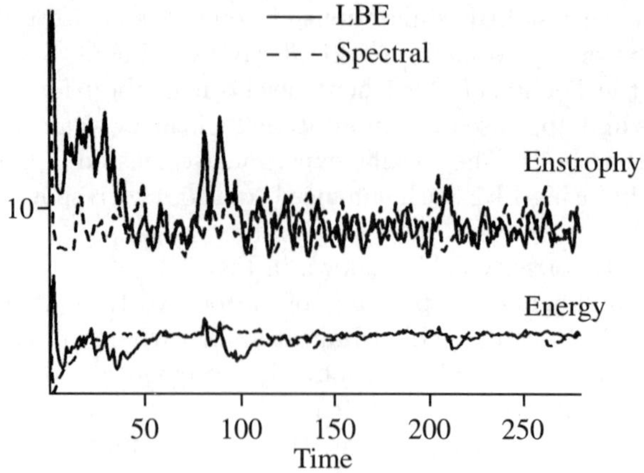

FIG. 9.3. Total energy $E(t)$ and enstrophy $\Omega(t)$ for the case of Fig. 9.2. Also shown are the same quantities for a corresponding spectral calculation (from [51]).

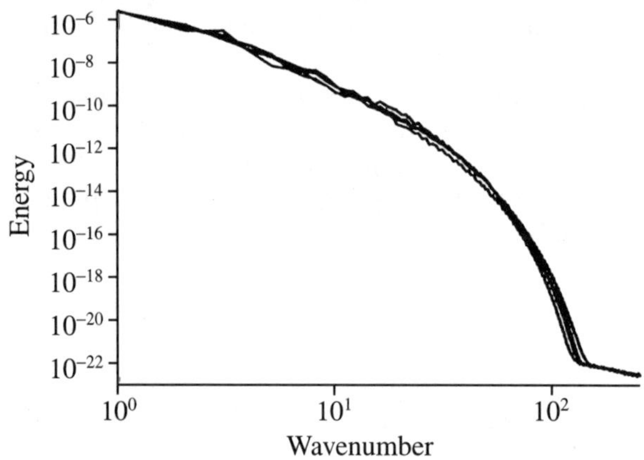

FIG. 9.4. Energy spectrum $E(k)$ for the case of Fig. 9.2 (from [51]).

small ones are sharply 'killed' as they cross the threshold wavenumber k_d. Indeed pseudospectral methods, especially in 2D, make heavy use of hyperviscous dissipation with fairly high exponents, $p = 8, \dots, 16$. This produces a twofold effect: first, the inertial range extends deeper down in the vicinity of l_k; second, energetic bursts are more likely to be absorbed, since they meet with a sharper dissipation barrier. While in the spectral formalism the implementation of hyperviscosity is fairly straightforward, just a power law, in the real space LBE formalism the same task is more cumbersome since it requires smoothing averages over remote

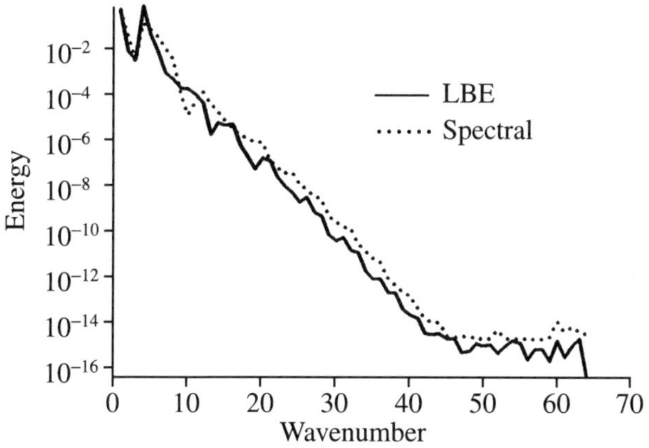

FIG. 9.5. Comparison of LBE and spectral energy at 128×128 grid resolution. Both spectra show underdamped tails, a sign that the grid could probably host a higher Reynolds number (from [51]).

sites (p sites apart) which hurt LBE locality. We may therefore conclude that as far as optimization of grid resolution is concerned, LBE lags behind the spectral method (no shame, the spectral method has asymptotic exponential accuracy!).

On the other hand, many researchers would contend that hyperviscosity is just a numerical trick. On such a premise, LBE could well be held as good as pseudospectral methods even on serial computers. This is, however, down-deep technical matter which is better left for the specialized literature.

9.2.1 *Seeing the invisible: sub-grid scales and numerical stability*

Let us discuss another practical aspect of hyperviscosity, namely the ability to protect against numerical instabilities caused by occasional burst of turbulence (*intermittency*). We have already mentioned that LBE was poised to a less ambitious task than LGCA *vis-à-vis* turbulence. Still, at its inception, there was some hope (at least as far as this author is concerned) that LBE could support a sort of built-in sub-grid scale modeling, namely the ability to describe the effect of unresolved scales ($l < 1$ in lattice units) on the resolved ones [104].

Such a property is readily tested by making the viscosity ν (via the eigenvalue λ) so small as to send the dissipative scale $l_d \sim l_k$ below the lattice spacing, i.e., $l_d < 1$ in lattice units. This means that the fluid flow is left with enough energy to do interesting things even below the smallest resolved scale.

The question is whether the numerical scheme can be instructed to dissipate this energy inflow in a physically plausible form.

Of course, this is asking a lot ...

Below the lattice scale, LBE still gives hydrodynamic behavior since the nominal mean free path can be made just a fraction of the lattice spacing.

Nevertheless, at these sub-grid scales it is the Taylor expansion of the streaming operator that becomes questionable because the expansion parameter $c\Delta t$ is by definition $O(1)$ at the scale of the lattice spacing.

In other words, at the scale of a single lattice site we are (probably) solving the Navier–Stokes equation no more, but Navier–Stokes plus HOT (higher-order terms) in the *lattice Knudsen number* $1/l_d$. In analogy with continuum kinetic theory, we shall dub Navier–Stokes + HOT as a *lattice Burnett* equation.

Lattice Burnett theory is still in its infancy, since it is difficult to carry the triple Knudsen–Mach–Taylor Chapman–Enskog expansion beyond second-order. Even in continuum kinetic theory, the resulting macroscopic equations (Burnett and super-Burnett equations), are notoriously unstable due to the presence of higher-order spatial derivatives.

In the absence of a well-established lattice Burnett equation, numerical experimentation proves the most direct probe of sub-grid behavior.

Several experiments in this direction demonstrated that once l_d goes below about 1.5 lattice units, destructive instabilities may develop in the form of negative populations, and the calculation goes wrong. This is the problem of nonlinear stability which will be addressed in more depth in the sequel to this book. The conclusion is that ν cannot be brought below a lower nonlinear threshold ν_{min} such that:

$$Re < Re_{\text{max}} \sim \left(\frac{L}{l_{\text{lbe}}}\right)^{1/m}, \tag{9.11}$$

where L is the linear size of the box in lattice units, m the scaling exponent of the spectrum, namely $m = 1/2$ in 2D and $m = 3/4$ in 3D and l_{lbe} some empirical LBE scale typically of the order of 1.5 lattice units. More sophisticated estimates based on the strength of the gradients are given in Qian and Orszag [105]. The above expression yields $Re_{\text{max}} \sim 10^4$ for $L = 150$ and $Re_{\text{max}} \sim 10^6$ for $L = 1500$ in two dimensions.

These figures significantly overestimate the current practice of LBE simulations. In fact, the estimate (9.11) does not take into account numerical prefactors, which cannot be neglected when it comes to pin down hard numbers. These factors, and to a weaker extent the scaling exponents as well, are sensitive to the phenomenon of *intermittency*, the ability of turbulent flows to generate occasional bursts of velocity fluctuations at scale l well in excess of the Kolmogorov estimate $\delta v(l) = (\epsilon l)^{1/3}$, where ϵ is the average dissipation in the fluid.

These bursts generate high local grid Reynolds numbers $Re_a = u_a a/\nu > 1$, a being the lattice spacing, which may drive the populations negative, thus triggering fatal numerical instabilities.

Let us consider a specific example. Suppose we aim at simulating $Re = 10^4$ on a 150^2 square grid. Since the Mach number must be small, a plausible set of parameters is $U = 0.1$ and $\nu = 0.001$. The condition for nonlinear stability reads, in lattice units: $\delta u_a < \nu$.

This means that any velocity fluctuation of more than 0.001, that is one per cent of the u_{rms}, between any two neighboring sites of the lattice, is very likely to produce local destructive instabilities. These instabilities materialize in the form of negative populations, and rapidly propagate around the system due to the action of the streaming operator.

Of course, a ten times finer lattice would tolerate ten times higher fluctuations, thus ensuring a roughly tenfold better stability.

The 'bare' LBE scheme has no means of protection against these short-scale fluctuations, probably because at the single lattice spacing, LBE is simulating some sort of Burnett rather than Navier–Stokes equations. Finding a discrete kinetic formulation ensuring stable and faithful sub-grid scale behavior would be a major, so far unachieved, breakthrough.

To date, we must be content with less ambitious recipes which work the other way around, namely import fluid dynamics techniques within the discrete kinetic language. Typical choices are smoothing averages, filters and stress-dependent eddy diffusivities to be described shortly. More details and refined comparative analyses of LBE versus pseudospectral methods in both two and three dimensions can be found in the original papers [106].

These authors converge basically to the same conclusions:

- *Excellent quantitative agreement on global observables (total energy, enstrophy as a function of time).*
- *Fairly good agreement on energy spectra, minor discrepancies on long time fine-grain morphological features such as the specific location of coherent vortices at late stages of the evolution.*
- *Finally, comparable efficiency on serial computers.*

Let us leave this 2D turbulence section with a few considerations on *computational efficiency*. We have already mentioned that LBE takes about $O(100)$ floating-point operations per site/step.

For the current grid sizes $L \sim 10^4$ grid points, this yields a performance pretty close to the one delivered by pseudospectral methods [51].

In view of what we learned in the previous chapter, namely that LBE does not need to solve the Poisson problem for the pressure field, the reader might wonder why LBE is not outperforming pseudospectral methods?

The point is that the spectral representation is even smarter than LBE in handling the Poisson problem: the Laplace operator reduces to a simple algebraic factor $\Delta = -k^2$, so that the Poisson problem is trivialized in a fairly effective way.

The next ace of the pseudospectral method is the underlying Fast Fourier Transform (FFT) algorithm which brings the originally N^2 complexity of the Fourier transformed convective term down to a mildly super-linear $N \log_2 N$ dependence [107]. For sizes of actual interest $N = O(10^4)$ the operation count of LBE and pseudospectral is quite comparable so that a similar performance results on serial computers.

The pseudospectral method shines in the computation of flows with *periodic* boundary conditions. On the other hand, as soon as periodicity is spoiled, as for many geometries of engineering interest, this nice property falls apart and much of the pseudospectral power with it.

LBE, on the other hand, has no problem in accommodating generic boundary conditions. Thus, homogeneous incompressible turbulence in a periodic box is possibly the *least* favorable test for LBE against pseudospectral methods. The fact that even this test results basically in a tie, must therefore be regarded as a very encouraging result for the use of LBE in computing incompressible turbulent flows.

9.3 Three-dimensional turbulence: parallel performance

The results discussed in the previous section show that LBE performs competitively with state of the art pseudospectral methods for the case of two-dimensional homogeneous incompressible turbulence.

The same conclusion applies in three dimensions.

From the methodological point of view, 3D LBE simulations proceed along the same line as two-dimensional ones. One feature which is however highlighted by the third dimension is the excellent LBE amenability to parallel computing. We have repeatedly stated that one of the major assets of LBE methods is their amenability to parallel computers. This is the right time to back up this statement with actual data. We refer to a series of three-dimensional benchmarks performed on the IBM SP-1 scalable parallel computer on a sequence of grids of increasing size.

These grid sizes were run on up to 128 processors of the IBM SP-1 parallel computer, using a 3D LBE code designed to study thermal turbulence [108]. The performance results are shown in Fig. 9.6.

From this figure we see that LBE achieves basically linear speed-up on all but the smallest configurations.[26] These data prove that the LBE amenability to parallel computing is not heralded in vain. Not only is parallel performance quite satisfactory but it is also achieved with relatively little programming effort, surely way less than pseudospectral codes. This is due to the space–time *locality* of the LBE algorithm, which makes it quite easy to partition the grid in subdomains and assign each subdomain to a different processor. A simple example of a two-dimensional partitioning of a two-dimensional domain is sketched in Fig. 9.7.

A detailed explanation of the parallel implementation can be found in [108, 109]. This 'much gain no pain' opportunity offered by LBE is probably one of the main assets for its future.

[26] It is well known to the specialists, and quite intuitive to everybody, that small grids offer less potential for parallel processing simply because by increasing the number of processors at a given size means that each processor gets less and less volume of computing and more interface communication. This implies a loss of parallel efficiency which shows up earlier for smaller grid sizes.

TABLE 9.1. Total memory requirements of the three-dimensional parallel LBE code as a function of the grid size.

Grid size	Total memory
$64 \times 96 \times 128$	72 Mbytes
$128 \times 192 \times 256$	576 Mbytes
$256 \times 192 \times 256$	1152 Mbytes
$256 \times 256 \times 256$	1536 Mbytes
$256 \times 512 \times 256$	3072 Mbytes

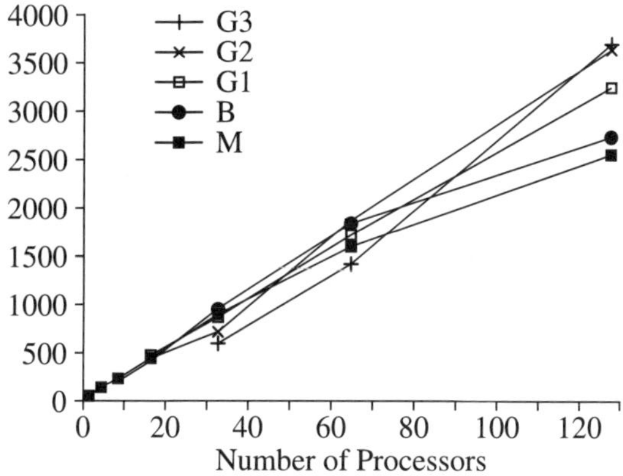

FIG. 9.6. Computational performance (Gigaflops) of a series of parallel LBE simulations on the IBM SP-1 parallel computer. Cases M to G3 range from 72 Mbytes to 3.072 Gbytes of central memory (from [108]).

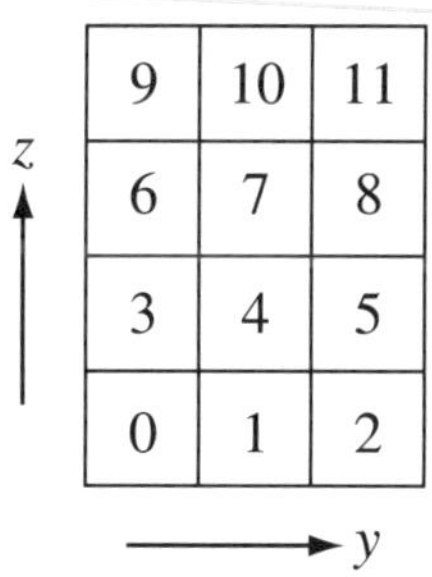

FIG. 9.7. A simple 3×4 two-dimensional domain decomposition (from [108]).

9.4 Three-dimensional channel flow turbulence

Always in the spirit of bringing solid evidence that LBE is a quite valuable method for large-scale parallel fluid dynamic computations, we discuss briefly another three-dimensional application on a completely different parallel computer. The application is three-dimensional turbulence in a square channel, on a Single Instruction Multiple Data (SIMD) computer [110], featuring up to 2 GB of central memory and at 25 Gflop peak processing speed. The machine named APE is a special purpose Italian computer, designed to perform numerical investigations of lattice quantum chromodynamics, the theory of strongly interacting elementary particles.

This computer is extremely well suited to any kind of 'crystal computation', i.e., computation on regular domains (regular lattices) and simple data structures. Much like quantum chromodynamics, lattice hydrodynamics has all that. And in fact excellent parallel performance has been obtained by running a three-dimensional channel flow version of LBE on a series of configurations.

So much for parallel computing.

The 3D LBE code is operated on the same principles exposed in the previous chapter: cyclic boundary conditions at inlet/outlet, no-slip boundary conditions on lateral walls. Turbulence is sustained by a volume force mimicking pressure gradient just the way we did for the Poiseuille flow.

The physics of wall turbulence is very rich, still pretty much under debate, and certainly not something to be squeezed into a few sentences. It is nonetheless possible to convey a flavor of the main aim behind these 3D channel flow LBE simulations. The idea is to explore the dynamics of coherent structures, typically vorticity 'blobs' ejected from the walls, and understand their effects on the statistical properties of the turbulent flow, primarily the way energy is transported

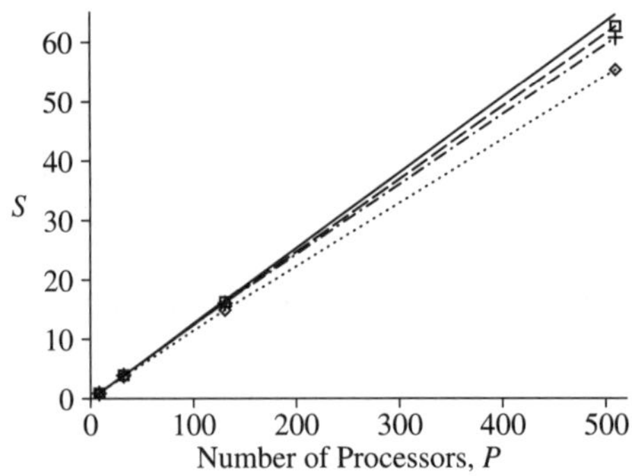

FIG. 9.8. Parallel speed-up of a 3D parallel LBE code on the 'APE' SIMD computer, as a function of the number of processors. The dotted, dash-dotted and dashed lines correspond to 25, 118 and 295 Mbytes of memory, respectively, and the solid line is the theoretical upper limit $S = P$ (from [151]).

from/to the bulk to/from the wall. This information is of great practical value as it dictates the effective drag exerted by the walls, hence the amount of energy required to push the fluid across the channel. A typical flow configuration showing the presence of these structures is reported in Fig. 9.9.

This picture reports isosurfaces of the magnitude of the mainstream flow $|u|(x, y, z) = \text{const}$. Coherent patches sprouting off the walls are clearly visible.

Just 'colorful fluid dynamics'? Not really.

After a series of validation tests, the parallel LBE code has spawned a variety of new physical results relating to the scaling properties of incompressible turbulent flows.

Among others, a new form of scaling known as extended self-similarity [111], which seem to extend beyond the range of validity of Kolmogorov's theory.

9.5 Sub-grid scale modeling

Most real-life flows of practical interest exhibit Reynolds numbers far too high to be amenable to direct simulation by any foreseeable (electronic) computer. This

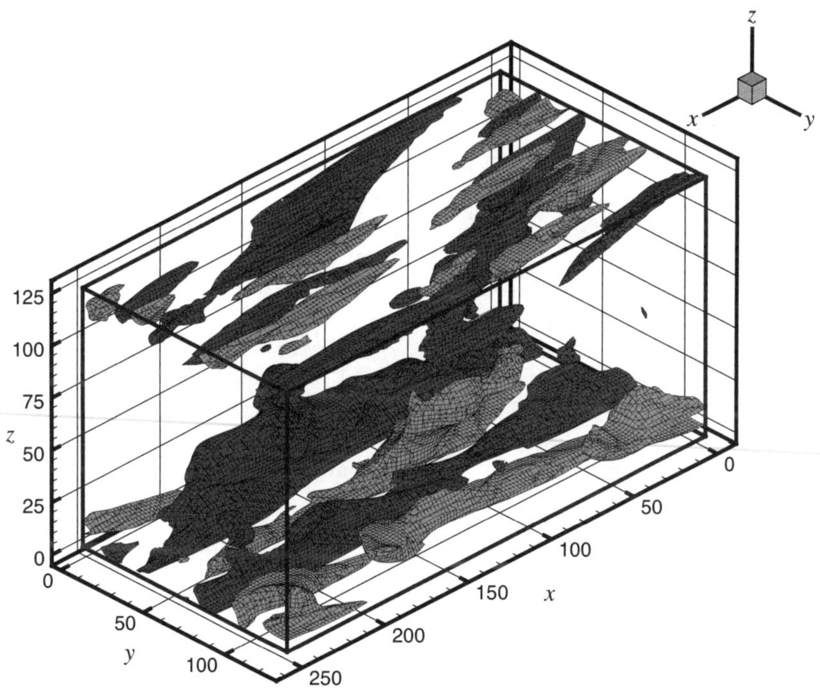

FIG. 9.9. Coherent structures in a 3D LBE channel flow simulation. Isosurfaces of streamwise flow speed magnitude. The turbulent activity in the near-wall region is easily visible. (Courtesy of G. Amati.)

raises the challenge of predicting the behavior of highly turbulent flows without ever directly simulating all scales of motion.

More specifically, one splits the actual velocity field into large-scale (resolved) and short-scale (unresolved) components:

$$u_a = U_a + u'_a \tag{9.12}$$

and seeks turbulent closures yielding the *Reynolds stress tensor*:

$$\sigma_{ab} = \rho\langle u'_a u'_b\rangle \tag{9.13}$$

in terms of resolved fields. Here brackets denote ensemble-averaging, typically replaced by space/time averaging on the grid. This task makes the object of an area of intense turbulence research known as turbulence modeling and Sub-Grid Scale (SGS) modeling.

One of the most powerful guiding ideas behind SGS is the concept of *eddy viscosity*. This idea, possibly the best contribution of kinetic theory to fluid turbulence, assumes that the effect of small scales on the large ones can be likened to a diffusive motion caused by random collisions. Small eddies are kinematically transported without distortion by the large ones, while these latter experience diffusive Brownian-like motion from jerky collisions with the small eddies. The simplest and most popular model in this class is due to Smagorinski [112]. This model is based on the following representation of the Reynolds stress tensor:

$$\sigma_{ab} = \nu_e\left(|S|\right) S_{ab}, \tag{9.14}$$

where ν_e is the effective eddy viscosity and $S_{ab} = (\partial_a U_b + \partial_b U_a)/2$ is the large-scale strain tensor of the incompressible fluid.

The specific expression of the Smagorinski eddy viscosity is

$$\nu_e = C_S^2 \Delta^2 |S|, \quad |S| = \left(\sum_{a,b} 2 S_{ab} S_{ab}\right)^{1/2}, \tag{9.15}$$

where C_S is an empirical constant of the order 0.1, and Δ is the mesh size of the numerical grid. The stabilizing role of the effective viscosity is obvious: whenever large strains develop, σ_{ab} becomes large and turbulent fluctuations are damped down much more effectively than by mere molecular viscosity (totally negligible effect at this point).

A critical assessment of the Smagorinski SGS model is out of context here; we shall simply point out that accommodating Smagorinski-like (*algebraic*) turbulence models of the form $\nu_e = \nu_e[\sigma]$ within the LBE formalism is fairly straightforward.

In fact due to the nice property that the stress tensor is locally available as a linear combination of the populations,

$$S_{ab} = \sum_i f_i^{\text{ne}} Q_{iab}, \tag{9.16}$$

the implementation is even simpler than in conventional CFD programs.

Once the stress tensor S_{ab} is available, all we need is to modify the relaxation parameter according to (9.15), so as to obtain the desired expression $\nu_e = C_S^2 \Delta^2 |\sigma|$.

Just recall the definition of the LBGK viscosity:

$$\nu = \frac{2\tau - 1}{6}, \quad \tau = \frac{1}{\omega}. \tag{9.17}$$

By inverting, we obtain $\tau = (6\nu + 1)/2$, and consequently

$$\tau_e(S) = \frac{6\,(\nu_e\,(S) + \nu_0) + 1}{2}, \tag{9.18}$$

where ν_0 is the bare, 'molecular' value.

This simple Smagorinski LBE was used to simulate 2D cavity flows at $Re = 10^6$ on a small 256^2 lattice [106]. Three-dimensional pipe flows were simulated by Somers at $Re = 50\,000$ on a 80^3 mesh. Eggels and Somers included the turbulent stress tensor σ_{ab} directly in the expression of the equilibrium distribution function $f_i^e = \rho(c_{ia}c_{ib}\sigma_{ab} - Ma^2\sigma_{cc}\delta_{ab})$.

Using this model they performed an impressive (parallel) large eddy simulation of a turbulent flow in a baffled stirred tank chemical reactor [113].

9.5.1 *Two-equation models*

The Smagorinski SGS model has the great virtue of simplicity: it is an algebraic model which does not bring *any* change in the mathematical structure of the Navier–Stokes equations. Unfortunately, life isn't that simple, and certainly not with turbulent flows. The next level of refinement in turbo-models is to link the eddy viscosity to the actual content of turbulent kinetic energy:

$$k = \frac{1}{2}\sum_a \langle u_a' u_a' \rangle \tag{9.19}$$

and dissipation

$$\epsilon = \frac{\nu}{2}\sum_{a,b} \langle (\partial_a u_b')^2 \rangle. \tag{9.20}$$

These quantities are postulated to obey transportive equations of the generic advection–diffusion–reaction form [114]:

$$\tilde{D}_t k - \nu_k \Delta k = S_k, \tag{9.21}$$

$$\tilde{D}_t \epsilon - \nu_\epsilon \Delta \epsilon = S_\epsilon, \tag{9.22}$$

where ν_k, ν_ϵ are effective viscosities for k and ϵ, respectively, and S_k, S_ϵ are local sources of k and ϵ representing the difference between production and dissipation [118]. Finally, $\tilde{D}_t \equiv \partial_t + \tilde{u}_a \partial_a$ with $\tilde{u}_a \equiv u_a - \partial_a \nu$, where $\nu \equiv \nu_k$ or $\nu \equiv \nu_\epsilon$ [115]. These equations constitute the well-known k–ϵ model of turbulence, still one of the most used turbulence models for industrial applications. It is not difficult to accommodate k–ϵ within the LBE harness by defining extra populations carrying k and ϵ as the respective densities [116].[27] Alternative routes, based on a mixed Lax–Wendroff LBE transportive algorithm [117] have been successfully demonstrated by Teixeira [118].

9.5.2 *Non-local eddy viscosity models*

Another class of turbulent models bearing a practical interest is based on the idea of a k space-dependent eddy viscosity. The basic idea is to account for space–time 'memory' effects of fluid turbulence. These memory effects translate into a non-local BGK collision integral of the form

$$C_i^{EV} = \int \omega\,(x-y)\,[f_i - f_i^{\mathrm{e}}]\,(y)\,\mathrm{d}y. \tag{9.23}$$

Here stabilization comes from space averaging, just as for hyperviscous dissipation as discussed previously. The implementation of a scale-dependent eddy viscosity is somewhat uneasy for LB since it entails a loss of locality. Notwithstanding this difficulty, Hayot and Wagner proposed an interesting extension of LB models based on the use of Levy flights to mimic the long distance momentum transfers occurring in turbulent flows [119]. Let us recall here that Levy flights are unnormalizable distributions where the probability of experiencing a jump of size l goes like $l^{-\alpha}$, where α is some positive exponent greater than one. They appear well suited to describe the sudden bursts of activity which characterize the dynamics of turbulent flows (intermittency) although (to the best of this author's knowledge) nobody has been able to put this link on a solid *ab-initio* basis; that is, starting from the equations of motion.

By suitably tuning the exponent α, Hayot and Wagner could reproduce satisfactory logarithmic profiles in a turbulent channel flow [119]. The long range nature of the Levy flight slightly impairs the computational performance of LBE, but only to a very minor extent since the Levy jumps take place much less frequently than ordinary LBE collisions.

The problem with the Levy approach is that it is difficult to generate the correct Reynolds stress tensor because short-scale fluctuations are smoothed out by space averaging. A promising variant is to use Levy distributions for sub-grid unresolved scales and represent their effects on the resolved ones. At the time of this writing, this is only a speculation which it would be interesting to check against actual calculations.

9.5.3 *Wall–turbulence interactions*

Before closing this chapter, we wish to emphasize that the real challenge for the successful implementation of <u>any</u> turbulence model is to devise appropriate boundary conditions describing wall–turbulence interactions. This subject

[27] By now, the reader should be willing to accept that any physical phenomenon governed by the general triad *convection–diffusion reaction* is amenable to LBE treatment!

is highly specialized and is still developing, hence, probably not yet ripe for a book. Surely it is a frontier of LBE research which will expand fast in the coming years, and one which is key to the successful penetration of LBE techniques in the commercial arena of computational fluid dynamics [120].

9.6 Summary

To sum up, LB methods prove good performers for both direct simulation and modeling of highly turbulent incompressible flows.

By sticking to the principle of space–time locality, LBE shows a considerable potential edge on parallel computers. Only the future can tell us whether this edge will make LBE a popular choice for the direct simulation of fluid turbulence. As for turbulence modeling, to date we have 'only' been able to adapt existing ideas to the LBE formalism. This can be done in simple and fairly efficient ways for most popular turbulence models currently in use. The dream is to generate innovative methods based upon genuine kinetic-theoretical knowledge. So far, this task remains unaccomplished, although preliminary efforts are starting to appear [121].

9.7 Exercises

1. Compute the cost of a physical degree of freedom in units of computational degrees of freedom for 2D turbulence. What do you conclude in the limit of $Re \to \infty$?
2. Prove that the vorticity source/sink term in the vorticity equation vanishes identically in 2D.
3. Draw a qualitative plot in k space of the hyperviscous operator $\nu_p \Delta^p$ for various values of p. What are the implications of raising p for near-dissipative eddies?
4. Write down the LBE equation of motion for the k_i, ϵ_i variables in the LBE k–ϵ turbulence model.
5. Using the sample code provided within this book, set up a 2D turbulent simulation at $Re = 10^3$.

10

OUT OF LEGOLAND: GEOFLEXIBLE LATTICE BOLTZMANN EQUATIONS

If we can't beat them, let's join them!
B. Alder

The LBEs discussed so far lag behind 'best in class' Computational Fluid Dynamics (CFD) methods for the simulation of fluid flows in realistically complicated geometries (think of most industrial devices, cars, airplanes and the like). This traces back to the constraint of working along the light-cones $\mathrm{d}\vec{x}/\mathrm{dt} = \vec{c}_i$. *Various methods have been proposed [122–126] to remedy this unsatisfactory state of affairs. The basic strategy is to borrow geomctrical freedom from well established techniques which can afford it, namely Finite Volumes (FV), Finite Differences (FD) and Finite Elements (FE). In this chapter we shall present the main ingredients behind the marriage between LBE and these grid techniques.*

10.1 Coarse-graining LBE

We shall handle the merging of LBE with finite 'X' (X = volumes, differences, elements) methods within the more general framework of *coarse-graining.* The idea is to gain geometrical flexibility by coarse-graining the information carried by the *differential* form of LBE, and define hydrodynamic observables on a coarse grid of virtually arbitrary shape. The basic idea is that the coarse grid *need not be tied down to the symmetries of the underlying fine grid* where the dynamics of the differential LBE takes place.

The starting point of the coarse-graining process is the differential form of LBE:

$$\boxed{\partial_t f_i + c_{ia}\partial_a f_i = C_i, \quad i = 1, \dots, b,} \tag{10.1}$$

where C_i denotes any suitable LBE collision operator, either in scattering or BGK form. The expression (10.1) represents a set of b hyperbolic Partial Differential Equations (PDEs), which are obviously amenable to standard finite difference/volume/element discretization. The result is a set of bN ordinary differential equations for the generic degree of freedom f_i^l representing the i-th population attached to the spatial node/cell/element labeled by l. These equations take the generic form:

$$\sum_{m=1}^{N}\left[A^{lm}\frac{\mathrm{d}f_i^m}{\mathrm{d}t} + B_i^{lm} f_i^m\right] = C_i^l, \quad i = 1, \dots, b, \quad l = 1, \dots, N, \tag{10.2}$$

where A^{lm}, B_i^{lm} are the grid representations of the identity and streaming operators, respectively.[28]

10.2 Finite volume LBE

In this section we shall substantiate the abstract eqn (10.2) to the specific case of Finite Volume (FV) discretization. Finite volumes provide a natural framework to carry on the information removal procedure underlying all LBE theory, in a way compliant with the basic conservation principles. This is because cell-centered variables change in time under the effect of boundary fluxes and, since fluxes are defined on cell boundaries, they automatically fulfil the congruence condition: 'outgoing flux from actual cell C to, say, right-neighbor cell $C+1 =$ incoming flux to cell $C+1$ from left-neighbor cell C.' This is nearly a tautology, but a very useful one since it secures automatic conservation (to machine round-off) of physically conserved quantities. For the sake of simplicity, we shall refer to a two-dimensional situation, although extensions to three dimensions are conceptually straightforward. We begin by considering a coarse cell in the form of a two-dimensional trapezoid (see Fig. 10.1).

The coarse-graining projection operator $\mathcal{P}$ is defined by a simple spatial integration over the cell Ω_C, and generates coarse-grained populations $F_i(C)$ at each cell center C:

$$F_i\,(C) = \frac{1}{V_C}\int_{\Omega_C} f_i\,(x,y)\,\mathrm{d}x\,\mathrm{d}y. \tag{10.3}$$

Application of $\mathcal{P}$, as combined with explicit Euler forward time marching in steps $\Delta t = h$, yields:

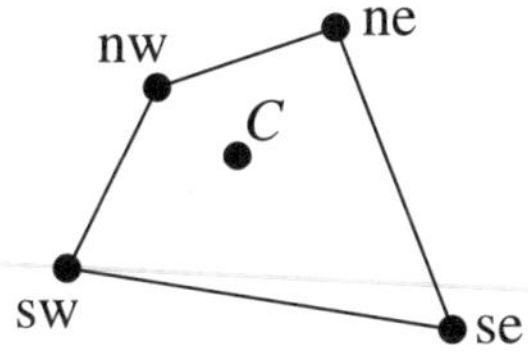

FIG. 10.1. A trapezoidal finite volume in 2D defined by the north-east (ne), north-west (nw), south-west (sw) and south-east (se) corners. Actual unknowns are located in the cell centroid C.

[28] We recall that the identity operator translates into a unit matrix only if the unknown function is projected upon an orthonormal set of basis functions. This is typically the case for finite differences and volumes but *not* for finite elements, since the latter are generally overlapping in space.

$$F_i(C,t+h) - F_i(C,t) + \sum_{s=1}^{4} \Phi_{is} = h \sum_j A_{ij} \left[F_j(C) - \bar{F}_j^{\mathrm{e}}(C)\right], \tag{10.4}$$

where

$$\Phi_{is} = \frac{h}{V_C} \vec{c}_i \cdot \int_{\partial\Omega_C} \vec{n}_s f_i(x,y)\, \mathrm{d}x\, \mathrm{d}y \tag{10.5}$$

are the fluxes across the four 'east, north, west, south' (index $s = 1,\ldots,4$) surfaces (lines) of the boundary $\partial\Omega_C$ surrounding the cell and

$$\bar{F}_i^{\mathrm{e}}(C) = \frac{1}{V_C} \int_{\Omega_C} f_i^{\mathrm{e}}\, \mathrm{d}x\, \mathrm{d}y \tag{10.6}$$

is the coarse-grained local equilibrium distribution.

Equation (10.4) is not yet operational because the integrals defining the coarse-grained streaming and collision operators involve the fine-grain distribution f.

A *reconstruction operator* (interpolation), $\mathcal{R}: F \to f$, mapping the coarse-to-fine populations is needed. Formally, $\mathcal{R} \equiv (1-\mathcal{P})\mathcal{P}^{-1}$, but this expression is ill-defined since the projection operator $\mathcal{P}$ cannot have a unique inverse $\mathcal{P}^{-1}$. The actual form of $\mathcal{R}$ depends on the specific interpolation scheme.

For the sake of simplicity, we shall restrict our analysis to the case of quadrilaterals of sides a and b.

10.2.1 *Piecewise-constant streaming*

The simplest reconstruction procedure consists of assigning the same value $F_i(C)$ to all points (x,y) in cell Ω_C. This corresponds to a (*piecewise-constant*) spatial representation and yields the following streaming expression (for rightward movers):

$$F_C(t+1) = (1-v)\, F_C(t) + vF_{\mathrm{W}}(t), \tag{10.7}$$

where $v = c\Delta t/\Delta x < 1$ is the Courant number, and W denotes the 'west' neighbor cell.

Unfortunately, this representation introduces severe numerical diffusion effects which make the piecewise-constant LBE scheme totally unviable for all but the lowest Reynolds flows. To appreciate this statement, let us Fourier analyze the above expression,

$$\mathrm{e}^{\mathrm{i}[k-\omega]} = 1 + (1-v)\left(\mathrm{e}^{\mathrm{i}k} - 1\right). \tag{10.8}$$

By writing $\omega = \omega_r + \mathrm{i}\gamma$, and acting upon the real and imaginary parts of the above relation (summing the squares and dividing imaginary over real components), we obtain:

$$\mathrm{e}^{2\gamma} = 1 + (1-v)^2 (2 - 2\cos k) + 2(1-v)(\cos k - 1), \tag{10.9}$$

$$tg\,(\omega_r - k) = \frac{(1-v)\sin k}{(1-v)\cos k + v}. \tag{10.10}$$

It is readily seen that in the limit $v \to 1$, the exact continuum dispersion relation $\omega_r = k$ and $\gamma = 0$ is recovered. Departures from these continuum relations are driven by a non-unit Courant number. In the limit of short wavelengths, $k \ll 1$, $\cos k \sim 1 - k^2/2$, the second of the above relations simplifies to

$$e^{2\gamma} = 1 - k^2 v\,(1 - v)\,. \tag{10.11}$$

In the limit $\gamma \ll 1$, this delivers

$$\gamma \sim -k^2 v\,(1 - v)\,, \tag{10.12}$$

namely a diffusive correction with a (positive) numerical diffusion coefficient $D_{\text{num}} = v(1-v)$. This numerical diffusion vanishes in the LBE limit $v = 1$, and it takes its maximum, $1/4$ at $v = 1/2$. In the limit $v \ll 1$, $D_{\text{num}} \sim v$, which represents quite a sizable effect since v cannot be made much smaller than 1 on pain of incurring severe numerical dispersion effects, as it is apparent from the second dispersion relation (10.9).

10.2.2 *Piecewise-linear streaming*

The natural move to improve on the unsatisfactory piecewise-constant representation is to go to a higher-order interpolation. The next level is (*piecewise-linear*) interpolation, namely (index i dropped for simplicity):

$$f\,(x, y) = F_C + F_1\,(x - x_C) + F_2\,(y - y_C) + F_3\,(x - x_C)\,(y - y_C)\,, \tag{10.13}$$

where the coefficients F_k, $k = 0, \ldots, 3$ must be expressed in terms of the nodal values F_P, $P = [C, \mathrm{E}, \mathrm{N}, \mathrm{W}, \mathrm{S}, \ldots]$.

This is a statement on the value of the function at the cell center C, together with its partial derivatives $\partial_x F$, $\partial_y F$ and $\partial_{xy} F$.

Upwind ideas are useful for this purpose. For instance, the x-derivative of the east movers on the east boundary reads as (see Fig. 10.2)

$$\partial_x F\,(\mathrm{e}) = \frac{F\,(\mathrm{E}) - F\,(C)}{x_{\mathrm{E}} - x_C}\,, \tag{10.14}$$

and similar expressions for other direct connections.

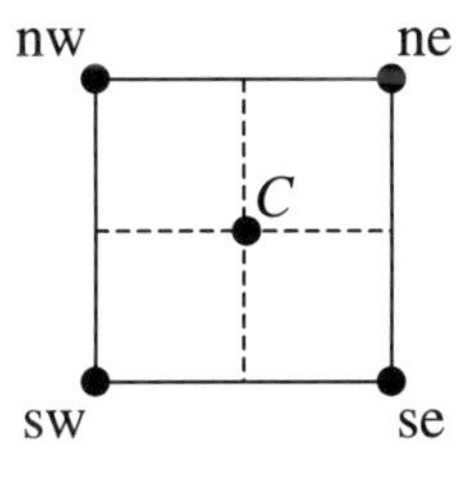

FIG. 10.2. Square cell centered in C and related areas as weighting factors.

Cross derivatives are slightly more complicated. The complete expression for north-east movers ($i = 2$) is:

$$f_2\left(\mathrm{e}\right) = p\left[F_2\left(C\right) + \frac{F_2\left(C\right) - F_2\left(\mathrm{W}\right)}{x_C - x_{\mathrm{W}}}\left(x^*_{\mathrm{e}} - x_C\right)\right] + \left(1 - p\right) F_2\left(\mathrm{S}\right), \tag{10.15}$$

where $p = 1 - 1/2a$ and the effective east location $x^*(\mathrm{e})$ can to some extent be treated as a free parameter to minimize numerical viscosity.

Some algebra [127] yields

$$x^*\left(\mathrm{e}\right) = x_C + \frac{a-1}{2p}, \tag{10.16}$$

which provides virtually zero numerical viscosity in a uniform mesh of size a.

Note that the location $x_C + (a-1)/2$ is geometrically transparent: it is the position of particles in the fine grid which cross the boundary of the macrocell at half integer times $t + 1/2$ (see Fig. 10.3).

The correction p has no transparent geometrical interpretation, at least not for this author. It is also worth observing that while p is irrelevant at large a, it becomes key at $a = 1$. In this limit, the expression (10.15) reduces to $f_2(\mathrm{e}) = (F_2(C) + F_2(\mathrm{S}))/2$, and it is left for the reader to prove that, by summing up all four contributions from east, north, west and south boundaries, the standard LBE expression for north-east propagation is recovered:

$$F_2\left(C, t+1\right) = F_2\left(\mathrm{SW}, t\right). \tag{10.17}$$

10.2.3 *Piecewise-linear collision operator*

Since the fluid field is a linear combination of the discrete populations, the piecewise-linear assumption on f carries over to the velocity field:

$$u_a = U_a + u'_a. \tag{10.18}$$

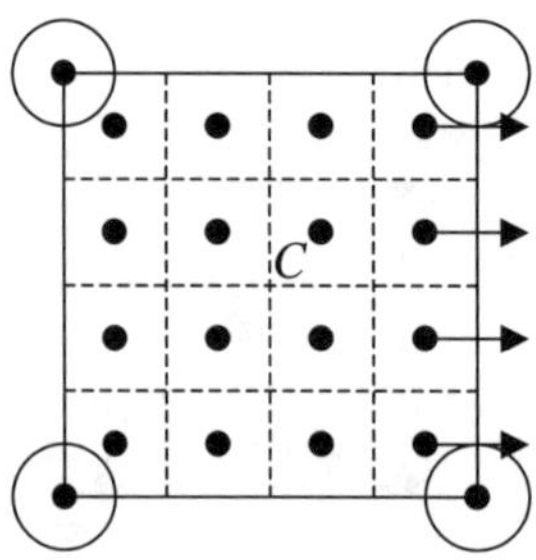

FIG. 10.3. Fine grid particle crossing the coarse cell boundary.

As a result, the contribution of sub-cell fluctuations takes the form

$$\mathcal{P}\, u'_a u'_b = \left\{ \int_{\Omega(C)} [x'_c - x_c(C)]\,[x_d - x_d(C)]\, \mathrm{d}x\, \mathrm{d}y \right\} \partial_c U_a \partial_d U_b. \tag{10.19}$$

The cell integration is easily performed analytically, and the partial derivatives of the coarse-grained field are obtained as linear combinations of the nodal values $F(C)$ via interpolated expressions such as (10.13). A detailed formulation of the piecewise-linear finite volume LBE in two and three dimensions can be found in [128]. The calculations are elementary but lengthy and will not be reported here. We shall simply remark that the piecewise-linear collision operator becomes weakly non-local through the partial derivatives of the coarse-grain velocity field. This is the price to pay for the significant drop of numerical viscosity achieved by piecewise-linear versus piecewise-constant (basically recovering the second-order accuracy of the original LBE scheme). This reduction was clearly demonstrated in [127] for a simple two-dimensional Poiseuille flow test with a two grid lattice with 32 grid points. The first 8 points starting from both walls are equispaced whereas the remaining 16, covering the central region, have doubled spacing $a = 2$, for a total channel width $W = 8 + 8 + 16 \times 2 = 48$ (fine) lattice units. Along the vertical direction (y) an equispaced lattice $b = a = 1$ is retained.

The pressure gradient (volume force) is adjusted in such a way as to produce a maximum speed $U_c = 0.02$ at the centerline of the channel, independently of the physical viscosity ν. This 'numerical viscosimeter' permits the computation of the additional viscosity introduced by the numerical scheme, by simply measuring the ratio $0.02/U_c$ where U_c is the value provided by the numerical simulation. Shown in Figs 10.4 and 10.5 are a series of simulations with varying viscosities using *piecewise-constant* and *piecewise-linear* interpolations, respectively.

From Fig. 10.4, we see that the analytical result is only attained for very large values of the physical viscosity, $\nu = O(1)$. This is consistent with the theoretical

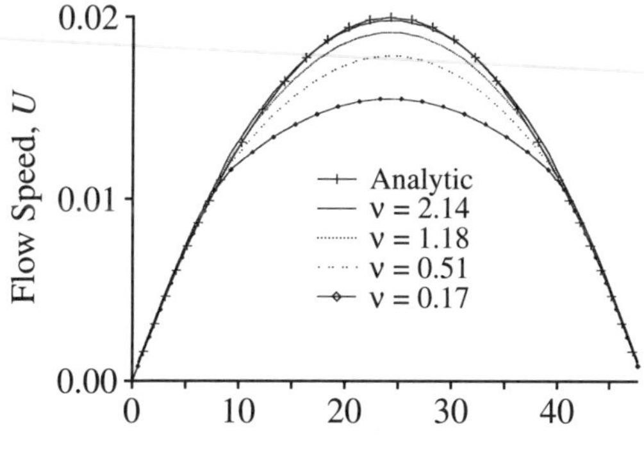

FIG. 10.4. Two-dimensional Poiseuille flow with piecewise-constant finite volume LBE at various viscosities ν (from [127]).

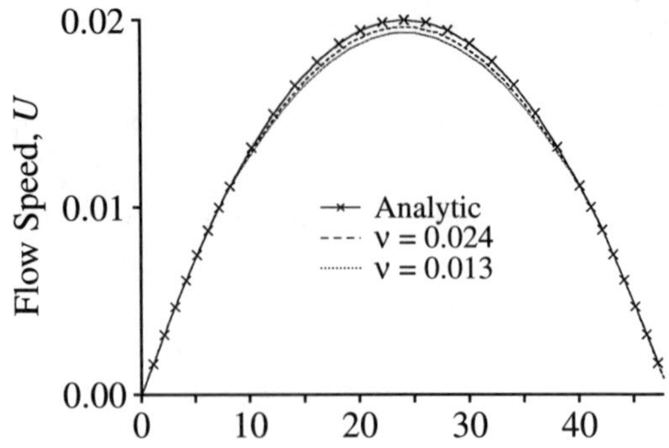

FIG. 10.5. The same case as in Fig. 10.4, with a piecewise-linear finite volume LBE (from [127]).

expectations and shows that the piecewise-constant finite volume LBE scheme (FVLBE) is inadequate for all but smallest Reynolds number simulations. Fig. 10.5 shows the dramatic improvement, nearly two orders of magnitude, resulting from piecewise-linear interpolation.

Nonetheless, the critical reader may wonder why, after having stated that the *piecewise-linear* version does not yield *any* numerical diffusion, do we still observe departures from analytical results for Poiseuille flow.

The point is that, while it is true that there is no numerical diffusion on each of the two separate uniform sublattices, at the interface there is a definite 'kink' in the slope of the velocity profile, which is responsible for numerical errors. This kink amplitude scales with the ratio $(b-a)/\nu$ and becomes negligible only at viscosities $\nu \gg (b-a)$.

Care must therefore be exercised in tailoring the mesh so as to minimize these interface kinks. The earliest finite volume LBE versions were limited to integer jumps of the lattice pitch a; subsequently [128] this limitation has been lifted, allowing higher flexibility in the grid construction and smoother control of the kink effects.

10.2.4 *Piecewise-parabolic interpolation*

The piecewise-linear finite volume LBE is good for low–moderate Reynolds flows, but still not entirely up to the expectations for cutting edge 3D direct numerical simulations of turbulent flows in the range $10^3 < Re < 10^4$.

The calculation is easy: for a 100^3 grid with $U = 0.1$ (in uniform LBE lattice units) the physical viscosity required to achieve $Re = 1000$ is around $\nu = 0.01$. With a local stretching $b/a = 1 + \epsilon$, the condition for interface kinks to be negligible reads as $\epsilon/\nu \ll 1$, which means $b/a < 1.01$, a very mild stretching which cannot yield major upsizing of the computational domain with a reasonable number of grid points.

The bottom line is that for state of the art direct finite volume LBE simulations, higher-order interpolates are needed [129]. In fact, second-order interpolation should restore second-order accuracy of the original LBE scheme.

Modern finite volume LBE formulations have been introduced recently [130], which claim great geometrical flexibility without compromising the simplicity of the LBE method. These methods are very elegant and perform convincingly on moderate Reynolds number flows. The future will tell whether these schemes can also deal with cutting edge turbulent flows at $Re \sim 10^4$.

10.3 Finite difference LBE

Once it is recognized that the differential form of LBE is nothing but a set of hyperbolic PDEs, it becomes natural to observe that both time and space derivatives can be discretized in many ways, not just the lock-step light-cone rule $\mathrm{d}\vec{x} = \vec{c}\,\mathrm{d}t$ characterizing LBE. With independent time and space discretizations the lattice is set free from the tyranny of symmetry requirements which tie down LBE dynamics and considerable geometrical flexibility can be bought thereof.[29]

This point of view has been endorsed by the Los Alamos–IBM T. J. Watson team (S. Chen and co-workers), who developed a number of Finite Difference LBE (FDLBE) methods based on higher-order (Runge–Kutta) time marching schemes, as combined with various spatial discretization schemes [124]. As a specific example, they use a central finite difference scheme for spatial discretization and a second-order Runge–Kutta (modified Euler) for time marching.

The resulting FDLBE schemes take the general form:

$$f_i\left(\vec{x}, t+\tfrac{1}{2}\right) = f_i\left(\vec{x}, t\right) - \tfrac{1}{2}\Delta t R_i\left(\vec{x}, t\right), \tag{10.20}$$

$$f_i\left(\vec{x}, t+1\right) = f_i\left(\vec{x}, t\right) - \Delta t\, R_i\left(\vec{x}, t+\tfrac{1}{2}\right), \tag{10.21}$$

where

$$R_i\left(\vec{x}, t\right) = -c_{ia} D_a f_i - \omega\left(f_i - f_i^{\mathrm{e}}\right) \tag{10.22}$$

and $D_a f_i = [f_i(\vec{x} + \Delta x_a) - f_i(\vec{x} - \Delta x_a)]/2\Delta x_a$ is the centered partial derivative with respect to x_a.

This FDLBE scheme achieves second-order accuracy in both space and time and has been successfully demonstrated for various flows in bounded geometries, such as two-dimensional Taylor vortex flow, Couette flow with temperature gradient between the walls, and others.

10.4 Interpolation-supplemented LBE

Another approach along the line of *grid freedom* is the so-called *ISLBE* (Interpolation-Supplemented Lattice Boltzmann Equation). The idea is still to move the discrete distributions along straight paths $\mathrm{d}x_{ia} = c_{ia}\mathrm{d}t$, with the important proviso of doing away with the constraint that endpoints of the flights should land on lattice sites.

[29] One more link with LGA theory being jettisoned ...

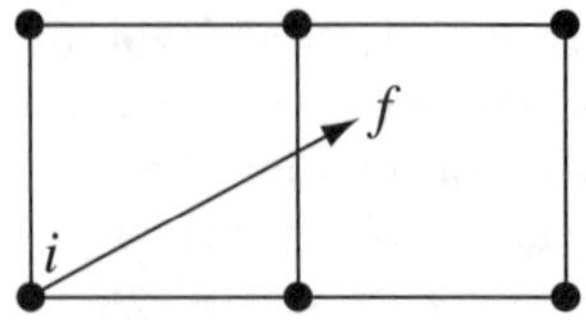

FIG. 10.6. An off-lattice particle flight starting at i and ending at f. The starting point i lies on the lattice, the ending point f does not.

This is yet another way of setting the control grid free from the symmetry constraints bearing on the microscopic particle motion. The price to pay for this geoflexibility is, as in finite volume LBE, the need of interpolating between the spatial grid and the particle positions.

It is clear that LBE becomes here a close relative of well-known Particle In Cell (PIC) methods [131, 132], with the LBE highlight that velocity space is represented by a fistful of discrete speeds. As in PIC methods, the geometry of the spatial control grid can be chosen at will and non-uniform or even non-Cartesian meshes can be adopted. Unlike PIC, however, the particle degrees of freedom carry quasihydrodynamic information which obviates the need for statistical averaging. Also, the asset of locality is preserved to a great extent, since particles contribute to the grid densities only at the nodes of the cells they belong to.

To minimize numerical viscosity, second-order interpolation is used.

He *et al.* prove the viability of ISLBE for a two-dimensional flow past a cylinder using cylindric coordinates [133], at both low and high Reynolds number (up to $Re = 10\,000$). Their results compare well with the wide body of existing data and support claims of substantial memory savings (more than an order of magnitude) with respect to standard uniform LBE.

10.5 Finite element LBE

Having discussed the merge of LBE with finite volume and finite difference methods, for the the sake of completeness, we shall also mention the possibility of marrying LBE with the third general grid method, namely Finite Elements (FELBE).

Consistent with the general finite element philosophy, we expand the discrete populations $f_i(x,t)$ onto a set of *localized* (finite support) basis functions $\psi_l(x)$ [134]:

$$f_i\left(x,t\right) = \sum_l f_{il}\left(t\right)\psi_l\left(x\right). \tag{10.23}$$

The basis functions ψ_l are generally piecewise polynomials functions restricted to some local neighborhood of the lattice node x_l (*finite elements*). Such an expansion, as applied to the differential LBE (10.1), delivers a set of ordinary differential equations like (10.2), where (in one dimension for simplicity):

$$A_{lm} = \langle \psi_l, \psi_m \rangle, \tag{10.24}$$

$$B_{lm} = \langle \psi_l, \frac{d\psi_m}{dx} \rangle \tag{10.25}$$

are the so-called 'mass' and 'force' matrices, brackets standing for scalar product $\langle f, g \rangle \equiv \int f(x) g(x) \, dx$.

To the best of this author's knowledge, such FELBE schemes have not been addressed by the published literature thus far. They might nonetheless represent an appealing alternative to finite volumes for very complex geometries requiring *unstructured* grids (grids with space-dependent connectivity).

10.6 Native LBE schemes on irregular grids

We began this chapter with the idea of buying geoflexibility to LBE schemes by marrying them with proved performers in the field, chiefly finite volumes. This is certainly wise, but more daring paths can be tried as well.

The question is: *can we formulate the native LBE microdynamics in an irregular space–time*? Irregular (random) lattices are no new faces in lattice field theory [135], where they are generally credited with a number of virtues, typically regularization of anomalies (the benefits of disorder ...).

The theory of random LBE is only just being formulated and is certainly not yet ripe for a book. Still the main idea is worth conveying to the interested reader. The main point is to define a set of space-dependent discrete speeds $c_{ia}(\vec{x})$ as a local deformation of some suitable standard uniform set c^0_{ia}. The new speeds can to all effect be regarded as a 'grid flow' with its own speed $c_a = \sum_i c_{ia}$ and deformation tensor $G_{ab} = \partial_a c_b$ (both these quantities vanish identically in a uniform mesh). The non-Euclidean metric gives rise to inertial forces, not at all desirable in a numerical scheme. The crucial idea is to reabsorb the inertial forces due to the space-dependence of the discrete speeds into properly generalized equilibria. At the time of this writing, it seems that such generalized equilibria not only do exist but they can even be found by a systematic procedure based upon entropy minimization principles [136].

Whether this procedure will ever lead to a practical alternative to finite volume or finite difference LBEs is a separate story. Still, the subject carries an interest on its own as a discrete kinetic model for disordered systems.

10.7 Implicit LBE schemes

All schemes discussed so far differentiate on the basis of the way the space variable is treated; in all cases the time variable has been treated via explicit discretization. This responds to the wise criterion of preserving causality in the numerical scheme.

On the other hand, once we are willing to forsake the light-cone space–time discretization, we might consider doing away with explicit time integration as well. The purpose is clear: *march in larger steps towards steady-state configurations.* More general LBE schemes, including *implicit* LBEs can be devised by starting from the Lagrangian integration of LBGK along the characteristic trajectories.

By formally rewriting LBGK as an ordinary differential equation

$$\frac{\mathrm{D}f_i}{\mathrm{D}t} + \omega f_i = \omega f_i^{\mathrm{e}}, \tag{10.26}$$

a straight quadrature along the trajectory yields:

$$f_i\left(\vec{x} + \vec{v}_i\tau, t + \tau\right) = \mathrm{e}^{-\omega\tau} f_i\left(\vec{x}, t\right) + \mathrm{e}^{-\omega\tau} \int_t^{t+\tau} \mathrm{e}^{-\omega\left(t'-t\right)} f_i^{\mathrm{e}}\left[\vec{x} + \vec{v}_i\left(t' - t\right), t'\right] \mathrm{d}t'. \tag{10.27}$$

It is readily checked that in the limit $\tau \to 0$ and by evaluating the collision integral explicitly, namely at $t' = t$, the above equation reproduces exactly the familiar LBGK form.

Any other choice would produce an implicit algorithm since both right- and left-hand sides contain the actual unknown at time $t + \tau$. The properties of implicit LBEs have not received much consideration thus far. This kind of study has an intrinsic interest for practical computations of steady-state flows. The chief interest of implicit solvers is to move to steady-state in large time-steps, happily unaware of Courant–Friedrichs–Lewy-like restrictions. The computational cost is significantly increased because each time-step requires the solution of a sparse linear system of size bN, b being the number of discrete speeds and N the size of the grid. Useful work along this direction is just starting to appear [137, 138].

10.8 Multiscale lattice Boltzmann scheme

Many phenomena of physical and engineering interest exhibit violent excursions over highly localized regions (boundary layers, shock fronts) which require a correspondingly highly clustered mesh. A popular response to this kind of need in modern computational fluid dynamics is provided by *unstructured* meshes, namely discrete grids where the number of neighbors of a given node may change from place to place. This allows much stronger distortions of the computational grid, but also requires a significantly more complex data structure. Another popular possibility is provided by *locally embedded* grids, namely grids in which the local connectivity (number of neighbors) is unchanged but the lattice spacing is refined or coarsened locally, typically in steps of two for practical purposes. Local embedding is a specific instance of a more general framework known as *multiscale algorithms.*

Multiscale LBE schemes have been pioneered by Filippova *et al.* [64], who tested and validated them for moderate Reynolds flows around cylinders and wings.

Technical details can be found in the original literature [64, 139].

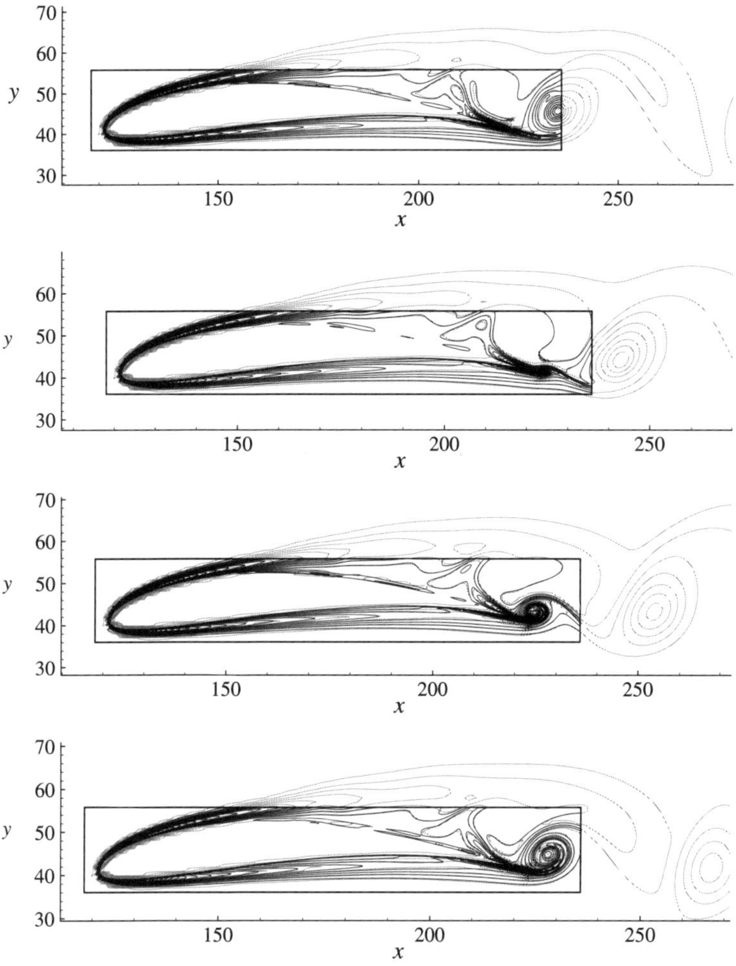

FIG. 10.7. Two-dimensional flow past an airfoil at $Re = 10^4$ as computed with a multiscale LBE scheme (courtesy of O. Filippova). The grid within the box is refined by a factor of 6 as compared to the exterior. The four figures represent the time evolution of the vorticity field.

10.9 Summary

To sum up, while the original LBE was severely grid-bound and awkward in front of realistically complex geometries, many variants are available today that cure the initial flaw. These variants (with the exception of the native formulation) take LBE closer and closer to the realm of computational fluid dynamics and their future success will critically depend on the extent to which they will manage to handle complex geometries without sacrificing too much the original LBE assets, simplicity and amenability, to parallel computing.

10.10 Exercises

1. Draw a graph of the coupling coefficients of the piecewise-constant finite volume LBE scheme as a function of the grid size h.
2. Estimate the numerical viscosity of piecewise-constant finite volume LBE.
3. Compute the piecewise-linear streaming coefficients for right movers.
4. Derive eqn (10.27).
5. Compute the mass and advection matrix of piecewise-linear finite elements.

11

LBE IN THE FRAMEWORK OF COMPUTATIONAL FLUID DYNAMICS

The purpose of computing is insight, not numbers
R. Hamming

The purpose of computing is insight, not pictures
L. N. Trefehen

And the purpose of insight is ... ?
Anonymous

The LBE lies at the crossroad between two distinct lines of thought, one that views it as an appealing method to compete with numerical fluid mechanics and the other that sees it rather as a sort of 'telescope' of molecular dynamics. This 'dual' nature of LBE is a potential source of richness and of some confusion too. Therefore, after having covered the basic theory of LBE and a set of representative fluid dynamics applications, it is now time to try to put LBE in a critical perspective within the broader framework of Computational Fluid Dynamics (CFD).

11.1 LBE and CFD

The LBE is a brainchild of computational statistical mechanics; this is the 'humus' from which it draws both its advantages and its disadvantages.

Indeed, the method really took off the day it was realized that, besides a number of beautiful properties, statistical mechanics was also placing needless restrictions on lattice fluids.

The most telling examples in point are the infamous anomalies discussed in Part I, lack of Galileo invariance, spurious invariants, low Mach numbers and others. Most of these disturbing flaws were lifted by cutting the umbilical cord with lattice gas cellular automata and regarding LBE as a self-standing kinetic model of the Navier–Stokes.

This 'information removal' procedure undoubtedly takes LBE closer and closer to computational fluid dynamics. This is only natural, since formally LBE is nothing but a set of finite difference equations that must obviously have its own entry in the ponderous list of numerical methods!

This is not the point, though. The real question is whether LBE represents something new, and not just '*yet another finite difference scheme*' for the Navier–Stokes equations ('Navier–Stokes in disguise').

In this chapter we shall bring evidence in support of the latter perspective.

To begin with, let us characterize the distinctive features of LBE, leaving apart kinetic theory and endorsing the language of numerical analysis instead [140].

The main properties of any numerical scheme can be classified as follows (the order corresponds to presentation convenience):

- *Causality*;
- *Accuracy*;
- *Stability*;
- *Consistency*;
- *Efficiency*;
- *Flexibility*.[30]

With respect to the above properties, LBE can be classified as an *explicit, Lagrangian, finite-hyperbolicity* approximation of the Navier–Stokes equations with the following highlights:

- *Causality*: Fully local in space and time.
- *Accuracy*: Second-order in space and time (likely with a very small prefactor).
- *Stability*: Unconditional linear stability, conditional nonlinear stability. Pointwise conservativeness (to machine round-off).
- *Consistency*: Marginal, conditioned to low Mach numbers.
- *Efficiency*: Good efficiency on serial computers (no need to solve the Poisson problem for pressure). Outstanding efficiency on parallel computers.
- *Flexibility*: Easy handling of grossly irregular boundary conditions. Poor natural adaptability to non-uniform grids (in its original version). Handy inclusion of mesoscopic physical effects.

11.1.1 *Causality*

Numerical schemes for time-dependent equations fall within two broad categories, depending on the way the time variable is discretized. These are:

- *Explicit*;
- *Implicit*.

In explicit schemes the current state of a given variable is computed in terms of a (usually small) number of neighbors at a preceding time. Thus, the current state is uniquely and explicitly specified in terms of the past state of the neighborhood. For a simple one-dimensional equation, a typical explicit scheme reads as follows:

[30] Flexibility is definitely not a canonical category of numerical analysis! It reflects however the daily experience that simple schemes tend often to outlive more sophisticated ones just because of better ease of use. This is not at all rigorous, but vital nonetheless to actual computational practice.

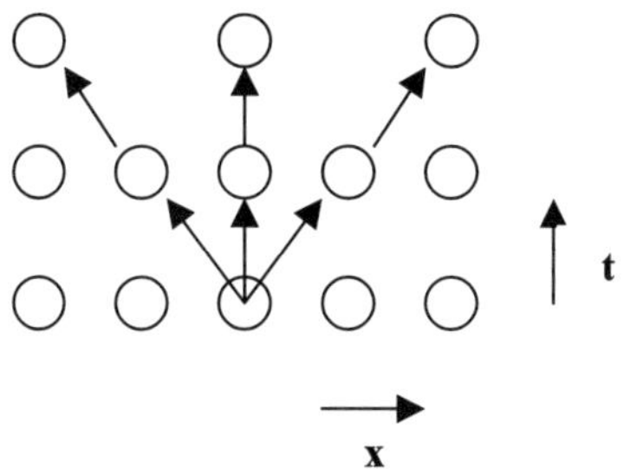

FIG. 11.1. Space–time diagram of an explicit method based on nearest-neighbor connections in one space dimension.

$$f(x_l, t_n) = \sum_{m,k<n} T_{lm,kn} f(x_m, t_k), \tag{11.1}$$

where the coefficients $T_{lm,kn}$ are the space–time connectors between the actual space–time location ($x_l = l\Delta x$, $t_n = n\Delta t$) and its neighborhood. Note that causality implies $t_n > t_k$, whereas purely spatial connections are generally two-sided.

In implicit schemes the current state of a given variable depends on the *simultaneous* state of its neighborhood:

$$f(x_l, t_n) = \sum_{m,k<n} T_{lm,kn} f(x_m, t_k) + \sum_{m} T_{lm,nn} f(x_m, t_n). \tag{11.2}$$

Simultaneity introduces the non-physical notion of 'action at distance' thus breaking the aforementioned causal link. The immediate operational effect is a *loss of locality*: in order to compute the actual value at (x_l, t_n), we need to solve the linear system (11.2) involving *all* space locations at time t_n. In fact, even if the spatial connection is local (that is, the matrix T_{lm} is sparse) with only a few nonzero entries around the diagonal, its inverse is generally full, which means that all space locations are simultaneously coupled. As a result, to advance in time, we must solve a linear system at each time-step, which makes implicit schemes much more demanding on a per time-step basis. The reward is larger time-steps without breaking the stability thresholds (beware of accuracy though ...).

LBE belongs to the family of *explicit, second-order* time-marching schemes.

To back up this statement, let us consider the differential form of the streaming operator:

$$D_t f \equiv (\partial_t + v_a \partial_a) f \tag{11.3}$$

and discretize it according to a first-order Lagrangian scheme along the characteristic $\mathrm{d}x_a = v_a\,\mathrm{d}t$.

The result is

$$\Delta_a f \equiv f\left(\vec{x}+v_a\,\mathrm{d}t, v_a, t+\mathrm{d}t\right)-f\left(\vec{x}, v_a, t\right)+O\left(\mathrm{d}t\right)=0. \tag{11.4}$$

Now specialize $v_a = c_{ia}$, $i = 1, \ldots, b$ to the set of discrete speeds and what we obtain is precisely the left-hand side of the lattice Boltzmann equation with $dt = 1$. The explicit nature of the discretization scheme, as well as first-order accuracy in time, is therefore apparent.

Nonetheless, we claimed <u>second</u>-order time accuracy. How is this?

The point is that even if the Lagrangian integration is first-order (in fact, it is a forward Euler), since each discrete characteristic comes with its mirror conjugate, the macroscopic hydrodynamics resulting from the kinetic scheme is second-order in time. On the other hand, this is consistent with the fact that the multiscale Taylor expansion upon which LBE is based implies second-order space and time derivatives (see Chapter 1). And also with the fact that space and time are tied up together by the light-cone marching condition $\mathrm{d}x_{ia} = c_{ia}\mathrm{d}t$.

11.1.2 *Accuracy*

Accuracy refers to the errors introduced by replacing differential operators with finite difference ones. If this error is zero for polynomials up to degree p, we say the discretization has p-th order accuracy.

For instance, a one-sided finite difference of the form $\left(f(x+\Delta x)-f(x)\right)/\Delta x$ reproduces exactly the derivative of a linear function, but not a quadratic (or higher) one, which means that accuracy is only first-order.

It can easily be checked that a centered difference such as $(f(x+\Delta x) - f(x-\Delta x))/2\Delta x$ is second-order accurate.

To discuss spatial accuracy of LBE it is expedient to consider a parity conjugate pair of discrete speeds $\vec{c}_i$ and $\vec{c_{i^*}} = -\vec{c}_i$. Now introduce the sum of the streaming operators $\Delta_i^+ \equiv \Delta_i + \Delta_i^*$ associated with free motion along the i-th direction.

This reads (index i omitted) $\Delta^+ f = \left(f(x+\Delta x) - 2f(x) + f(x-\Delta x)\right)/2\Delta x$ where $\Delta x = c\Delta t$. It is immediately checked that this is precisely the centered discretization of the second-order derivative $\partial_{xx} f$, which is known to be second-order accurate.

Now consider the difference $\Delta_i^- \equiv \Delta_i - \Delta_i^*$. This reads as $\Delta^- f = (f(x+\Delta x) - f(x-\Delta x))/2\Delta x$, which is again a second-order accurate discretization, this time of the first-order derivative $\partial_x f$. Since it is precisely these sum and difference of streaming operators that ultimately enter the LBE discretization of the Navier–Stokes equations, the above expressions tell us a story of second-order accuracy.

Note that we have chosen the propagation direction along x for the sake of the argument. Since the set of b discrete speeds splits naturally into $b/2$ conjugate pairs, every propagation direction is equivalent. Second-order spatial accuracy is an expected result since the existence of a mirror conjugate for every discrete speed $\vec{c}_i$ implies a centered finite differencing scheme.

Second-order accuracy applies to the bulk of the computational domain, and might eventually degrade to first-order at the boundaries. For instance, the bounce-back procedure used to impose no-slip boundaries is apparently only first-order because it breaks the space-centered nature of the streaming operator. Ways to restore second-order accuracy such as mid-grid placement of the boundary have been described in a previous chapter (including a cautionary footnote).

The accuracy of the method could be improved by adding further discrete speeds. This is akin to higher-order finite difference schemes involving a larger number of neighbors [141, 142]. We shall briefly touch upon these higher-order schemes in connection with thermal extensions of the LBE method.

Before closing this section, we would like to mention that very often LBE methods are empirically observed to provide higher quality results than second-order finite difference schemes. This is definitely the case for the turbulence computations discussed in a previous chapter, where LBE was shown to provide results of quasispectral accuracy (asymptotically *exponential* accuracy!).

This may be tentatively attributed to the *exact conservativeness* (to machine round-off) of the LBE scheme: the streaming operator is exactly conservative (no floating-point operation), while the collision is also conservative to machine round-off. It is quite likely that such properties make the prefactor in front of the quadratic decay of the numerical error with grid resolution very small indeed.[31]

11.1.3 *Stability*

Stability is a key property of any numerical scheme since it helps to protect against numerical runaways due to cumulative error build-up or other generic sources of inaccuracy. For explicit schemes, the basic notion is that the lattice is a discrete world which can only support signals with a finite propagation speed. Once this is recognized, the necessary criterion for stability is simply that *physical information should not travel faster than the fastest speed supported by the lattice.* This is the physical content of the well-known Courant–Friedrichs–Lewy (CFL) conditions.

For a simple one-dimensional convective equation of the form

$$\partial_t f + U \partial_x f = 0, \tag{11.5}$$

the CFL condition reads:

$$C = \frac{U \Delta t}{\Delta x} < 1, \tag{11.6}$$

where C is the Courant number of the scheme.

The physical interpretation of this inequality is that the physical speed U should not exceed the highest speed supported by the discrete grid $U_g \equiv \Delta x / \Delta t$, on pain of putting the numerical results in jeopardy. It is a simple matter to notice

[31] This author is grateful to Dr Hudong Chen for bringing this point up, and to Dr Irina Ginzbourg and D. d'Humières for further elucidations.

that LBE light-cone marching $c_{ia}\Delta t = \Delta x_i$ implies $C = Ma \ll 1$, so that LBE is well within the CFL condition.

The CFL condition alone is no guarantee of *practical* stability since numerical debris due to imperfect conservations may build-up systematically and ruin the calculation. Before entering this discussion, which touches upon the thorny issue of nonlinear stability, let us address the sweet cake first, linear stability analysis.

Linear stability Let us consider the *linearized* BGK equation recast in the following form:

$$(\Delta_i + \omega)\left(f_i - f_i^0\right) = 0, \tag{11.7}$$

where f_i^0 is the uniform (no flow) global equilibrium and Δ_i is the discretized streaming operator. Since $\Delta_i f_i^0 = 0$, this term can safely be added to the standard LBGK equation to obtain (11.7).

A Fourier transform of the eqn (11.7) delivers the following dispersion relation:

$$\prod_i (z_i - 1 + \omega) = 0, \tag{11.8}$$

where $z_i = \mathrm{e}^{j(k_a c_{ia} - \omega)}$, and j denotes the imaginary unit to avoid confusion with the discrete speed index i.

The above equation has a single pole of multiplicity b (the number of discrete speeds):

$$z_i = 1 - \omega. \tag{11.9}$$

Upon splitting the frequency ω into real and imaginary parts, Ω and γ, respectively, eqn (11.9) delivers two equalities

$$\mathrm{e}^{-\gamma}\cos\theta_i = 1 - \Omega, \tag{11.10}$$

$$\mathrm{e}^{-\gamma}\sin\theta_i = 0, \tag{11.11}$$

where $\theta_i = (k_a c_{ia} - \Omega)$ is the phase of the Fourier modes propagating along the i-th direction.

The latter equation delivers $\theta_i = 0$ (that is, $\Omega = k_a c_{ia}$) which is precisely the dispersion relation of free traveling waves in the continuum.

This tells us that *the lattice free-streaming operator preserves spatial coherence* since it does not introduce any distortion of the spatial profile (*dispersion-freedom*). This nice property is strictly related to the light-cone discretization.

By squaring both eqns (11.10) and (11.11), and summing them up, we obtain

$$\mathrm{e}^{-2\gamma} = (1 - \Omega)^2. \tag{11.12}$$

Linear stability implies $\gamma < 0$, which in turns requires $|1 - \Omega| < 1$, namely the condition

$$0 < \Omega < 2 \tag{11.13}$$

we met many times earlier in this book when discussing the LBE viscosity.

This analysis shows that LBGK is *unconditionally linearly* stable, provided the relaxation parameter obeys the above inequality.

In view of the expression of the fluid viscosity ($\omega \equiv \Omega$):

$$\nu \sim \frac{1}{\Omega} - \frac{1}{2}, \tag{11.14}$$

we recognize that (11.13) is precisely the condition for the fluid viscosity to be positive, modulo the factor $-1/2$ coming from second-order terms in the Taylor expansion of the streaming operator. This matches the intuitive notion of negative viscosity as a byword of physical instability.

Both limits are, so to say, 'virtual'. The limit $\Omega \to 0$ (infinite viscosity) breaks the adiabaticity assumption that begets the convergence of a hyperbolic equation like LBE to a diffusive equation like the Navier–Stokes equation. The other extreme, $\Omega \to 2$, (zero viscosity) cannot be reached either, because it gives rise to sub-grid scales (shorter than the lattice spacing) which the scheme is not able to dissipate, thereby leading to an ultraviolet catastrophe. This takes us back again to the sensitive issue of nonlinear stability.

Nonlinear stability A rigorous, exact, theoretical analysis of nonlinear stability of LBGK schemes is impossible, for it would amount to solving the LBE itself!

However, a number of general guiding criteria prove fairly useful [143].

One of these guiding criteria is the concept of *conservativeness* of a numerical scheme, namely the ability to ensure that physically conserved quantities remain conserved *exactly* in the lattice as well.

The actual value of the conserved quantity might not be exactly the same as in continuum, but if it does not change in time, we are at least guaranteed that the numerical trajectory remains adhered to an invariant (hyper)surface in phase-space and consequently it cannot run astray. A property which is paramount for very long simulations.

LGCA are ideally placed to guarantee this (remember the excitement of 'exact computing' in the early LGCA days). Less so for LBE because in any floating-point computation conservativeness can only be guaranteed up to machine round-off, namely 14 digits on 64 bit word computer, at best. Still, as compared to many floating-point techniques, LBE is not badly placed at all.

The streaming operator is perfectly conservative since it implies a plain transfer from one site to another of the same distribution function, *no floating-point* involved. The collision operator is also conservative, to machine round-off, owing to conservation laws explicitly encoded either in the scattering matrix or in the local equilibria. This makes LBE an 'exactly' conserving numerical scheme. Does this automatically protect against numerical blow-ups in the actual simulations?

Unfortunately not, or, better said: yes, but only in the linear regime.

By linear, we mean hydrodynamically linear, no $u\nabla u$ terms and the attendant nonlinear mode–mode coupling. This nonlinear mode–mode coupling can be triggered by either defective transportiveness or genuine high Reynolds flow physics. Either way, this takes us to a more detailed discussion of nonlinear stability and its connections with consistency requirements.[32]

11.1.4 *Consistency*

As mentioned in the chapter devoted to turbulent flows, the nonlinear regime of the Navier–Stokes equations implies that short-scale motion is excited and a nonlinearly stable numerical scheme should be able to ease numerical drifts resulting from the approximate representation of these short scales. A well-known manifestation of short-scale under-representation is the so-called Gibbs phenomenon; that is, spurious oscillations [141] that threaten the *positivity* of the distribution function:

$$f > 0. \tag{11.15}$$

This consistency requirement is sometimes called a *realizability* constraint, in the sense that negative values eventually produced by the numerics are not realizable in the physical world as we know it. (For the sake of completeness, the full realizability constraint should read as $0 < f < 1$, but the troublemaker is typically the lower side of the bound.)

Realizability is neither a sufficient nor (probably) a necessary condition for nonlinear stability, but it is generally understood that if f goes negative the computation should not be trusted even in the absence of numerical blow-up.

To analyze the realizability condition in the nonlinear regime, let us recast LBE in the form of a pseudo-ordinary differential equation as follows:

$$\partial_t f = -\omega f + \Phi, \tag{11.16}$$

where the non-equilibrium flux Φ is given by

$$\Phi = \omega f^{\mathrm{e}} - v_a \partial_a f^{\mathrm{e}}. \tag{11.17}$$

If Φ is slow compared to f, the formal solution of eqn (11.16) reads

$$f = f_0 \mathrm{e}^{-\omega t} + \frac{\Phi\left(1 - \mathrm{e}^{-\omega t}\right)}{\omega}. \tag{11.18}$$

If the equivalent system is linearly stable ($\omega > 0$ in the continuum, $0 < \omega < 2$ in the lattice) this solution relaxes to a renormalized equilibrium

$$\tilde{f}^{\mathrm{e}} = f^{\mathrm{e}} - \frac{v_a \partial_a f^{\mathrm{e}}}{\omega} \tag{11.19}$$

in a time lapse $1/\omega$.

[32] Before doing so, let us take an exception. When we talk about LBE, what we mean are the quasilinear versions described in Part I, not the full nonlinear McNamara–Zanetti version. This latter has an H-theorem securing positivity even in the nonlinear regime and, consequently, probably also a better stability. Unfortunately, as we have seen, it is not practical for three-dimensional high Reynolds computations and therefore it will not be discussed any further.

Consider now smooth equilibria, such that $|v_a \partial_a f^{\mathrm{e}}| \ll |\omega f^{\mathrm{e}}|$ (adiabatic approximation). For these smooth equilibria, the positivity constraint translates into an equivalent (but much more controllable) positivity constraint on the local equilibrium:

$$f^{\mathrm{e}} > 0. \tag{11.20}$$

This requirement is met by construction in continuum kinetic theory, but in the discrete world it cannot be taken for granted. For low Mach non-thermal flows the condition (11.20) is met for all but the highest (supersonic) speeds, so that the situation is not bad. However, we ought to bear in mind that discrete lattice equilibria are *not* positive-definite for all values of the thermohydrodynamic variables. For the sake of concreteness, let us refer to the simple D1Q3 BGK equilibria:

$$f_1^{\mathrm{e}} = \frac{1}{6}\left(1 + 3u + 3u^2\right), \tag{11.21}$$

$$f_2^{\mathrm{e}} = \frac{1}{6}\left(1 - 3u + 3u^2\right), \tag{11.22}$$

$$f_0^{\mathrm{e}} = \frac{4}{6}\left(1 - \frac{3}{2}u^2\right), \tag{11.23}$$

where all speeds are measured in units of the sound speed $c_s = 1/\sqrt{3}$. Consider a region of the flow where $u > 0$. The 'endangered' species is f_0, which goes negative for $u > \sqrt{2/3}$. This is significantly above the weakly compressible regime, and therefore it does not constitute any problem.

For this model, the nonlinear stability limit is clearly dictated by non-smooth renormalized equilibria, in which the gradient part plays a driving role. The non-equilibrium component is easily evaluated by using the definition (11.19): $f_0^{\mathrm{ne}} = -2u_x/\omega$, so that we obtain ($u_x \equiv \partial_x u$):

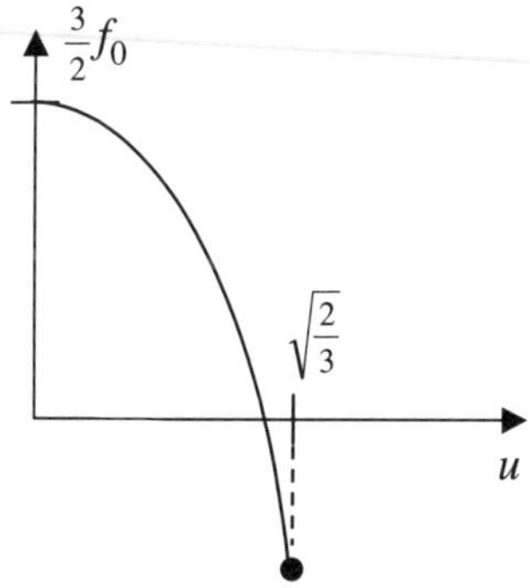

FIG. 11.2. Endangered species going negative beyond a given flow speed.

$$\tilde{f}_0^{\mathrm{e}} = f_0^{\mathrm{e}} - 2\frac{u_x}{\omega}. \tag{11.24}$$

The positivity constraint modifies into

$$\frac{u^2}{2} + \frac{u_x}{\omega} < \frac{1}{3}, \tag{11.25}$$

neatly showing that high local speeds (convective transport) *and* high local gradients (diffusive transport, since $1/\omega \sim \nu$) both conspire to endanger realizability.

The H-theorem The issue of nonlinear stability is intimately related to the existence of an H-theorem for LBE.

Let us first consider the H-theorem for the BGK equation in the continuum. The simplest candidate as a BGK H-function is the 'Euclidean distance':

$$h[f] = f^2. \tag{11.26}$$

Upon multiplying by $-f$ and integrating over velocity space, the BGK equation delivers

$$\partial_t s + \partial_a S_a = \omega\left[(f, f) - (f, g)\right], \tag{11.27}$$

where

$$s(x,t) = -\frac{1}{2}(f,f) \equiv \int_v \frac{1}{2} f^2 \, \mathrm{d}v, \tag{11.28}$$

$$S_a(x,t) = -\int_v v_a f^2 \, \mathrm{d}v \tag{11.29}$$

are the local entropy density and the corresponding entropy flux, respectively.

Here g is a shorthand for the *attractor* of the collisional operator. We should not hasten to identify g with a local Maxwellian at this stage, but rather regard it as a generic (nonlinear) map $g = M[f]$, whose *fixed points* $f = M[f]$ deliver the local equilibria.

As a result, we shall refer to g as a *target equilibrium* [144].

The existence of a local H-theorem implies that the local entropy production

$$\sigma \equiv \omega\left[(f,f) - (f,g)\right] \tag{11.30}$$

is positive for any (positive-definite) functions f and g.

The first term on the left-hand side serves us well, since it is clearly positive-definite. The second term, on the other hand, has no single sign *a priori* and prevents general conclusions on the existence of an H-theorem. A notable exception is the case in which the target equilibrium coincides with the global uniform equilibrium $g = \rho/b$. In this case, g reduces to the first term of the Hermite expansion $g = \rho h_0(v)$, $h_0(v) = 1$, and since the Hermite basis is orthogonal, it follows that f and g fulfil the following relation:

$$(f, g) = (g, g). \tag{11.31}$$

On account of (11.31), the entropy production reads simply $\sigma = +\omega(f - g, f - g)$, which is clearly positive-definite.

Can we extend the idea to more general equilibria?

A simple calculation shows the way. Suppose the target equilibrium consists of two terms in the Hermite expansion, i.e., $g = g_0 h_0(v) + g_1 h_1(v)$ and $f = f_0 h_0(v) + f_1 h_1(v) + \ldots$, where dots stand for the (orthogonal) remainder of the expansion. The scalar products yield $(g, g) = g_0^2 + g_1^2$ and $(f, g) = f_0 g_0 + f_1 g_1$ thus proving that the condition (11.31) is fulfilled if and only if $f_0 = g_0$ and $f_1 = g_1$. On the other hand, this is precisely the usual requirement that the actual distribution and the target equilibria share the same density ρ and current density ρu_a (note that we say nothing on quadratic terms, which means that these target equilibria do not capture hydrodynamic advection). This observation makes contact with the standard low Mach expansions of Maxwellian equilibria discussed in Part I. The progress is that empirical discrete equilibria can be related to a formal H-theorem. A constructive procedure for the systematic identification of target equilibria with a local H-theorem has recently been proposed in the literature [144]. This is a rather technical and still moving subject, so that the interested reader is best directed to the original reference.

We now move on to the discrete LBGK world.

In the lattice, things are slightly more complicated because f_i and $\hat{f}_i \equiv f(\vec{x} + \vec{c}_i, t+1)$ are two distinct objects which contribute separately to the local entropy production. Upon multiplying (11.7) once by f_i and once by $\hat{f}_i$ and summing up over all discrete speeds *and* lattice sites, we obtain the following expression for the global entropy production:

$$\Delta S \equiv -\left(H_2(t+1) - H_2(t)\right) = 2\omega \sum_x \sum_i 2\left(f_i - g_i\right)\left(f_i + \hat{f}_i\right). \tag{11.32}$$

We have assumed that boundary conditions are such as to allow the replacement $\hat{H}_2 \equiv \sum_x \sum_i \hat{f}_i^2 = H_2 \equiv \sum_x \sum_i f_i^2$. Typically, periodicity would do.

The global entropy production term on the right-hand side of (11.32) has the same sign indeterminacy of its continuum analogue, plus the additional problem of the presence of the hat distribution $\hat{f}_i$. The latter can be disposed of simply by rewriting LBGK in the form

$$\hat{f}_i = (1 - \omega) f_i + \omega g_i. \tag{11.33}$$

Substituting this in (11.32) yields

$$\Delta S = -\omega \sum_x \sum_i \left[\omega \left(f_i - g_i\right)^2 - 2 f_i g_i\right]. \tag{11.34}$$

Invoking again the 'cylinder condition' $(f_i, g_i) = (g_i, g_i)$, the above expression simplifies to

$$\Delta S = \omega\left(\omega - 2\right) \sum_x \sum_i \left(f_i - g_i\right)^2, \tag{11.35}$$

which is the direct transcription of the continuum counterpart with the renormalization $\omega \to \omega\left(\omega - 2\right)$. The expression (11.35) shows that discrete target equilibria fulfil a global H-theorem provided ω lies in the range $0 < \omega < 2$, the same stability range of linear stability.

Target equilibria with underlying H-theorem are obviously much prized since they are likely to exhibit better nonlinear stability.

It is worth pointing out that 'cylinder equilibria' are special and in general the nonlinear H-theorem cannot be extended to the *over-relaxation* regime $1 < \omega < 2$, because in this range a positive pair f_i, g_i does *not* imply that the shifted population $\hat{f}_i$ is positive as well. This follows directly from eqn (11.33).

It is important to note that *it is precisely this kinetic over-relaxation regime which is mostly relevant for turbulent flows*! With $\omega = 1$, the lattice viscosity is in fact $\nu = c_s^2/2$, hence well above 0.1 for most lattices, too large a value to reach high Reynolds numbers on affordable grid resolutions.

Nonlinear stability holds in a restricted strip $0 < \omega < \omega_{NL}[f] < 2$, where the threshold $\omega_{NL}[f]$ is an unknown functional of the solution f via the flow itself and its gradients. As we said already, nailing down the exact form of this functional is basically (almost) as difficult as solving the LBE itself! Common practice suggests $\omega_{NL} \sim 2 - O(Kn)$. Nonetheless, very recent developments [52, 145, 146] suggest that the actual value of $\omega_{NL}[f]$ can be computed by requiring that a given H-function, or, more generally, Lyapunov functional $\Lambda[f]$, should not decrease under the effect of collisions. This suggests a typical *over-relaxation* scenario $(1 < \omega < 2)$ in which the local equilibrium is attained through a sequence of overshoots rather than by monotonic convergence.

The actual value of the relaxation parameter is obtained by solving the (non-linear) algebraic problem

$$\Lambda[f] = \Lambda[f^*], \tag{11.36}$$

where star labels the post-collisional state such that entropy is conserved in a collision.

The post-collisional state f' can then be obtained by a linear interpolation between the isentropic partners f and f^*:

$$f' = \beta f + (1 - \beta) f^*, \tag{11.37}$$

where $0 < \beta < 1$ controls the viscosity (virtually zero for $\beta = 0$).

From a geometrical point of view, this is quite a different move as compared to a standard LBGK relaxation. The latter always proceeds along the segment joining f and f^e, whereas the entropic update proceeds along the segment connecting f and f^*. Far from equilibrium, these two paths may diverge significantly (see Fig. 11.3).

The major progress with respect to a normal LBGK update is that since Λ is a true Lyapunov functional, entropy is bound not to decrease in a collision step, so that better stability should result. These so-called *entropic* LBE models hold some renewed promise to revive the 'mirage' of zero viscosity discussed earlier in this book. Whether such an exciting opportunity holds true in actual simulations remains to be proven at the time of this writing, but prospects look fairly interesting.

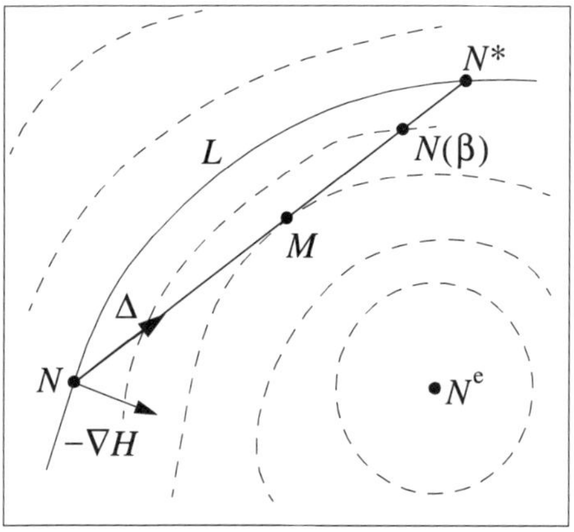

FIG. 11.3. Entropic LBE update. The solid line L is the isentropic curve where $\Lambda(f) = \Lambda(f^*)$ ($N \equiv f$ in this figure). The vector Δ represents the collision operator and the angle between Δ and ∇H is the entropy production inequality ($H \equiv \Lambda$). Point M is the minimum entropy along the segment joining the isentropic partners. The result of the update (11.37) is represented by point $N(\beta)$ (the case shown is over-relaxed). A normal BGK update would be represented by a point along the segment joining N to the local equilibrium N^e. (From [52], courtesy of I. Karlin.)

Entropy minimization strategies We conclude this section with a few comments on the interrelation between the existence of an H-theorem (nonlinear stability) and the absence of lattice anomalies (consistency).

We have learned previously that Maxwell equilibria necessarily lead to breaking of Galilean invariance. We also learned that Galilean invariance can be restored by replacing Maxwellian (Fermi–Dirac) distributions with suitable second-order polynomials in the Mach number. The question is whether any of such polynomial equilibria can be equipped with an H-theorem.

This type of question is best addressed within the general framework of entropy minimization principles [144]. In principle, one may wonder whether anything can be gained by considering alternative functional forms of the microscopic H-function, other than f^2, as for BGK, or $f \ln f$ for the continuum Boltzmann equation [147].

The strategy is to minimize the following constrained functional (energy is not included because thermal LBE schemes have yet to be discussed):

$$H_f = \sum_i h(f_i) + A\left[\sum_i f_i - \rho\right] + B_a\left[\sum_i f_i c_{ia} - J_a\right], \qquad (11.38)$$

where $h(\cdot)$ is a generic *convex, positive-definite* functional and A, B_a are the Lagrangian parameters.[33]

If such a convex functional can be found, the *local* H-theorem follows straight from eqn (11.33):

$$H_f(t+1) = (1-\omega)\, H_f(t) + \omega H_g(t), \tag{11.39}$$

which, upon invoking convexity of the H-functional, as well as the minimum property of H_g, yields

$$H_f(t+1) - H_f(t) \leqslant \omega\, [H_g(t) - H_f(t)] \leqslant 0, \tag{11.40}$$

thus proving the discrete H-theorem in local form.

The global H-theorem follows straight from the local one (with suitable boundary conditions, of course) upon recognizing that the streaming term cannot alter the entropy inside the fluid volume.

Let us now put the entropy minimization procedure to work to identify the corresponding discrete equilibria:

The variational requirement $\min\{H\}$ implies

$$\frac{\delta H}{\delta f_i} = 0 \quad \to \quad h'[f_i] = -I_i \equiv -\left(A + B_a c_{ia}\right), \tag{11.41}$$

where the prime means derivative with respect to f_i and '$\to$' means logical implication.

For the specific case of the LBGK operator, it proves convenient to inspect the following family of polynomial H-functions:[34]

$$h_p[f] = \frac{f^{p+1} - f}{p}, \quad p > 0. \tag{11.42}$$

Equation (11.41) delivers a whole family of solutions of the form:

$$g_i = \frac{1}{p+1}\left(1 + pI_i\right)^{1/p}. \tag{11.43}$$

It can readily be checked that Boltzmann entropy is recovered as a special case $p \to 0$.

For any integer $p \geqslant 1$, the equilibria (11.43) represent irrational functions of the argument I_i, which cannot be captured by any finite degree polynomial

[33] A function $f(x)$ is said to be strictly convex if it satisfies the following inequality: $f(ax+by) \leqslant af(x) + bf(y)$, with $0 < a, b < 1$, $a + b = 1$. The geometrical meaning is that the segment joining any two points of the graph of f lies below the graph.

[34] This is a truly glorious functional in modern statistical physics, one that relates to many important subjects such as multifractal processes, the replica trick in spin-glass theory and, more recently, Tsallis' non-extensive thermodynamics.

expansion. The author [147] concludes that anomaly-free LBGK polynomial equilibria of any finite degree cannot possess an H-theorem.

This gloomy conclusion refers to an exact version of the H-theorem, one applying independently of the lattice spacing. Restricted versions, where compliance with H-theorem is only requested to the order of accuracy of the LBE method (a fairly reasonable request after all), have indeed been found by Karlin *et al.* [52] by extending the list of constraints with the momentum flux tensor $\rho u_a u_b + P\delta_{ab}$ as well. Some of the resulting equilibria, like the D2Q9 model, are recognizable, some others are not.

It is important to realize that most of these entropies are difficult to analyze because the actual match of the hydrodynamic constraints generally results in a highly nonlinear multi-dimensional algebraic problem for the Lagrangian parameters, which does not generally deliver closed analytical solutions of these parameters as a function of the hydrodynamic fields.

This is a rather serious snag, hindering further progress in the field. New hopes in this direction have been provided by very recent work by Boghosian and co-workers, which is based on a very elegant geometrical decomposition of kinetic space into hydrodynamic and kinetic *polytopes* [145]. As pointed out by L. S. Luo in a recent informative review [148], the practical effectiveness of these latest developments remains to be proven by actual numerical simulations.

It is certain that there will be more to learn on this matter in the forthcoming years [149, 150].

11.1.5 *Efficiency*

By now we have learned that LBE is *more* than a Navier–Stokes solver.

Together with density, flow and pressure fields it also tracks the momentum flux tensor and a bag of additional fields, the ghosts. This makes LBE memory and CPU intensive as compared to most explicit methods.

The extra information brings some rewards as well.

First, as we have seen, *Poisson-freedom*, no need to solve the Poisson equation for the pressure field.

Second, *no second-order derivatives to describe diffusion*: diffusion results implicitly from adiabatic relaxation of the traceless stress tensor to local equilibrium (finite-hyperbolicity). This is an important property, since it turns the CFL diffusive constraint $\nu\Delta t < \Delta x^2$, the most restrictive one in standard explicit finite difference methods, into the purely convective condition $\Delta t < \Delta x/c$. Both properties make of LBE a very attractive explicit finite difference solver.

The situation turns even for better on parallel computers.

As described in the Appendix C, in order to qualify for efficient parallel execution, a numerical scheme should feature the following highlights:

- *High fraction of parallel content (Amdahl's law).*
- *Low Communication to Computation Ratio (CCR).*
- *Even distribution of the computational workload across computational space–time (load balance).*

It is readily checked that LBE has all that.

A high fraction of parallel content is guaranteed by the local nature of the method: by far most of the CPU time of any LBE application is taken by the collision step, literally a communication-free process (ideal efficiency). This is a gracious inheritance of LGCA, a powerful gift not lost along the way.

Typical parallel performance of LBE codes has been described earlier in this book [151], and more can be found in the current literature [152, 153]. Optimal load balance is guaranteed by the uniform nature of the computation: the same operation repeated over and over again on all sites. This is a property shared by all CFD schemes working on regular data structures. Eventually, a minor source of load unbalance comes from boundary conditions. By using buffer nodes, boundary conditions can be embedded into the bulk computation so that no special handling is required.

Finally, low CCR is secured again by the dominance of collision versus streaming steps of the algorithm, typical value being CCR $= O(10)$.

A word of caution is in order: LBE is computationally slightly more intensive than a normal explicit finite difference method, simply because there are more populations than hydrodynamic fields to keep track of. Typical CCRs for a plain explicit finite difference code is $O(1)$, which explains why these algorithms do not always easily achieve very high parallel efficiency.

A just critique needs be forestalled here.

Job turnaround, elapsed time to solve the problem, not parallel efficiency in itself, is the real measure of success. There is little point in heralding a high parallel efficiency obtained by raising the computational share instead of reducing the communication costs!

The fact remains that modern computer architectures put a high reward on the policy of '*data reusability*', namely once information is being read off from memory to the local buffer (high-speed buffer, or simply 'cache' in computer jargon), processing it with two floating-point operations costs less than twice a single operation, just because data transfer from/to memory to/from CPU is an expensive operation.

Bottom line: the computational overload of LBE does not really reflect into a corresponding toll on execution time just because LBE takes full advantage of the data reusability principle fostered by cache-based modern computer architectures.

This is even more true of parallel computers.

Serial and, more so, parallel efficiency, as combined with user-friendliness, are the basic assets of LBE for the years to come.

11.1.6 *Flexibility*

We discuss two distinct instances of flexibility: (i) geometrical and (ii) physical.

Geometrical flexibility Having assessed that LBE corresponds to a second-order Lagrangian integration along a set of constant characteristic speeds, we can proceed to examine further consequences of this peculiar feature. The main point is that space and time are *not* treated independently, but they are coupled instead via the light-cone condition $dx_{ia} - c_{ia}dt = 0$.

This light-cone discretization reflects the covariant, hyperbolic nature of LBE in which space and time are treated on the same footing. This is what confers on LBE its character of *finite-hyperbolicity* approximation (recall the adiabatic approximation of Part I). Within this picture, the molecules are *grid-bound*, i.e., they move in lock-step mode from one lattice site to another never stopping in between. They cannot live other than on lattice sites.

This property, a strict legacy of LGCA, yields important formal and practical advantages. For instance, the fact that the characteristics are constant may constitute a significant advantage for compressible applications because it trivializes the Riemann problem.

It also makes the scheme very simple in terms of data structure organization and permits the discretization of the streaming operator without any floating-point operation. It implies some limitations as well, most notably the inability to apply plain LBE to the more general case of non-uniform grids.

As discussed in the previous chapter, only recently, researchers in the field have come up with practical recipes to do away with this constraint. A much heralded virtue of LB methods is the ability to easily accommodate grossly irregular boundary conditions such as those encountered in flows through porous media. This emphasis is certainly well deserved, for it is true that accounting for whatever irregular distribution of solid obstacles within the flow is *conceptually* straightforward. Credit for this goes directly to the dual 'fluid–particle' nature of the LB scheme: move 'swarms' of particles along simple particle trajectories. Slightly underemphasized, perhaps, is the fact that the actual implementation of these boundary conditions is somewhat painful due to the relatively high number of populations, often to be treated differently for different propagation directions (outward, inward movers, corners being the typical nightmare of the hasty programmer). With this little practical caveat in mind, the fact remains that flows through porous media, and generalizations thereof, prove one of the most successful applications of LBE methods. A few comparative studies report that LBE superiority versus finite element methods surfaces more and more neatly as the 'randomness' of the media is increased [154]. More recent results on the comparison of LBE versus finite elements are provided by Kandhai *et al.* [155].

Physical flexibility and user-friendliness Ease of use and a gentle learning curve is another unquestionable asset of LB methods. Again, this is not surprising, since we are talking a simple explicit method on a regular uniform space–time grid.

But this is exactly the point!

Explicit finite difference methods are 'vanilla CFD', and as such, they were (and possibly still are) regarded as just too expensive (too many time-steps, too many grid points) to come anywhere near in competition with well rugged implicit solvers.

The dramatic growth of computing power (and the attendant drop of computing costs) is changing all that: quoting J. Boris [142], '*we are reaching the point of diminishing returns in terms of trading off computational cost for accuracy.*' And Boris goes further, '*It is now more effective to increase the number of grid points to improve spatial resolution and hence accuracy than to seek greater accuracy through high-order algorithms.*' Moral: if you have a simple, low-order, fast explicit solver just use it! This is even more true as parallel computing enters the scene. Also, as already noted, the hyperbolic nature of LBE implies $\Delta t \sim \Delta x$ rather than $\Delta t \sim \Delta x^2$, which means that time-steps do not collapse quadratically with decreasing mesh size [156].

Besides computational efficiency, ease of use brings other important bonuses along. A major one is the ability to embody mesoscopic physics via external effective potentials and/or generalized equilibria. We have noted previously that the whole nonlinear physics of the Navier–Stokes equations translates into a few more lines coding quadratic terms in the local equilibrium! The good news is that the same idea extends to generalized hydrodynamics.

Substantial new physics can be added within a few lines of extra code!

Whenever additional meso/macroscopic physics can be moulded into an effective force of the form $F(x, t; M(x, t), \nabla M)$, where M denotes any macro/mesoscopic field, this physics can also be encoded within a few lines into the very same LBE harness doing plain hydrodynamics. The power of discrete kinetic theory in action! This is the leading theme of the 'beyond' part of this book to be described shortly.

11.2 Link to fully Lagrangian schemes

As already mentioned, there are mixed opinions as to whether the physical roots of LB methods should be regarded as an asset rather than a liability. Work on the former line has been mentioned earlier in this book, when it was shown that LB methods can be viewed as specific instances of discretization schemes for the continuum BGK equation using Hermite expansions and numerical Gaussian quadrature for the actual evaluation of moments of the distribution function [157].

The conclusion drawn by the authors was that '*by realizing that LBE models are merely simple and rather primitive finite difference representations of the discrete BGK equation, we can employ more sophisticated numerical techniques in solving these equations with better efficiency, stability and flexibility.*'

In a rather remarkable paper, Ancona puts forward a rather opposite point of view [158], namely that, '*the lack of separation of physics and numerics by conventional kinetic theory approach is a major drawback when the purpose is solving partial differential equations.*'

To substantiate this statement, he presents an in-depth analysis of LBE schemes and points out their relation to a wider class of so-called Fully Lagrangian (FL) methods for partial differential equations. He shows that LB methods are a subclass of FL methods characterized by the following distinctive properties:

- *Pointwise conservation of physical quantities.*
- *Perturbative expansion of the Lagrangian variables.*
- *Use of discretization errors to represent physical effects.*

It is instructive to analyze these points in a little more detail.

First, let us clarify what we mean by fully Lagrangian methods.

The crucial idea is to transform Eulerian fields, such as density, flow, pressure as a function of x and t into a set of 'moving fields', the Lagrangian variables in point. Of course the way to achieve this is by no means unique. For instance, in the simple case of a 1D fluid, density ρ and speed u would be turned into a pair of right and left propagating fields $r(x,t)$, $l(x,t)$ obeying the following relations:

$$\rho = \frac{r+l}{2}, \tag{11.44}$$

$$\rho u = c\frac{r-l}{2}, \tag{11.45}$$

where c is the common speed magnitude of the Lagrangian movers ('particles'). By replacing (11.44) into the continuity and Navier–Stokes equations in 1D, we readily obtain the evolution equations for the Lagrangian movers:

$$D_t^+ r = S, \tag{11.46}$$

$$D_t^- l = -S, \tag{11.47}$$

where $D_t^{\pm} \equiv \partial_t \pm c\partial_x$ are the left/right Lagrangian derivatives and

$$S = 2c\partial_x \frac{rl}{r+l} + \frac{\partial_x P}{2c} \tag{11.48}$$

is the source term mediating the interaction between the two Lagrangian fields.

Finally, $P = -p + 2\mu\partial_x u$ is the one-dimensional stress tensor. It is immediately seen that the Lagrangian formulation discloses a beautiful duality which was left hidden in the Eulerian version. The transformation

$$c \to -c, \tag{11.49}$$

$$(r,l) \to (l,r) \tag{11.50}$$

leaves the eqns (11.46) and (11.47) invariant (in the limit $\mu \to 0$).

It is not hard to see that LB falls within this kind of formulation. To this purpose all we need is to march in time with a first-order Euler forward scheme and discretize space with a so-called *upwind* technique, namely $\partial_x r = r(x+\Delta x) - r(x)$ and $\partial_x l = l(x-\Delta x) - l(x)$.

Clearly, LBE is just a specific FL representation, many other, for better or for worse, being certainly conceivable.

In particular, coming to the first point of the list, it is easy to show that not all FL formulations lead to *pointwise* conservation of mass and momentum on a node-by-node basis. LBE does (by construction) and this surely marks a point in its favor.

The second point, perturbative treatment of the Lagrangian variables, is also easily illustrated. The source term S is a nonlinear (quadratic since we only have two fields) function of the Lagrangian fields. The way to discretize this term is again a matter of choice, each spawning its own numerical scheme. There is a clear link to hyperbolic methods with sources [159] and kinetic flux splitting methods in gas dynamics [160]. The qualifying LBE feature is that these nonlinear source terms are expanded *perturbatively* around local equilibria, thus making the scheme fully explicit.

This is a strictly physically inspired move, which buys a great computational simplification. Ancona points out that the replacement of the nonlinear source S with the BGK relaxation term $\omega(r - r^{\mathrm{e}})$ with a constant ω, makes the method only *conditionally consistent.* Recovery of the original continuum PDEs limit is no longer guaranteed for any generic value of ρ and u, but becomes conditioned on the small Mach number assumption. This is basically the positivity issue discussed previously in this chapter.

Helas, a weakness of LBE.

The final point, using discretization errors to model physical effects, refers to the idea of retaining second-order terms in the Taylor expansion of the discrete free-streaming operator in powers of the differential one D_i: $\Delta_i \sim D_i + D_i^2/2$.

As we learned in Part I, this leads to the so-called propagation viscosity, $\nu_p = -1/2$ (in lattice units) which adds to the collisional one to produce an effective viscosity $\nu = \nu_c - 1/2$. This is no coincidence and responds to the physical arguments behind the multiscale procedure.

Always in the spirit of locating LBE within the framework of numerical analysis, Ancona shows that the LBE discretization of the 1D Navier–Stokes equation, better known as Burger's equation, is *exactly* equivalent to the Dufort–Frankel scheme.[35] Indeed, the Dufort–Frankel scheme shares many general features in common with LBE, explicitness, second-order accuracy in space, unconditional stability (for diffusion equations) and conditional consistency.

On the other hand, the author himself recognizes that this identity between LBE and DF does not apply generically, which means that LBE is not just a matter of 'reinventing the wheel'. In fact, very recent work from the numerical analysis/applied mathematics front seems to reveal that LBE is a kind of *multi-difference* method in which propagation along different directions (nearest-neighbor, next-to-nearest neighbor) are subject to distinct type of finite differ-

[35]For a 1D diffusion equation $\partial_t f - D\partial_x^2 f = 0$, the Dufort–Frankel scheme reads as follows:

$$f(i,n+1) - f(i,n-1) = \frac{2D\Delta t}{\Delta_x^2}\left[f(i+1,n) - (f(i,n-1) + f(i,n+1)) + f(i-1,n)\right].$$

encing [161]. It is probably this multi-difference nature which makes of LBE more than a pre-existing finite difference scheme in disguise.

Both Ancona's and He–Luo's analyses show that, while it is true that LBE can be derived by combining ideas drawn from other numerical methods, it is no less true that this combination does not follow straight from any single other existing method.

On the other hand, as pointed out in [162], heralding the physics (as opposed to numerics) oriented character of LBE as a decided source of superiority might also be misleading. For instance, LBE schemes for complex geometries are heavily leaning on finite difference and finite volume ideas. A balanced and critical view is probably the healthiest attitude towards a fair appreciation of the real power and limits of the method.

11.3 LBE in a nutshell

We summarize the main LBE properties as follows:

- *Mathematical*

  Hyperbolic superset of the Navier–Stokes equations.
  Linear streaming operator, as opposed to nonlinear $u\nabla u$ Navier–Stokes advection.
  Nonlinearity hidden in a *local* collision operator, and formally separated from streaming.

- *Physical*

  Sound, as it is derived entirely within the statistical mechanics framework.
  Simple, as it is based on clear-cut stream–collide dynamics.
  Flexible towards additional (mesoscopic) physics.

- *Numerical*

  Explicit, synchronous.
  Second-order space–time accurate.[36]
  Advection-limited time-step.
  Unconditionally linearly stable.
  Conditionally nonlinearly stable.

- *Computational*

  Efficient on serial computers.
  Outstanding on parallel ones.

[36] Likely with a very small prefactor.

11.4 Exercises

1. Find the CFL condition for the 1D diffusion equation.
2. Write down an implicit LBE scheme: analyze the matrix structure of the resulting linear system.
3. Write down a 1D LBE scheme with 5 speeds: discuss spatial accuracy.
4. Prove that the $\sum_i f_i^2$ is an H-function for the linearized BGK equation.
5. Write the Dufort–Frankel scheme for the 1D diffusion equation in LBE form.

Part III

Beyond fluid dynamics

The ideas and applications discussed thus far have been restricted to flows of 'simple' fluids that can be described in terms of the Navier–Stokes equations alone. Modern 'do better–do faster' high-tech applications are placing a pressing demand of quantitative understanding of situations in which even the micro-physical world is not simple any more. Flows with chemical reactions, flows with phase transitions, multiphase flows, suspensions of colloidal particles, non-Newtonian flows, polymer melts, to cite but a few. This emerging sector of modern science, often referred to as 'complex flows', or more trendily, 'soft matter physics', portrays a multidisciplinary situation where fluid dynamics makes contact with other disciplines, primarily material science. There is a growing consensus that LBE, and extensions thereof, holds good promise as a computational tool for soft matter research. Ideally, LBE would fill the gap between fluid dynamics and molecular dynamics, namely the huge and all-important region where fluid dynamics breaks down and molecular dynamics is not yet ready to take over for lack of compute power. LBE is a good candidate to fill this gap because of its ability to seamlessly incorporate micro/mesoscopic details into the kinetic theory formalism via suitable external fields and/or equivalent generalizations of local hydrodynamic equilibria: a microscope for fluid mechanics, a telescope for molecular dynamics.

The final part of this book is therefore devoted to a cursory illustration of LBE applications in the direction of soft matter research. Because this is an open research front, the material is largely qualitative in character, the main task being again not to provide hard-core details, but rather to stimulate new ideas and further activity in this exciting and rapidly evolving frontier of LBE research.

12

LBE SCHEMES FOR COMPLEX FLUIDS

12.1 LBE theory for generalized hydrodynamics

The LBEs discussed so far revolve around two basic ideas:

1. *Construction of suitable discrete local equilibria, consistent with hydrodynamic conservation laws.*
2. *Collisional relaxation to these local equilibria, possibly in compliance with the second principle of thermodynamics.*

This is all we need to generate faithful dynamics of 'simple'[37] *isothermal, quasi-uniform, incompressible* fluids.

Many applications in modern science and technology need to address significantly more complicated problems, such as fast flows with substantial heat transfer, compressibility, chemical reactions, phase transitions and other assorted complexities. To handle this much broader context a new generation of LBEs is called for.

A very elegant way of generalizing LBE along the above direction is to convey the new physical effects into a generalized force term of the form:[38]

$$\frac{\vec{F}}{m} \cdot \frac{\partial f}{\partial \vec{v}}, \tag{12.1}$$

where the generalized force accounts for *both* external fields (such as gravity or electric fields) and *self-consistent* interaction forces associated with intermolecular interactions or any other effective interaction we wish to include.

The corresponding generalized LBE reads as follows:

$$\Delta_i f_i = \sum_j A_{ij} \left[f_j - f_j^{\mathrm{e}} \right] + S_i, \tag{12.2}$$

where the source term S_i can be formally interpreted as the discrete velocity representation of (12.1).

Operationally, S_i can be regarded as source of mass, momentum, energy or any other macroscopic observable of the form:

$$\Phi\left(\vec{x}, t\right) \equiv \sum_i \phi\left(\vec{c}_i, \vec{x}\right) f_i\left(\vec{x}, t\right), \tag{12.3}$$

where $\phi(\vec{c}_i, \vec{x})$ is a generic phase-space dependent microscopic quantity.

[37] Beware: simple fluids does *not* mean simple flows (turbulence docet!).

[38] By allowing a velocity dependence, $\vec{F} = \vec{F}(\vec{v})$, such a 'force' can also model mass conversion phenomena.

The rate of change in time of the observable Φ due to the source S_i is given by:

$$\frac{\delta\Phi\left(\vec{x},t\right)}{\delta t}=\sum_i\phi\left(\vec{c}_i,\vec{x}\right)S_i. \tag{12.4}$$

Depending on its strength and shape in momentum space, this term can result in dramatic changes of the flow dynamics, such as sharp interfaces, phase transitions, chemical reactions and other assorted complexities. In the simplest instance, S_i could just be a set of constants representing the effect of a gravitational field or a constant pressure gradient as discussed earlier in this book. More general situations are readily generated by letting S_i carry a generic space–time dependence either explicitly and/or in the form of a self-consistent *mean-field* coupling to macroscopic fields and/or their derivatives:

$$S_i\left[\vec{x},t;\Phi\left(\vec{x},t\right),\nabla\Phi,\ldots\right]. \tag{12.5}$$

Expressions like (12.5) are well suited to encode phenomenological knowledge about the micro–mesoscopic behavior of the system.

Regardless of the complexities which may arise in the fluid flow configuration, we shall require that the system keeps being described by *weak departures from generalized local equilibria*, so that a (generalized) hydrodynamic picture still holds true. These generalized local equilibria are naturally defined from eqn (12.2) as:

$$f_i^{\mathrm{e}}\rightarrow\tilde{f}_i^{\mathrm{e}}\sim f_i^{\mathrm{e}}-A_{ij}^{-1}S_j, \tag{12.6}$$

where A_{ij}^{-1} is the inverse scattering matrix.

This suggests a projection of the source term onto the set of eigenvectors of the scattering matrix:

$$S_i=S_0+S_1^a c_{ia}+S_2^{ab}Q_{iab}+\ldots, \tag{12.7}$$

where S_0 is the contribution to the density, S_1^a to the flow speed, S_2^{ab} to the (traceless) momentum flux tensor and so on for all higher moments. Upon tuning the moments of the source term, we can selectively inject mass, momentum and energy into the fluid in a fairly general and flexible way.

The discrete equilibria discussed in Parts I and II are given by an explicit function of the conserved quantities, the fluid density ρ and speed $\vec{u}$:

$$f_i^{\mathrm{e}}=F_i\left[\rho,\vec{u}\right], \tag{12.8}$$

where F_i is linear in the density and quadratic in the flow field.

Depending on the form of the source term, it might or might *not* be possible to express the generalized equilibria as the *same* map of a correspondingly generalized flow and density fields:

$$\tilde{f}_i^{\mathrm{e}}=F_i\left[\tilde{\rho},\tilde{\vec{u}}\right], \tag{12.9}$$

where

$$\tilde{\rho} \equiv \rho - \sum_{i,j} A_{ij}^{-1} S_j, \tag{12.10}$$

$$\tilde{\rho}\tilde{\vec{u}} \equiv \rho\vec{u} - \sum_{i,j} \vec{c}_i A_{ij}^{-1} S_j. \tag{12.11}$$

Even when this is possible, the *nonlinear* behavior of fluctuations around the generalized equilibria $\tilde{f}_i^{\rm e}$ might differ considerably from the corresponding behavior around 'bare' hydrodynamic equilibria. In particular, the realizability constraints $0 < f_i < 1$ and $0 < \tilde{f}_i^{\rm e} < 1$ surely restrict the choice of possible source terms compatible with generalized hydrodynamics.

Since we are going to abandon the comfortable situation of low speed, isothermal flows with minor density changes, it is clear that higher-order terms in the Chapman–Enskog analysis, which laid dormant so far, need a careful reconsideration. The 'danger zones' are regions with sharp density contrasts and vicinities of thin interfaces. For instance, in the presence of a strong density contrast, mass conservation across a two fluid interface, $\rho_1 u_1 = \rho_2 u_2$, enforces large speeds on the light fluid side which can deteriorate the realizability/stability of the scheme.

The careful understanding of such delicate effects sets a pointer to a higher-order LBE theory (*Lattice Burnett Equation*) which is, however, still in its infancy [163, 164].

With these caveats in mind, we now proceed to illustrate some selected applications of GLBE, namely:

- *Reactive flows*;
- *Multiphase flows*;
- *Flows with suspended objects*.

This list is by no means exhaustive of modern LBE applications, nor does it imply any priority judgement. It only serves the purpose of fixing the aforementioned ideas to a concrete backdrop. Also, we shall not delve into any specific details of the applications for at least three good reasons: (i) because they are mostly in flux; (ii) because this author is not a pan-expert in all of them; (iii) because, even if he was, the treatment would take another couple of books! For a timely update of the current state of the art the reader is referred to the recent literature [165–167].

12.2 LBE schemes for reactive flows

Reactive flow dynamics lies at the heart of several important applications such as combustion, heterogeneous catalysis, pollutant conversion, pattern formation in biology, to name but a few [168]. A beautiful account of reactive lattice gas cellular automata can be found in [169]. In general, LGCA are extremely well suited to describe reaction–diffusion applications but they do not easily extend to the case of fast flowing reactants. This task is easily undertaken instead by LBE.

The scope of this section is to provide a simple guideline to the inclusion of reactive phenomena within the LBE formalism.

In order to describe a reactive flow mixture of N_s miscible species (components) we shall introduce a *multicomponent distribution function* f_{is} where the index s runs over the species $s = 1, \ldots, N_s$. Chemical reactions are represented by a mass source term R_{is} expressing the mass change of species s propagating along direction i. The corresponding reactive LBE equation then takes the form:

$$\Delta_i f_{is} = -\omega_s \left(f_{is} - f_{is}^{\mathrm{e}}\right) + R_{is}. \tag{12.12}$$

The reactive term R_{is} must be designed so as to reproduce the correct density change rates $\dot{\rho}_s$ and energy input $\dot{Q}_s$ for each given species s.

In equations:

$$m_s \sum_i R_{is} = \dot{\rho}_s, \tag{12.13}$$

$$m_s \sum_i R_{is} c_{ia} = 0, \tag{12.14}$$

$$m_s \sum_i R_{is} \frac{c_i^2}{2} = \dot{Q}_s, \tag{12.15}$$

where m_s is the mass of the generic s-th species.

Moreover, since overall mass is conserved, additional constraints apply:

$$\sum_s m_s \sum_i R_{is} = 0, \tag{12.16}$$

$$\sum_s m_s \sum_i R_{is} \frac{c_i^2}{2} = \dot{Q}, \tag{12.17}$$

where $\dot{Q}$ is the total heat release/absorption per unit volume in the reactive flow. It should be noted that the density change rates $\dot{\rho}_s$ and the heat input $\dot{Q}$ are very sensitive functions of the fluid temperature T, typically in Arrhenius form $\mathrm{e}^{-T_{\mathrm{att}}/T}$, where T_{att} is the activation temperature above which chemistry takes on.

The unknown R_{is} can be sought in the form:

$$R_{is} = A_s + B_s \left(c_i^2 - k\right), \tag{12.18}$$

where $k = \sum_i c_i^3 / \sum_i c_i^2$ serves the purpose of orthogonalizing the basis $E_i^{(0)} = 1$, $E_i^{(a)} = c_{ia}$, $E_i^{(2)} = c_i^2 - k$, and solving the system (12.13) to yield:

$$A_s = \frac{\dot{\rho}_s}{E_1}, \quad B_s = \dot{Q}_s - \frac{\dot{\rho}_s}{E_2}, \tag{12.19}$$

where $E_a \equiv \sum_i |E_i^{(a)}|^2$ is the square norm of the a-th eigenvector $E_i^{(a)}$. The relations (12.19) specify in complete detail the reactive LBE scheme.[39]

A few comments are in order.

First, since $\dot{\rho}_s$ only depends on the various densities $\rho_1, \ldots, \rho_{N_s}$, the reactive LBE scheme preserves in full the much prized locality in configuration space which makes it so valuable for parallel computers. A further comment regards numerical stability: in lattice units ($\Delta t = 1$), the requirement is:

$$0 < R_{is} < r < 1, \tag{12.20}$$

where r, basically the ratio of the lattice time-step to the fastest chemical scale τ_c, must ensure realizability of the scheme. This means that the lattice time-step Δt cannot exceed a fraction of the shortest chemical time scale τ_c. Securing this inequality under all thermodynamic conditions is virtually impossible because of the dramatic dependence of the reaction rates on the temperature which makes $\tau_c \to 0$ in the limit $T \gg T_{\text{att}}$. A 'back of the envelope' estimate yields $\tau_c \sim 1\,\mu$s as a typical value for computationally feasible problems on present day computers. It is perhaps instructive to derive this figure explicitly.

Consider a reactive fluid in a box 1 m in side, and suppose we cover the box with 10^9 lattice nodes (thousand per site). This yields $\Delta x = 0.1$ cm. Take $c_s \sim 300$ m/s as a sound speed, the resulting $\Delta t \sim 0.1/(3 \times 10^4) \sim 3 \times 10^{-6}$ s. Such a computation requires approximately $24 \times 4 \times 10^9$ bytes of memory (FCHC lattice, single precision), namely 96 gigabytes of storage. Assuming 100 floating operations (flops) per site at each time-step, a computer delivering 1 Gigaflop would take 100 CPU seconds per step, namely 10^5 CPU seconds, roughly a day, for a 1000 step long simulation (a millisecond in real time). These simple considerations suggest that fast chemistry can be handled 'head-on' only for very small systems.

A final comment concerns the need of efficient multicomponent transport algorithms. Naive extensions of LBE schemes with $2D$ discrete speeds per species in D spatial dimensions are ruled out by memory considerations. As of today, the best option appears to be a blend between LBE and an extended version of the Lax–Wendroff scheme [117].

12.2.1 *Reactive LBE applications*

Reactive LBEs have been applied to a number of reactive flow simulations, including:

1. *Isothermal reaction–diffusion–advection with no heat release (Selkov model)* [170].
2. *Two-dimensional methane laminar flame with instantaneous chemistry and significant heat release* [171].
3. *Low Mach laminar combustion* [172].

[39] *Caveat*: the author is only aware of applications with $B_s = 0$ in which mass is equidistributed among all different directions.

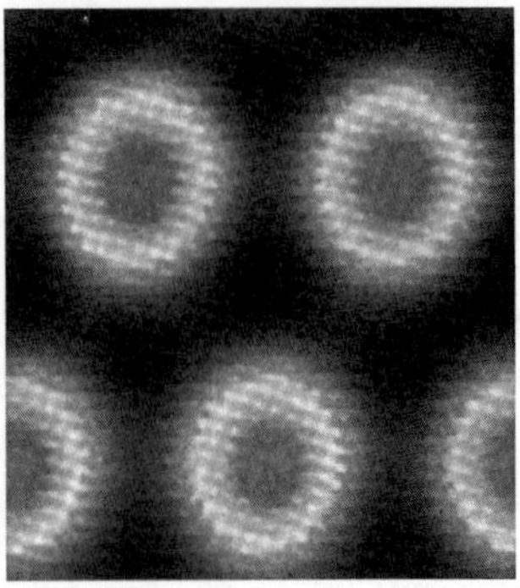

FIG. 12.1. The formation of hexagonal density patterns in an LBE simulation of the Selkov model (from [170]).

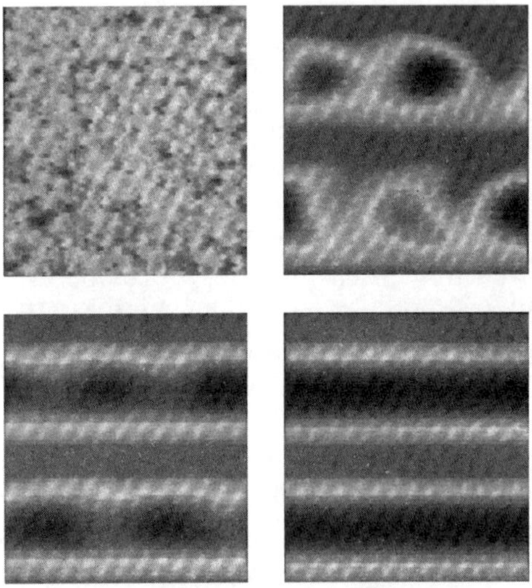

FIG. 12.2. The formation of lamellar density patterns in an LBE simulation of the Selkov model with fluid flow (from [170]).

The former case is relevant to biology-oriented applications, whereas the latter two relate to combustion engineering issues. The Selkov model is addressed by means of a reactive LBGK with an equidistributed mass source of the form (12.18) and temperature treated like a freely tunable parameter. The authors investigate several pattern formation scenarios with various diffusion coefficients with and without flowing solvent, always reporting satisfactory agreement with literature data.

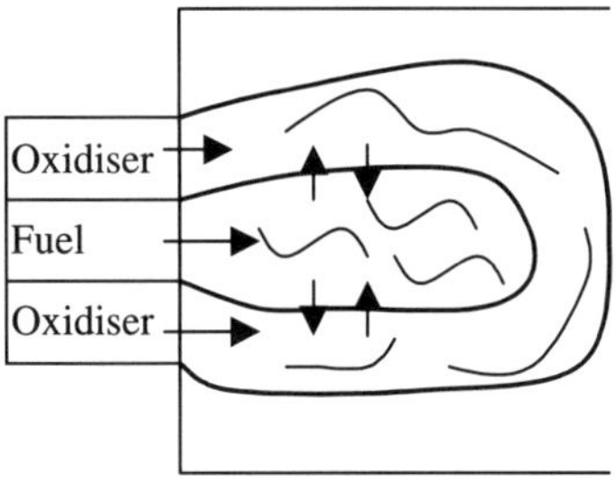

FIG. 12.3. Schematic picture of a diffusion flame. The central region is mostly filled with fresh unburned fuel, while the exterior one is mostly populated by the oxidizer. Upon diffusing at the interface, fuel mixes with the oxidizer and chemical reactions take place.

The laminar methane flame application was addressed with a four-dimensional FCHC scheme in which two dimensions are compacted to produce two additional scalars, the temperature T and the fuel/oxidizer mixture fraction Z, dynamically coupled to the two-dimensional reactive flow. The scheme is thermodynamically consistent only in the limit of infinitely fast chemistry where the reaction front becomes infinitely thin and can therefore be described by a single scalar function contour $Z(x, y, t) =$ constant associated with the geometrical location of the reaction front.

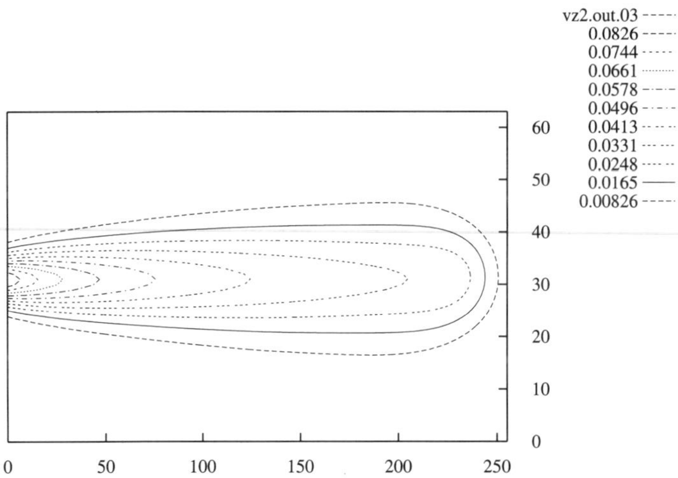

FIG. 12.4. Fuel mixture fraction $Z(x, y)$ contours of a diffusion laminar flame (after [171]).

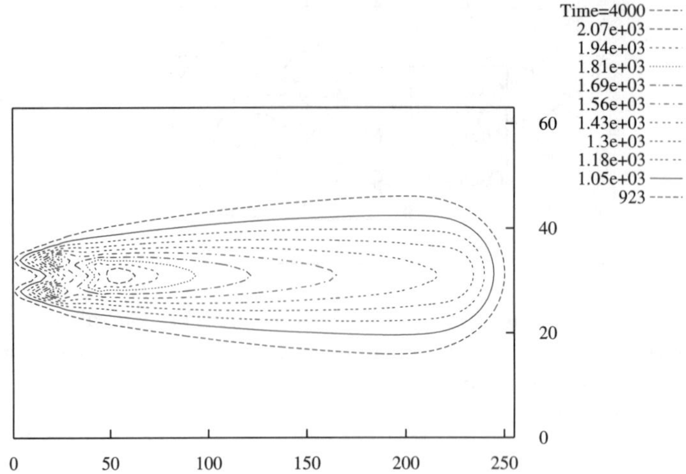

FIG. 12.5. Temperature $T(x, y)$ contours of the laminar flame (after [171]).

The third application refers to a weakly compressible low Mach flow with temperature dynamics. Finite time chemistry is accommodated via a suitable generalization of lattice BGK equilibria incorporating the effects of pressure fluctuations. The scheme is enriched with smart grid refinement techniques which raise significantly its computational competitivity.

Besides the applications discussed above, another area of great potential interest is catalytic conversion, namely low Reynolds flows with heterogeneous surface reactions [173]. These applications are well suited to reactive LBE schemes because they need not handle large systems and can easily accommodate surface irregularities thanks to the geometrical flexibility of the method. As we speak, the situation is less promising for combustion applications with significant heat release and compressibility effects, typically engine combustion. This is not due to chemistry in itself, but rather to the relative immaturity of LBE schemes in representing the wide temperature and density excursions occurring in compressible/compressed flows (see Chapter 14 on thermal LBE).

12.3 LBE schemes for multiphase flows

The dynamics of multiphase flows plays an important role in many areas of applied science and engineering, including oil–water flow in porous media, boiling fluids, liquid metal melting and solidification, to cite but a few.

The numerical simulation of multiphase flows is a challenging subject because, in addition to the usual difficulties associated with single phase motion, it also requires the tracking in time of the interfaces (moving boundaries) between the different fluids [174, 175]. A thorough description of multiphase LGCA models can be found in [176]. In the sequel we shall present a cursory view of the main

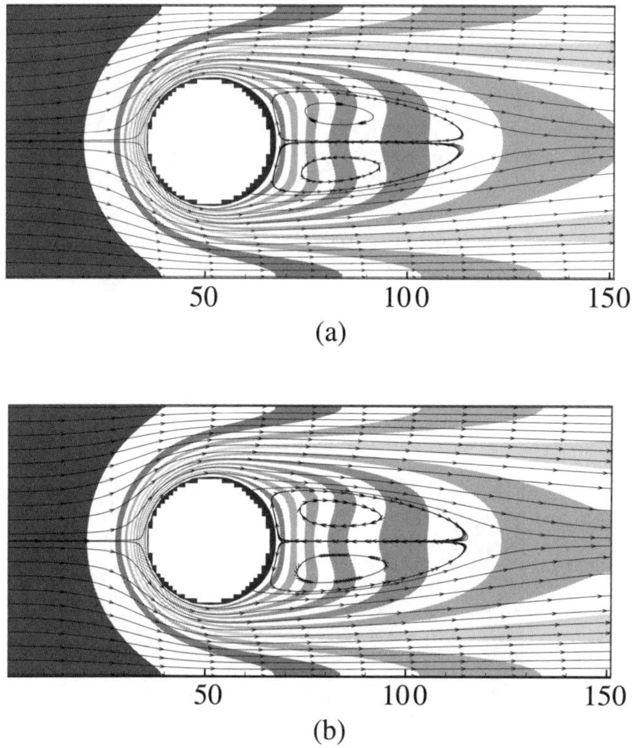

FIG. 12.6. Low Mach reactive flow LBGK calculation. Velocity streamlines and temperature contours past a cylinder. Case (a) finite difference solution of the Navier–Stokes equations with a pressure relaxation method. Case (b) LBGK solution. (After [172], courtesy of O. Filippova.)

LBE formulations for immiscible multiphase flows, namely:

- *Chromodynamic models*;
- *Pseudo-potential models*;
- *Free-energy models*;
- *Finite density models.*

Such a long list clearly indicates that no general consensus has emerged as yet on which one of these methods should be recommended as the best LBE solution for multiphase problems. In fact, all of them are still affected by a certain degree of inevitable empiricism due to the fact that, *even in the continuum*, a fully-fledged kinetic theory of complex fluid flow is still missing. In principle, we are faced here with an exciting opportunity to promote truly fundamental research work in a difficult sector of non-equilibrium statistical mechanics. On the other

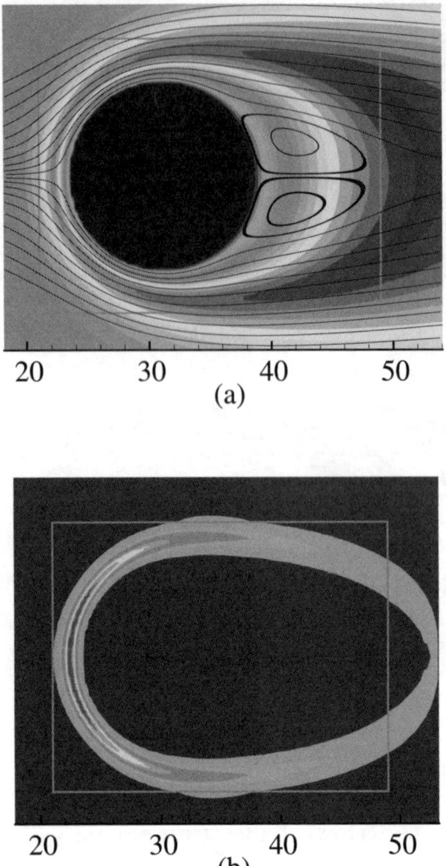

FIG. 12.7. Low Mach reactive flow LBGK calculation. Close-up of the temperature and reaction rates. (After [172], courtesy of O. Filippova.)

hand, the pathway to realize such an opportunity is littered with technical and conceptual problems related to the discrete nature of the lattice. Turning around these problems is the major challenge facing LBE research in this field.

12.3.1 *Surface tension and interface dynamics*

Before plunging into the description of the various LBE models a few basic notions on the physics of multiphase flows and the associated numerics are worth calling to mind.

The key to the physics of multiphase flows is the notion of surface tension. From a macroscopic point of view, surface tension is defined as the reversible work per unit surface needed to increase the area A of surface Σ by an amount

ΔA:

$$\Delta W = \sigma \, \Delta A. \tag{12.21}$$

Consider for clarity a spherical liquid droplet of radius R at a pressure P_l, immersed in its vapor at a pressure $P_v > P_l$. The question is how much work is to be spent on the vapor–liquid system to expand the radius of the liquid droplet from R to $R + \Delta R$ (condensation). This work is given by $\Delta W = (P_v - P_l)\Delta V$, where $\Delta V = 4\pi R^2 \Delta R$ is the volume change of the droplet. This is the energy supply from the exterior needed to win the action of surface tension, whose 'task' is to withstand the growth of the liquid surface. In equations:

$$(P_v - P_l) \, \Delta V = \sigma \, \Delta A. \tag{12.22}$$

Upon spelling out the elementary geometrical factors, we obtain the celebrated Laplace relation:

$$\Delta P = \frac{2\sigma}{R}. \tag{12.23}$$

Note that the surface tension dominates at large curvatures of the interface and its effects vanish on a flat surface. This is at the heart of capillary motion, namely the penetration of fluids in small pores where surface effects play a leading role.

Microscopically, surface tension is related to intermolecular interactions. With reference again to our liquid droplet, let us consider a molecule sitting right on the droplet interface. Since the liquid is denser than the vapor, this boundary molecule interacts with more 'liquid' molecules than 'vapor' ones and since the intermolecular potential is attractive (on scales above the hard-core repulsion region, typically a few Ångströms), the net result is that the boundary molecule naturally tends to be pulled back into the liquid region. This contrasts with the condition of an internal molecule which, being surrounded by an equal number of molecules in all directions, does not experience any net force.

The conclusion is that a surface molecule has an excess of energy (surface energy) with respect to an internal one, the difference representing the work needed to extract the internal molecule and 'peel it off' the surface. The same energy must be supplied to push a vapor molecule inside the liquid droplet (ticket to gain 'liquid status') which is another way of saying that surface growth involves an energy toll. Surface tension is a decreasing function of temperature and vanishes at the critical liquid–vapor point, where the two phases become virtually indistinguishable. A derivation of this dependence from first principles implies detailed studies of molecular dynamics.

As usual, these are of no direct concern to us. For our purposes, surface tension can be regarded as a given input parameter in the same league as viscosity, conductivity and other transport properties. The central difference, though, is that while viscosity and conductivity are well-defined concepts also for an ideal gas, surface tension commands a non-ideal equation of state describing *phase-transition* phenomena. This is a major feature that needs to be addressed by all numerical methods for multiphase flows, including of course LBE.

12.3.2 *Numerical methods for flows with interfaces*

The numerical simulation of multiphase flows is a challenging subject because, in addition to the usual difficulties associated to single phase motion, it requires the tracking in time of the interfaces (moving boundaries) between the different fluids, whose shape can become fairly complex as time unfolds. The existing methods to deal with moving interfaces split into basically two distinct categories:

- *Front-Tracking (FT);*
- *Front-Capturing (FC).*

In FT methods, dynamic degrees of freedom (markers) are attached to the moving interface, and their dynamics are explicitly designed to follow the interface evolution. By doing so, the interface can be tracked fairly accurately, as long as the topology remains sufficiently smooth. If, however, the interface topology becomes very tortuous or undergoes qualitative changes, such as break-up and reconnection, FT run into the typical difficulty of Lagrangian methods, namely ill-conditioning and singularities due to markers coming too close to each other.

Front-capturing methods solve this problem by defining a data structure throughout the entire computational domain (Eulerian approach). The interface is located there where discontinuities take place. These methods side-step all problems related to large distortions of the interface but, in revenge, they suffer from rather severe numerical diffusion effects which tend to smear out the interface in the course of the computation.

Many variants of both front-tracking and front-capturing methods have been proposed in the recent literature [174, 175], including hybrid methods combining the best of the two. This is a very active and fast moving 'front' (forgive the pun) of research which goes beyond way the scope of the present chapter.

So, let us now go back to LBE models for non-ideal fluids.

12.3.3 *Chromodynamic models*

The first multiphase LBE model was introduced by Gunstensen *et al.* (1991) [177], based upon the two component lattice gas model developed by Rothman and Keller [178].

In these models one begins by introducing two-particle distributions, say Red and Blue, f_{iR} and f_{iB}, for the two different fluids. Of course, color is just a mnemonic for different species, say water and oil, or any other attribute allowing to tell the two fluids apart. Each phase obeys its own LBGK equation:

$$\Delta_i f_{is} = -\omega_s \left(f_{is} - f_{is}^{\mathrm{e}}\right) + S_{is}, \quad i = 1, \ldots, b, \quad s = \text{Red, Blue}. \tag{12.24}$$

The source term S_{is} represents the mesoscopic interaction between the two phases and it is therefore in charge of describing phase separation via surface tension effects. The central quantities in Gunstensen, Rothman, Zaleski and Zanetti treatment are the *color current* K_a and *color gradient* G_a defined as follows:

$$K_a\left(\vec{x}, t\right) = \sum_i c_{ia}\left(f_{iB} - f_{iR}\right) = K_{aB} - K_{aR} \tag{12.25}$$

and

$$G_a(\vec{x},t) = \sum_i c_{ia}\left[\rho_B(\vec{x}+c_{ia}) - \rho_R(\vec{x}+c_{ia})\right]. \tag{12.26}$$

Note that both quantities vanish in color-blind homogeneous regions ($f_{iR} = f_{iB}$) where the two phases are in a perfect balance (perfectly miscible fluids). It can also be shown that single phase regions ($f_{iB} \cdot f_{iR} = 0$) are color gradient free if both fluids are incompressible. It is natural to define a *color potential* $V \sim G^2$ such that the gradient G_a points in the direction normal to the Blue–Red interface (see Fig. 12.8). Gunstensen *et al.* choose the source term in the heuristic form:

$$S_i = A\frac{G_i^2 - kG^2}{G}, \tag{12.27}$$

where $G_i = G_a c_{ia}$ is the projection of the color gradient along the i-th direction, $k = bc^2/D$ and A is a free parameter controlling the actual value of surface tension. Since this term does conserve blue and red densities separately, it can only set the stage for phase segregation but not promote it in actual terms. For this purpose Gunstensen *et al.* introduce an additional *color redistribution step*, which consists of (counter)aligning the color current K_a with the color gradient G_a in such a way as to minimize the 'color energy' (see Fig. 12.8):

$$W = \sum_a K_a G_a. \tag{12.28}$$

This rule lies at the heart of phase-segregating behavior.

In passing, we note that such a type of 'smell and go' dynamics is commonplace in many other sectors of complex fluid dynamics, including polar fluids and biological flows.

The Gunstensen *et al.* multiphase LBE opened the way to a series of applications such as two- and three-dimensional simulations of multiphase flows in porous media.

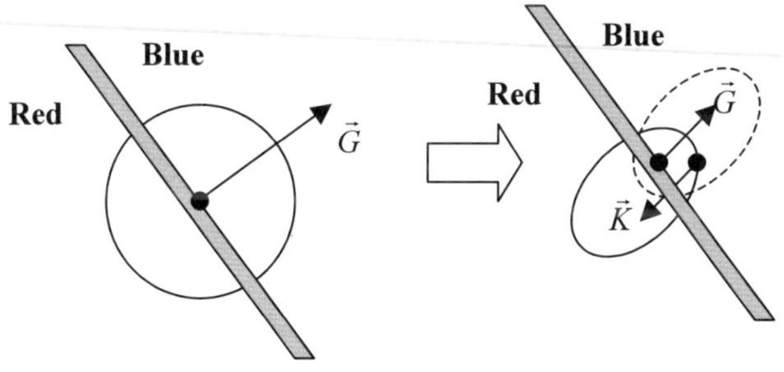

FIG. 12.8. Effect of the multiphase collision rule. The color gradient breaks the Red–Blue internal symmetry (circle) and introduces a preferential direction for Red or Blue particles to move across the interface (ellipses).

A more in-depth analysis revealed a few weak sides, however, in particular a certain degree of anisotropy of the surface tension, producing a spurious dependence on the interface orientation. In addition, near-interface parasitic microcurrents were reported which stem again from spurious invariants (subsequently, these parasitic currents were also discovered in other numerical methods!). In addition, the Gunstensen *et al.* scheme is relatively heavy from the computational point of view because of the variational minimization required by the recoloring step. This spurred the development of further models as discussed in the following.

12.3.4 *The pseudo-potential approach*

The chromodynamic approach of Gunstensen *et al.* builds on a significant abstraction of the physical reality; in fact, the 'color force' is nothing but the logical statement that molecules sitting at the interface between, say, dense and light fluid, experience a net force driven by the different values of the average intermolecular distance in the two fluids. It is therefore natural to look for more physically-oriented representations, in which these forces are directly encoded as the result of pairwise molecular interactions.

This is the pseudo-potential approach introduced by X. Shan and H. Chen [179]. These authors represent the source term in (12.24) as follows:

$$S_{is} = \vec{F}_s \cdot \vec{c}_i \equiv F_{sa} c_{ia}, \tag{12.29}$$

where the force F_{sa} comes from the following pairwise interaction potential:

$$V_s(\vec{x}) = \sum_{s'} \sum_i V_{ss'}(\vec{x}, \vec{x} + \vec{c}_i). \tag{12.30}$$

The physical meaning is clear: at each lattice site $\vec{x}$, the net force experienced by a particle of species s is the sum of the momentum exchanges with particles of all other species in the neighborhood $\vec{x}_i = \vec{x} + \vec{c}_i$ of the site $\vec{x}$.

The pseudo-potential matrix $V_{ss'}(\vec{x}, \vec{y})$ is chosen in the typical propagator form:

$$V_{ss'}(\vec{x}, \vec{y}) = \Psi_s(\vec{x})\, G_{ss'}(\vec{x}, \vec{y})\, \Psi_{s'}(\vec{y}), \tag{12.31}$$

where $G_{ss'}(\vec{x}, \vec{y})$ is the Green's function expressing the pairwise interaction between species s and s' at sites $\vec{x}$ and $\vec{y}$, respectively, and $\Psi_s(\vec{x}) \equiv \Psi[\rho_s(\vec{x})]$ is a kind of free energy of the system.

For most practical purposes, this interaction is taken in the 'corral' form:

$$G_{ss'}(\vec{x}, \vec{y}) = \begin{cases} G_{ss'} & |\vec{y} - \vec{x}| = |\vec{c}_i|, \\ 0 & \text{otherwise.} \end{cases} \tag{12.32}$$

The strength of the interaction is controlled by the amplitude parameters $G_{ss'}$, playing the role of effective (inverse) temperatures. Note that positive $G_{ss'}$ means attraction and negative $G_{ss'}$ repulsion. As a result, phase separation is encoded

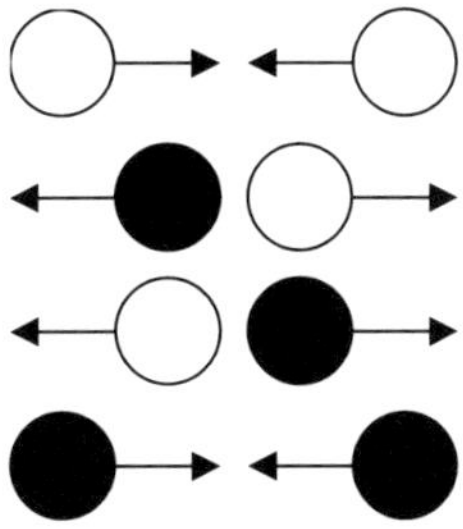

FIG. 12.9. The meaning of the pseudo-potential force. Each lattice population experiences an attractive force from color-mates neighbors and a repulsive one from the opposite color. The resulting force is the total sum over the neighborhood.

by positive entries in the diagonal of the interaction matrix $G_{ss'}$ and negative ones on the off-diagonal. Similar models for liquid–gas phase transitions have been used by Qian *et al.* [180].

A further simplification due to Shan and Chen permits the simulation of binary fluids with a single species only, i.e., $G_{ss'} = G$, whose density automatically identifies the light and heavy phases [181]. In this case, the transition between heavy and light fluid is controlled by the function $\Psi[\rho]$, which serves as an effective free energy of the model.

This function is typically taken in a switch-like form with a sharp transition between the dense and light phases,

$$\Psi(\vec{x}) = \Psi[\rho(\vec{x})] = \Psi_0 \left(1 - \mathrm{e}^{-\rho/\rho_0}\right), \tag{12.33}$$

where ρ_0 is a reference density marking the border $\Psi = 0/1$ between light/heavy fluids, respectively.

The force associated with this pseudo-potential is given by

$$F_a(\vec{x}, t) = -G\Psi[\rho(\vec{x})] \sum_i \Psi[\rho(\vec{x}_i, t)]\, c_{ia} \tag{12.34}$$

and it is easily checked to conserve the total momentum in the system, but not the local one on a site-by-site basis.

Thus this term contributes a source term ρF_a to the right-hand side of the Navier–Stokes equations of each species separately. Of course, the sum over all species yields zero since the momentum lost on one species must necessarily be gained by another one. This is an important difference with respect to the Gunstensen *et al.* model in which momentum was conserved locally for each species.

This force contributes an extra term to the momentum flux tensor

$$P_{ab} \to P_{ab}^{\rm NS} + \frac{1}{2} G \Psi \left[\rho\left(\vec{x}\right)\right] \sum_i \Psi\left[\rho\left(\vec{x}_i\right)\right] c_{ia} c_{ib}, \tag{12.35}$$

where, again, the superscript 'NS' denotes the ordinary single phase Navier–Stokes expression. By letting Ψ depend on the fluid density ρ, generic non-ideal gas equations of state $P = P(\rho)$ are obtained, including Van der Waals-like expressions $(P - a/V^2)(V - b) = $ constant, $V = m/\rho$ being the volume of the system.

The attractive feature of this model is that phase separation takes place spontaneously whenever the interaction strength G exceeds a critical threshold G_c, thus fitting naturally the physical notion of G as an inverse effective temperature of the system. Such a property is indeed beautifully confirmed by numerical simulations of binary fluid separation. Also, the model is really easy to use, which explains why it is probably the most popular choice to date.

Unfortunately, mechanical stability of a plane interface, as expressed by the differential condition $\partial_b P_{ab} = 0$, shows that $\Psi = \Psi_0 \mathrm{e}^{-\rho_0/\rho}$ is the *only* functional form compatible with continuum thermodynamics. This limitation reflects the lack of energy conservation.

12.3.5 *The free energy approach*

A step forward in the direction of thermodynamic consistency was taken by Swift, Osborne and Yeomans [182]. These authors introduce the equilibrium pressure tensor for a non-ideal fluid directly into an extended form of the collision operator. The idea is that, by doing so, the fluid is instructed to reach the right thermodynamic equilibrium directly under the effect of the correct equation of state. The Swift–Osborne–Yeomans method builds on the Van der Waals formulation of a two component isothermal fluid. The basic object of the theory is the free energy density functional Ψ, (free energy per unit volume) defined as

$$\Psi[\rho] = \int \frac{1}{2} k \left[(\nabla \rho)^2 + \psi(\rho) \right] \mathrm{d}V, \tag{12.36}$$

where the first term is the energy penalty paid to build density gradients, whereas the second term is the bulk free energy. The non-local pressure relates to Ψ through the following expression:[40]

$$P = \rho \frac{\mathrm{d}\Psi}{\mathrm{d}\rho} - \Psi = P_0 - k\rho\nabla^2\rho^2 - \frac{1}{2} k \left|\nabla\rho\right|^2, \tag{12.37}$$

where

$$P_0 = \rho\Psi' - \Psi \tag{12.38}$$

[40] We recall that the Helmholtz free energy F of a fluid of volume V is defined as $F = U - TS$ where U is the internal energy and S the entropy. The Gibbs free energy of the same fluid is given by $G = F + PV$. For an isothermal process ($\mathrm{d}T = 0$), we have $\mathrm{d}F = -P\,\mathrm{d}V$ and $\mathrm{d}G = V\,\mathrm{d}P$.

is the equation of state of the fluid (prime stands for derivative with respect to density). The full pressure tensor in a non-uniform fluid includes an off-diagonal component

$$P_{ab} = P\delta_{ab} + k\partial_a \rho \partial_b \rho, \tag{12.39}$$

where the second term is related to interfacial surface tension effects.

How do we encode this pressure tensor in the equilibrium distribution?

The recipe is to add (weakly) non-local terms to the discrete equilibria. In particular, for a seven state FHP lattice, Swift–Osborne–Yeomans propose the following expression:

$$f_{ie} = A + Bc_{ia}u_a + Cu^2 + Dc_{ia}c_{ib}u_a u_b + F_a c_{ia} + G_{ab}c_{ia}c_{ib}. \tag{12.40}$$

The Lagrangian parameters A, B, C, D, F_a, G_{ab} are prescribed by the usual conservation of mass, momentum, momentum flux tensor constraints, with P_{ab} given by (12.39). These relations are sufficient to compute the parameters as a function ρ *and its spatial derivatives*, thus solving the problem of identifying a proper free energy for the multiphase LB system.

Swift–Osborne–Yeomans demonstrate their model with 2D simulations based on the Van der Waals fluid free energy density $\Psi_{\mathrm{VdW}} = \rho T \ln \rho/(1-\rho^b) - a\rho^2$.

Like their predecessors, they test their scheme against Laplace's law:

$$P_{\mathrm{in}} - P_{\mathrm{out}} = \frac{\sigma}{R}, \tag{12.41}$$

where 'in/out' refer to inner/outer pressure of a bubble of radius R (Fig. 12.10). Another test refers to the dispersion relation of capillary waves (Fig. 12.11).

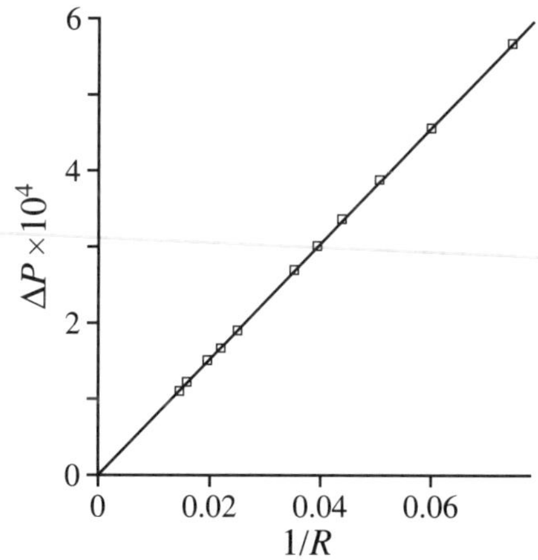

FIG. 12.10. Pressure difference between the inside and the outside of a spherical domain of radius R, as a function of $1/R$ (test of Laplace's law). The solid line is the exact result for a flat interface ($R \to \infty$) (from [182]).

They also compute the coexistence curve between the two phases for several values of the (static) fluid temperature, reporting excellent agreement with thermodynamic theory (Fig. 12.12).

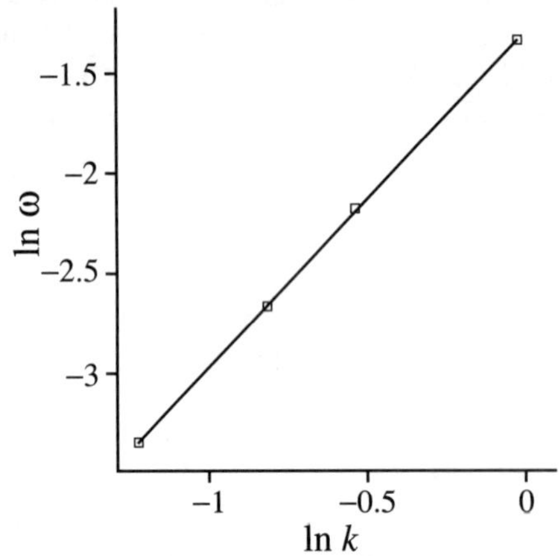

FIG. 12.11. The dispersion curve for capillary waves on an interface. The best fit line has a slope 1.6 ± 0.05, slightly in excess of the theoretical value $3/2$. This departure is attributed to curvature effects (from [182]).

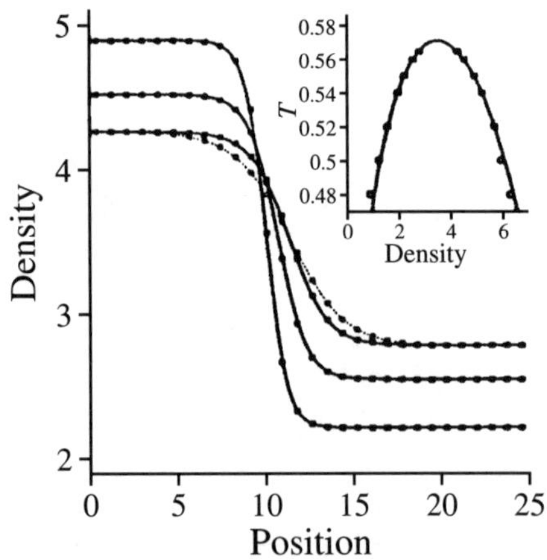

FIG. 12.12. Equilibrium density profiles normal to a flat interface for a Van der Waals fluid for the three highest values of temperature shown on the coexistence curve (inset). The solid lines are numerical solutions of the continuum thermodynamic equations, while the points are from the lattice Boltzmann simulations (from [182]).

The authors advocate significant reduction of the velocity fluctuations near the interface and good isotropy of the surface tension.

As a further bonus, they also propose to study wetting phenomena by including a suitable external chemical potential $\mu(\vec{x})$.[41] Gradients of the chemical potential act as an effective thermodynamic force $F_t = -\rho\nabla\mu/3$ and can be readily included as an extra source term in the right-hand side of LBE. By letting a nonzero μ only at the walls with a different intensity for the two species, the affinity of the wall to the two fluids can be varied in a simple and physically appealing way. In particular, by tuning the range of the chemical potential, one can construct diffuse boundaries along the lines of molecular dynamics. According to the authors, these diffuse boundaries significantly soften the pathological orientation issues associated with sharp crystallographic boundaries.

Potential liabilities to be signaled are: (i) density gradients in the local equilibria open up an exposure to higher-order lattice artifacts close to sharp interfaces; (ii) analysis of the energy equation reveals that the condition of a constant temperature cannot be satisfied, so that, again, thermodynamic consistency is not achieved.

The actual impact of these liabilities is likely to be problem-dependent and hard to judge *a priori*, but the model certainly looks interesting.

12.3.6 *Finite density models*

All multiphase models presented so far include temperature only as a static parameter, no self-consistent dynamics is allowed, on pain of incurring thermodynamic inconsistencies. This difficulty is not merely a technical one. The point is that the self-consistent inclusion of energy conservation necessarily involves the two-body radial distribution function, $\rho_{12} = \int f_{12}\, d\vec{v}_1\, d\vec{v}_2$:

$$E = E_{\text{kin}} + E_{\text{pot}} = \int f\,(x, v)\,\frac{1}{2}v^2\, d\vec{v} + \int \rho_{12}\,(x, x + r)\, V\,(r)\, dr. \tag{12.42}$$

This is a daunting task since the two-body distribution function carries a lot of information, thirteen independent variables in three dimensions!

Until very recently, this was regarded as *the* outstanding, to some perhaps even insurmountable [183], problem in multiphase LB research.

In a recent paper, however, L. S. Luo suggests that most problems fade away *by going back to the Boltzmann equation for dense (non-dilute) gases*, the time-honored Enskog equation [184].

Luo's point is that there is no consistent way of accounting for non-local interactions without including finite size (hence finite density) effects as well. His proposal is then to solve the Boltzmann–Enskog equation exactly the same way one does with the Boltzmann equation for dilute gases, namely by projecting onto a Hermite basis and subsequent evaluation of the kinetic moments by numerical quadrature [185].

[41] The chemical potential is a measure of the energy change with the number of molecules in the system, namely $\mu = \partial U/\partial N$.

Let us explore things in somewhat more detail.

As is well known, the standard Boltzmann equation does not apply to dense gases or liquids, but only to rarefied gases characterized by vanishingly small values of the inverse gassosity parameter s/l_μ, where s is the typical size of the molecule (range of its interaction potential) and l_μ its mean free path.

More specifically, the Boltzmann equation corresponds to the limit $n \to \infty$, $s \to 0$, $ns^2 \to$ constant, where n is the fluid number density and the resulting constant is basically the inverse mean free path. It is clear that this limit runs over any excluded volume effects (the finite size of the molecules, term $(V-b)$ in the Van der Waals equation), since the volume fraction, ns^3/N, goes to zero. No chance to capture non-ideal equations of state in a physically consistent way. Therefore, the key to achieve thermodynamic consistency is to explicitly acknowledge excluded volume effects; that is, finite size molecules. Finite size molecules call for finite density corrections to the Boltzmann equation, a classical, and difficult, topic of kinetic theory [186].

The Enskog equation reads [187]:

$$\partial_t f + v_a \partial_{x_a} f + F_a \partial_{v_a} f = J(f,f) \tag{12.43}$$

with the collision integral:

$$\begin{aligned} J(f,f) = \int [f(\vec{x},\vec{v}')\, f(\vec{x}+2\vec{s},\vec{v}_1')\, g(\vec{x}+\vec{s}) \\ - f(\vec{x},\vec{v})\, f(\vec{x}-2\vec{s},\vec{v}_1)\, g(\vec{x}-\vec{s})]\, \mathrm{d}\vec{v}_1 \mathrm{d}\vec{v}_1', \end{aligned} \tag{12.44}$$

where g is the *reduced radial distribution*:

$$g = \frac{f_{12}}{f_1 f_2} \tag{12.45}$$

describing two-body correlations ($g \to 1$ for an ideal gas).

It can be shown that $J(f,f;s)$ goes exactly into the usual quadratic Boltzmann collision operator $Q(f,f)$ in the limit $s \to 0$, so that finite size corrections to the Boltzmann equation can be represented by the operator $J_1(f,f) = J(f,f;s) - Q(f,f)$, arguably a linear quantity in the particle size s.

The key ingredient of the Enskog operator is the radial distribution g, manifestly an unknown, for it includes two-body information not available at the Boltzmann level.

Not surprisingly, this is the tunable knob of Luo's theory. Consistency with the Boltzmann equation requires any discretization of $J(f,f)$ to yield the corresponding discretization of $Q(f,f)$ in the limit $g=1$. Next, Luo replaces $Q(f,f)$ with a BGK relaxation term, and upon expanding about $\vec{x}$, to the first-order in s, he obtains the following finite-size BGK equation:

$$\partial_t f + v_a \partial_{x_a} f + F_a \partial_{v_a} f = -g\omega\,(f - f^{\mathrm{e}}) + (g-1)\, J', \tag{12.46}$$

where f^{e} is the usual quadratic local equilibrium, whereas the finite size remainder of the Enskog collision operator reads as follows:

$$J' = -b_2\,(g-1)\,f^{\mathrm{e}}\rho\,(\vec{v}-\vec{u})\cdot\nabla\ln\left(g\rho^2\right), \tag{12.47}$$

where b_2 is the second virial coefficient of the equation of state [187].[42] The last step to drive the point home is to discretize the finite size BGK eqn (12.46) according to the usual LBE practice. The finite size term J' does not raise any formal problem and can be treated the way we always did with mesoscopic sources, namely by local evaluation at $\vec{v}=\vec{c}_i$:

$$J'_i = -b_2\,(g-1)\,f^{\mathrm{e}}_i\rho\,(\vec{v}_i-\vec{u})\cdot\nabla\ln\left(g\rho^2\right). \tag{12.48}$$

Formally, this is a mesoscopic term, with a weak non-locality (density gradient) and a nonlinear dependence on the flow density, via the function $g(\rho)$ (beware the symbols, nothing to do with Galilean invariance!).

The corresponding hydrodynamic limit yields the Navier–Stokes equations with a viscosity

$$\nu = \frac{1}{3g\omega}-\frac{1}{6} \tag{12.49}$$

and the following equation of state:

$$P = \rho T\left[1+b_2\,(g-1)\,\rho\right]. \tag{12.50}$$

The standard ideal gas situation is recovered in the limit $g\to 1$, $b_2\to 0$. Given the equation of state, the Helmholtz free energy density is readily derived by the standard thermodynamic relation $H=U+PV$:

$$\frac{H(\rho)}{V} = -V^{-1}\int P\,\mathrm{d}V = \rho\int\frac{P}{\rho^2}\,\mathrm{d}\rho. \tag{12.51}$$

The free energy spelt out, the rest is easy. All relevant thermodynamic constructs, such as the Maxwell procedure to derive liquid–gas coexistence curves, can be performed with no concern of consistency.

Of course, since the Enskog equation itself is only an approximation (hard spheres, instantaneous collisions) of the true dense gas picture, Luo's theory cannot be criticism-free either. In addition its practical value remains to be demonstrated (no simulation results available to the best of the author's knowledge).

This theory nonetheless has an interest on its own, for it shows a possible way of analyzing the macroscopic limit of LBEs for non-ideal fluids on more systematic and mathematically well-defined grounds.

12.3.7 *Miscellaneous multiphase LBE applications*

Because LB methods do not track interfaces, but rather let them spontaneously emerge from the underlying microdynamics, they are well positioned to study

[42] The virial coefficient of order k, b_k, is defined as the coefficient of the k-th order term in the polynomial expression of the pressure as a function of the density $P/T=\sum_{k=1} b_k\rho^k$.

fluid flows with complex interface dynamics. Among others, successful multiphase LBE applications include:

- Spinodal decomposition of binary and ternary mixtures [188, 189].
- Liquid–gas two domain growth [190].
- Rupture and coalescence of fluid interfaces [191].

Typical phase separation patterns are shown in Fig. 12.13.

The reported advantages of the LB approach for this type of problem are:

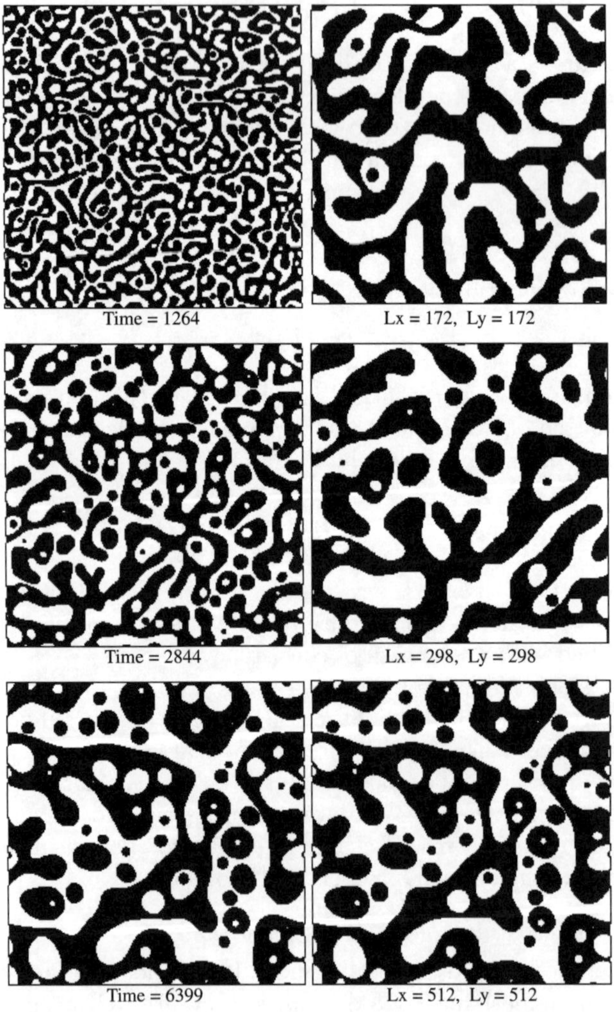

FIG. 12.13. Phase separation in a binary LBE fluid. (Courtesy of J. Yeomans.)

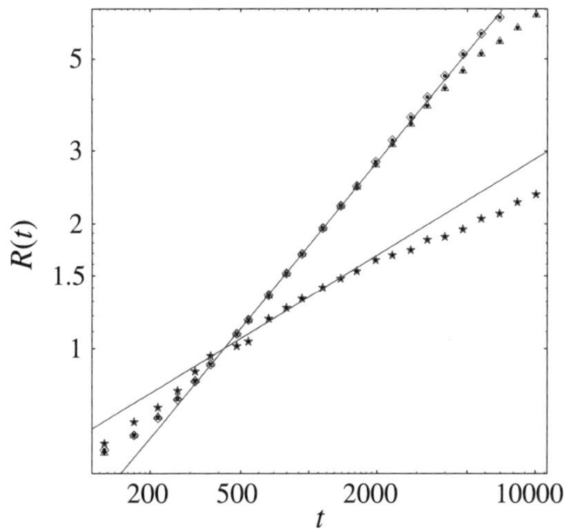

FIG. 12.14. Mean domain size of the two separated phases as a function of time. (Courtesy of J. Yeomans.)

- *Access to mesohydrodynamic scales virtually off-limits to molecular dynamics.*
- *Handy fine-tuning of the free energy to pick up the desired equilibrium as an attractor of the relaxation dynamics.*

Particularly promising is the combination of LBE and LGA methods for the study of interacting multiphase amphiphilic fluids [192], namely fluids consisting of two (or more) immiscible phases, such as oil and water, and an amphiphilic species (surfactant). In this case, fluctuations of the surfactant monolayer are often desirable since they permit the investigation of the effects of statistical noise on the complex pattern formation phenomena that characterize the physics of these complex fluids. As usual, a lot of careful comparative work is needed to establish the method on quantitative grounds, but the prospects look fairly stimulating.

12.4 LBE schemes for flows with moving objects

There is a common perception that complex flows with moving objects, such as suspensions, or polymer solutions, represent one of the most promising areas of application for LBE. In the sequel, we shall describe how such situations can be formulated by appropriately generalized lattice Boltzmann techniques.

12.5 Colloidal flows

The motion of suspended objects in fluid flows plays a major role in many important technological applications such as chemical reactors, paints and cements fabrication, pharmaceutical, ceramics processing, to name but a few [193, 194]. Numerical simulations accounting explicitly for hydrodynamic interactions between the suspended particles represent a very valuable tool to gain insight into the physics of suspensions. The numerical simulation of suspensions is traditionally handled by mesoscopic stochastic particle methods, such as the Brownian–Stokesian dynamics, in which the effects of microscopic scales on the mesoscopic ones are represented via stochastic sources (Langevin equations). The suspended particle–particle interactions take place via the intermediate of the surrounding fluid and give rise to long range forces which make the problem computationally very intensive (N-body problem). These long range forces result from the coherent superposition of short range interactions between the suspended macroparticles and the fluid molecules. However, a truly molecular treatment is ruled out by the huge scale separation between fluid molecules and the suspended particles. Hence, an appetizing opportunity for mesoscopic LBE techniques.

To seize these opportunities, however, two substantial extensions of the LBE methodology are called for:

1. *Incorporation of fluctuating forces.*
2. *Handling of solid–fluid moving boundaries.*

Let us examine the former (and simpler) item first.

12.5.1 *The fluctuating LBE*

The fluctuating LBE belongs to the general framework of *fluctuating hydrodynamics*. The basic idea of fluctuating hydrodynamics [195, 196] is that at mesoscopic scales, intermediate between molecular and hydrodynamic ones, thermally induced fluctuations can be reduced to random fluctuations in the fluxes of the conserved quantities. We shall be interested in the momentum flux tensor P_{ab}, whose divergence yields the force experienced by the fluid element. Since this tensor is explicitly tracked by LBE (see Part II), it is simpler to implement fluctuations directly at the level of this tensor rather than as volume forces. Mathematically, this is once again a supplementary source term $\tilde{S}_i$ (the tilde stands for fluctuating) on the right-hand side of LBE:

$$\boxed{\Delta_i f_i = A_{ij}\left(f_j - f_j^{\mathrm{e}}\right) + \tilde{S}_i.} \tag{12.52}$$

Since $\tilde{S}_i$ contributes only to the stress tensor, using the orthogonality of the FCHC scattering matrix, one comes up with the following expression:

$$\tilde{S}_i = A\,Q_{iab}\,S_{ab}, \tag{12.53}$$

where the (stochastic) amplitude A is pinned down by the requirement of Gaussian statistics of the two time correlator:

$$\langle \tilde{S}_{ab}\left(\vec{x}_1, t_1\right) \tilde{S}_{cd}\left(\vec{x}_2, t_2\right)\rangle = A\delta\left(\vec{x}_1 - \vec{x}_2\right)\delta\left(t_1 - t_2\right)\left[\delta_{ac}\delta_{bd} + \delta_{ad}\delta_{bc} - \frac{2}{3}\delta_{ab}\delta_{cd}\right]. \tag{12.54}$$

The variance is adjusted in such a way as to define the temperature of the system via the *fluctuation–dissipation* relation:

$$A = \frac{1}{3}\rho k_B T \left[1 - (1+\lambda)^2\right] = \frac{2\eta k_B T}{\lambda^2}, \tag{12.55}$$

where λ is the leading nonzero eigenvalue of the scattering matrix and η the fluid viscosity. This latter expression is not straightforward and follows from the solution of a discrete Langevin equation, details of which can be found in Ladd's original paper [197].

Ladd brings solid evidence that both over- and under-relaxation, $-1 < \lambda+1 < 0$ and $0 < \lambda + 1 < 1$, respectively, yield excellent agreement between theory and numerical experiment. The expression (12.55) fills the first task of fluctuating LBE, namely an explicit expression of the fluctuating stress tensor.

12.5.2 *Solid–fluid moving boundaries*

In order to account for the hydrodynamic interaction between solid particles, the fluctuating LBE must incorporate moving particle–fluid boundary conditions This task is technically thick, but given its practical and conceptual importance, it cannot be passed over.

The solid particle is represented by a closed surface $\mathcal{S}$, for simplicity a rigid sphere in the following. When painted in the lattice, the surface $\mathcal{S}$ gets replaced by a staircase approximation Σ defined by the set of lattice links cut by $\mathcal{S}$. As a result, each particle p is associated with a staircased sphere of surface Σ_p (see Fig. 12.15).

The accuracy of this representation is only $O(1/a)$, a being the radius of the sphere in lattice units, so that we are actually dealing with rough spheres.

Ladd suggests that both exterior and interior lattice volume be filled with fluid. This simplifies the technical procedure while leaving the physics basically unaffected as long as the *solid fraction*

$$\phi = \frac{4\pi \sum_p a_p^3}{3L^3} \tag{12.56}$$

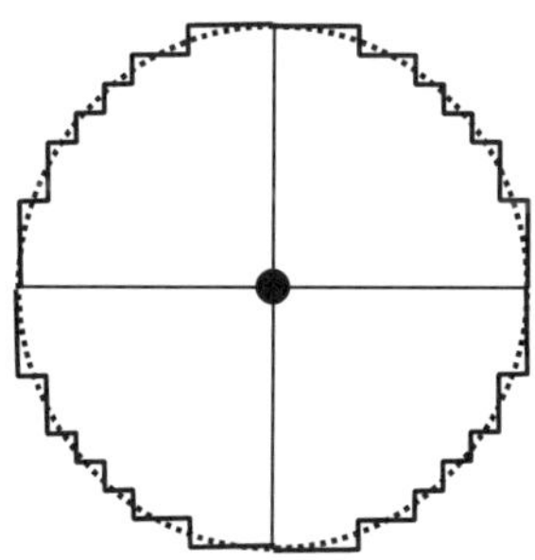

FIG. 12.15. Staircase approximation (solid) of a spherical particle (dotted).

remains negligible (diluted suspensions). In the above expression, L denotes the side of a cubic box where fluid motion takes place.

The solid–fluid interaction proceeds as follows:

Having placed solid nodes in the midpoint of a lattice link, we identify mirror conjugate pairs of directions say (i, j) colliding head-on along the cut link.

For the case of a solid at rest, one simply bounces back the conjugate populations just as we did for porous media.

If the solid site belongs to a particle p moving at speed $\vec{u}_p$ and angular speed $\vec{\Omega}_p$, then its speed is given by:

$$\vec{U}_p = \vec{u}_p + \left(\vec{r} + \frac{1}{2}\vec{c}_i\right) \times \vec{\Omega}_p, \tag{12.57}$$

where $\vec{r}$ is the coordinate of the lattice site relative to the center of the particle.

This speed sets the bias between colliding pairs:

$$f_i\left(\vec{r} + \vec{c}_i, t + 1\right) = f_j\left(\vec{r} + \vec{c}_i, t^+\right) + 2a\rho\vec{U}_p \cdot \vec{c}_i, \tag{12.58}$$

$$f_j\left(\vec{r}, t + 1\right) = f_i\left(\vec{r}, t^+\right) - 2a\rho\vec{U}_p \cdot \vec{c}_i, \tag{12.59}$$

where the notation t^+ denotes post-collisional states.

Note that these rules reduce to the usual bounce-back conditions in the case of a solid at rest, $\vec{U}_p = 0$.

These collision rules produce a net momentum transfer between the fluid and the solid site:

$$\vec{F}_r = \left[f_i\left(\vec{r} + \vec{c}_i, t + 1\right) + f_j\left(\vec{r}, t + 1\right) - f_j\left(\vec{r}, t + 1\right) - f_i\left(\vec{r}, t^+\right)\right]\vec{c}_i. \tag{12.60}$$

By summing all over the sites $\vec{r}$ belonging to Σ_p, we obtain the total force acting upon particle p:

$$\vec{F}_p = \sum_{\vec{r} \in \Sigma_p} \vec{F}_r. \tag{12.61}$$

Similarly, the total torque $\vec{T}_p$ is computed as

$$\vec{T}_p = \sum_{\vec{r}_i \in \Sigma_p} \vec{F}_r \times \vec{r}_i, \tag{12.62}$$

where $\vec{r}_i = \vec{r} + \frac{1}{2}\vec{c}_i$ (the position of square blocks in Fig. 12.16).

We are now in a position to advance the particle speed $\vec{U}_p$ and angular momentum $\vec{\Omega}_p$ according to the equations of motion:

$$M_p \frac{\mathrm{d}\vec{U}_p}{\mathrm{d}t} = \vec{F}_p, \tag{12.63}$$

$$I_p \frac{\mathrm{d}\vec{\Omega}_p}{\mathrm{d}t} = \vec{T}_p, \tag{12.64}$$

where M_p and I_p are the particle mass and inertial momentum, respectively, two input parameters.

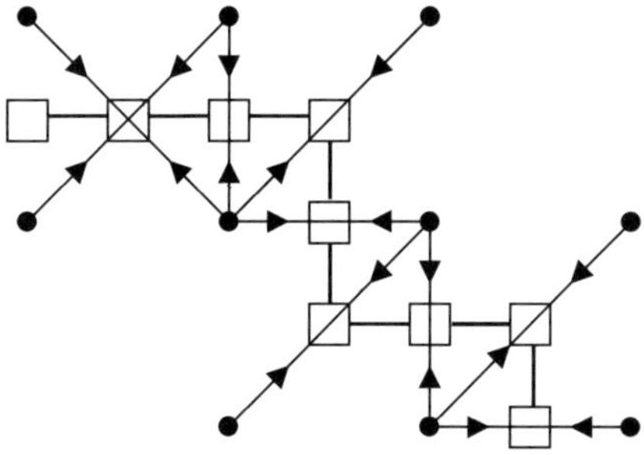

FIG. 12.16. Head-on collisions cutting through the surface. The surface is represented by the square blocks.

For the purpose of stability, Ladd finds it convenient to advance in time according to a leap-frog scheme:

$$\vec{U}_p(t+1) = \vec{U}_p(t-1) + \frac{2}{M_p}\vec{F}_p(t), \tag{12.65}$$

$$\vec{\Omega}_p(t+1) = \vec{\Omega}_p(t-1) + \frac{2}{I_p}\vec{T}_p(t), \tag{12.66}$$

where $\vec{F}(t) = [\vec{F}(t-1/2)+\vec{F}(t+1/2)]/2$ and $\vec{T}_p(t) = [\vec{T}_p(t-1/2)+\vec{T}_p(t+1/2)]/2$, consistent with the fact that solid–fluid collisions take place at half-integer times.

This smoothing average is key to damp out short scale torque oscillations triggered by the conservation of staggered invariants at the particle–fluid interface. By a similar token the fluid speed is smoothed out by a three point average:

$$\vec{u}(r,t) = \frac{\vec{u}(r,t-1) + 2\vec{u}(r,t) + \vec{u}(r,t+1)}{4}. \tag{12.67}$$

This implies some memory burden because three time levels need to be kept track of, but according to Ladd's recommendation, it cannot be helped. This set of equations takes into account the full many-body hydrodynamic interactions, since the forces and torques are computed with the actual flow configuration as dictated by the presence of all N_P particles simultaneously.

This is the promised set of equations involving only local operations, hence achieving *linear* computational complexity.

12.5.3 *Numerical tests*

In a series of thorough papers, Ladd performs an array of validation tests, including:

- Flow past a row of cylinders;
- Flow past a cubic lattice of spheres;
- Time-dependent hydrodynamic interactions (isolated moving sphere);

- Short time velocity autocorrelation;
- Short time dynamics, prior to onset of Brownian motion.

A detailed description of these tests can be found in the original papers. Here we just report some pictorial evidences.

These careful studies show excellent agreement with theoretical and previous numerical results. However, surface roughness may introduce slight deviations from continuum results (smooth surfaces) which fade away approximately like $1/a$.

A further comment on the stochastic fluctuating LBE dynamics is in order. A key task of fluctuating LBE is to (partly) regain non-Boltzmann correlations which fell alongside when moving from LGCA to LBE. A probing test for the validity of this approach is therefore to measure the decay of the linear and angular velocity autocorrelation functions:

$$A_u(t) = \langle u_p(t)\, u_p(0) \rangle, \tag{12.68}$$
$$A_\omega(t) = \langle \omega(t)\, \omega(0) \rangle. \tag{12.69}$$

As is well known from the famous Alder–Wainwright molecular dynamics experiment [198], the hydrodynamic interaction between the moving particle and the surrounding flow induced by the particle motion itself promotes the persistence of the particle trajectory along the initial direction, hence a memory effect which eludes Boltzmann's assumption of molecular chaos.

Mathematically, these non-Boltzmann memory effects manifest themselves into an algebraic decay of fluctuations, as opposed to purely exponential Boltzmann decay. In particular, in three dimensions, $A_u(t)$ is expected to decay like

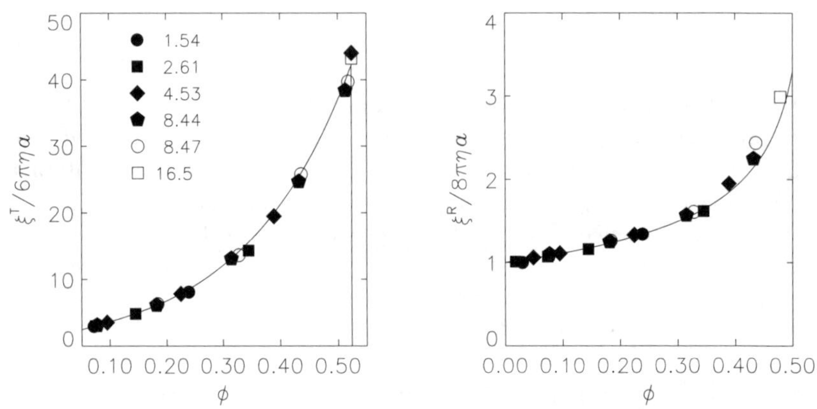

FIG. 12.17. Longitudinal and transversal drag of a cubic lattice of spheres as a function of the volume fraction. The solid lines are accurate solutions of the Stokes equations. (After [200], courtesy of A. Ladd.)

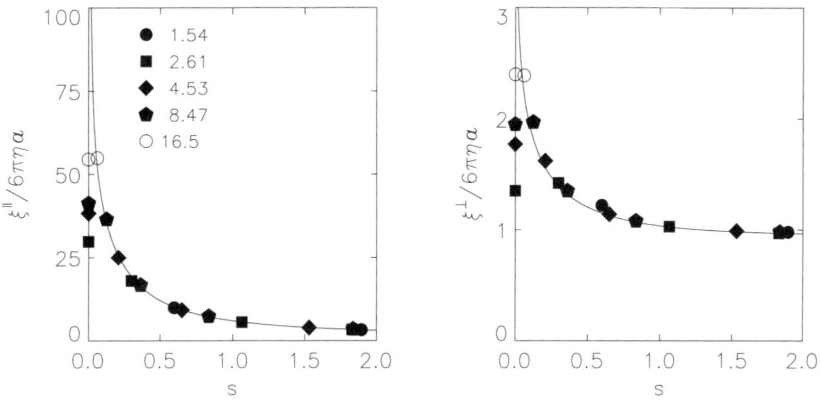

FIG. 12.18. Longitudinal and transversal drag between a pair of moving spheres as a function of the separation s. The solid lines are solutions of the Stokes equation in the same geometry. (After [200], courtesy of A. Ladd.)

$t^{-3/2}$. This algebraic decay and the ensuing long tail memory effects have indeed be reproduced in numerical experiments by Ladd [200] (the effect is measured at sufficiently short times, so as to rule out finite size effects). The simulations also show that the normalized autocorrelation falls off in time exactly like the steady decay of the translational and rotational velocities of the sphere, thus proving that the fluctuating LBE obeys the fluctuation–dissipation theorem and

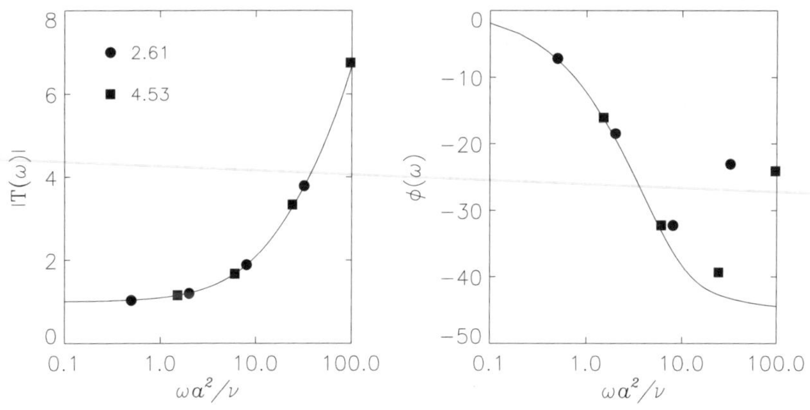

FIG. 12.19. Torque and phase-lag of a rotating sphere as a function of the reduced frequency $\omega a^2/\nu$. Solid lines are taken from expressions given in Landau–Lifshitz. (After [200], courtesy of A. Ladd.)

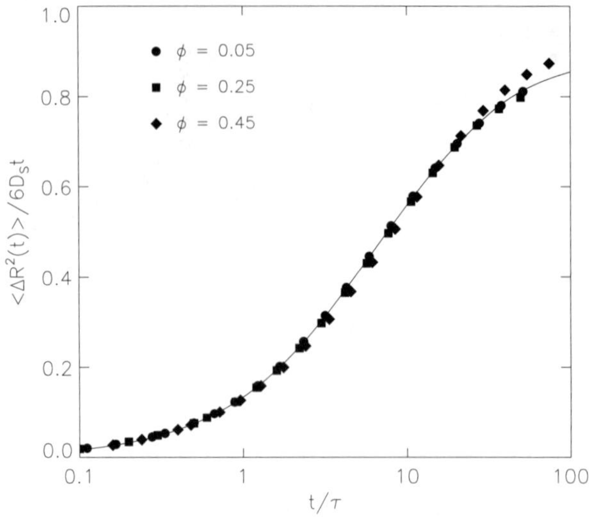

FIG. 12.20. Scaled mean square displacement $\Delta a^2/6D_s t$ at short pre-Brownian times versus reduced time $t^* \equiv t/\tau(\phi)$, where $\tau(\phi)$ is a universal function of the solid fraction. Simulation for 128 spheres (solid symbols) are shown at various packing fractions Φ. (After [200], courtesy of A. Ladd.)

reinstates (at least part of) the non-Boltzmann physics inaccessible to standard LBEs.

12.5.4 *Computational cost of the fluctuating LBE*

The fluctuating LBE is significantly more demanding than its forerunners, and it is therefore of some interest to estimate its computational cost.

Let $a = 5$ be a typical particle size in lattice units, and $\nu = 0.1$ the typical momentum diffusivity, also in lattice units. A typical simulation involving length scales of, say, 100 radii would require about 25 000 time-steps, given the fact that $Re_p \equiv u_p a/\nu = 1$ implies $u_p = 0.02$. Assuming about 1000 operations per particle update per time-step, this gives 25×10^6 floating-point operations, i.e., 25 Mflop/particle over the complete run. Following, say, thousands of particles, takes therefore 2500 Mflop $=$ 2.5 Gflop, which means that a mid-range workstation delivering 100 Mflops would complete the case in about 250 seconds, namely 4 minutes. Given the high amenability to parallel computing of the present scheme, simulations with up to hundreds of thousands of particles, and more, should be possible on present day supercomputers. Indeed, Ladd reports simulations with 64 000 particles on mid-range workstations, significantly higher than typical values for Stokesian dynamics simulations.

In summary, the combination of solid particle tracking with a fluctuating LBE gives rise to a fairly flexible and efficient tool for the simulation of the dynamics of diluted and moderately dense suspensions. Particle size, flow geometry, Péclet[43] and Reynolds numbers can all be changed independently without compromising the linear complexity of the numerical scheme and its outstanding amenability to parallel computing.

12.5.5 *Thermal equilibrium*

The fluctuating LBE simulations show that the translational and rotational degrees of freedom of the solid suspension come only into an approximate equilibrium with the underlying fluid. The culprit is again lack of energy conservation which prevents complete equipartition between the modes of the system.

It is found that the mean translational and rotational energy are almost equal and typically about twenty per cent lower than the fluid fluctuations. This lack of equipartition is disturbing as it hinders the measurement of temperature dependent transport parameters such as the viscosity of the suspension. Therefore, as for multiphase LBEs, the hottest open issue is again thermodynamic consistency.

12.6 Merging LBE with molecular dynamics: polymers in LBE flows

Ladd's fluctuating LBE is the trail-blazer of a series of schemes tracing extended objects within LBE flows. Besides spheres, another example of great practical importance is flexible strings and their immediate application to polymer-solvent solutions. In a recent paper, Ahlrichs and Dunweg, investigate a new method for simulating polymer-solvent systems combining LBE with a continuum molecular dynamics model for the polymer chain [201]. The two schemes are coupled via a frictional force proportional to the speed of the monomer relative to the local fluid speed. The friction coefficient is treated like a free tunable parameter. The polymer is assumed to be formed by a chain of monomers interacting via some potential $V(r)$, typically Lennard–Jones-like for free monomers, plus a spring-like potential for the chained ones:

$$V_{\mathrm{LJ}}(r) = V_0 \left[\left(\frac{s}{r} \right)^{12} - \left(\frac{s}{r} \right)^{6} \right], \tag{12.70}$$

$$V_K = \frac{1}{2} K \left(r_{p+1}^2 - r_{p-1}^2 \right), \tag{12.71}$$

where r is the generic separation between any two free monomers of radius s, $r_{p\pm1}$ is the separation along the polymer chain of the two neighbors of the p-th bound monomer and K is the spring stiffness.

Each monomer obeys the Newton–Hamilton equations for *point-like* particles

$$\frac{\mathrm{d}\vec{X}_p}{\mathrm{d}t} = \frac{\vec{P}_p}{m}, \tag{12.72}$$

$$\frac{\mathrm{d}\vec{P}_p}{\mathrm{d}t} = -\nabla_p V + \vec{F}_p. \tag{12.73}$$

[43]The Péclet number is defined as $Pe = UL/D$, where U is a typical flow speed, L is a typical length and D is the mass diffusivity.

These can be integrated in time with any of the efficient methods available from the molecular dynamics literature [202, 203]. The key issue, of course, is how to evaluate the force $\vec{F}_p$ exerted by the fluid on the p-th monomer.

The simplest recipe is to assume linear proportionality with the monomer–fluid relative speed, namely

$$\vec{F}_p = -\eta \left(\vec{u}_p - \vec{U}_p \right). \tag{12.74}$$

The local flow speed $\vec{U}_p$ at the monomer position can be evaluated by a simple interpolation from the grid points:

$$U_a^P = C^{\text{ne}} U_a^{\text{ne}} + C^{\text{nw}} U_a^{\text{nw}} + C^{\text{sw}} U_a^{\text{sw}} + C^{\text{se}} U_a^{\text{se}}, \quad a = x, y, \tag{12.75}$$

where the interpolation coefficients C^{ne}, C^{nw}, C^{sw}, C^{se} represent the area of the blocks identified by P and the generic grid point G, over the total area of the cell surrounding the monomer located at P. To come full circle, all we need is to compute the back-reaction on the fluid. A straightforward procedure is to interpolate this force on each of the grid nodes surrounding the monomer and bias the corresponding populations according to the usual scheme we used many times in this book to impart momentum to the LBE populations.

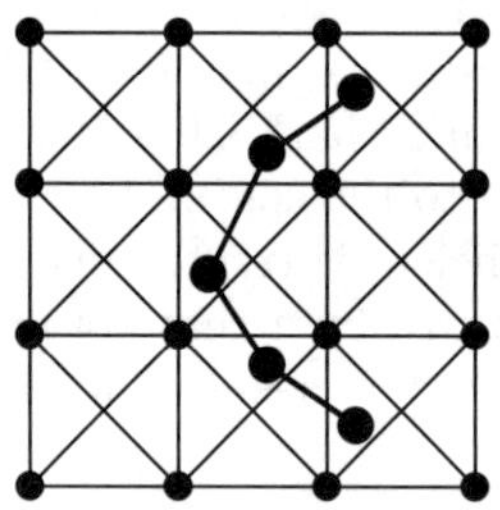

FIG. 12.21. A polymer chain swimming in a LBE flow.

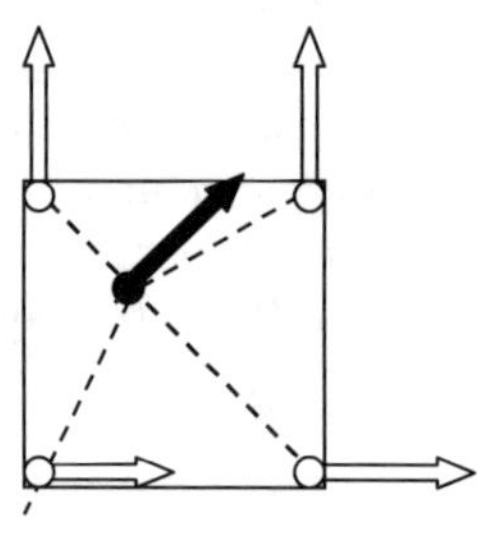

FIG. 12.22. Interpolating the force acting on the single monomer sitting at location P (black dot) within the LBE cell identified by the four north-east, north-west, south-west, south-east corners.

In equations:

$$F_a^G = C^{G'} F_a^P, \quad G = \text{ne, nw, sw, se}, \quad G' = \text{sw, se, ne, nw}, \quad a = x, y. \tag{12.76}$$

Once the force on the generic node is defined, the directional share on the i-th population is computed according to a simple projection:

$$F_i^G = \frac{c_{ia} F_a^G}{2b_a}, \tag{12.77}$$

where b_a is the number of discrete speeds projecting upon direction x_a. The update on the generic node G proceeds as usual

$$f_i^G \leftarrow f_i^G + F_i^G. \tag{12.78}$$

The combination of LBE plus molecular dynamics, supplemented with the interpolation steps (12.75)–(12.77), yields a self-consistent and seamless procedure to evolve the polymer within the LBE flow.

Life is easy because monomers are point-like molecules. Important physical effects, such as excluded volume (finite size of the monomers) are nonetheless retained via the use of a Lennard–Jones potential.

Ahlrichs and Dunweg demonstrate their method for a series of deterministic and stochastic experiments. The latter reveal a basic failure of the above LBE–MD scheme to recover the fluctuation dissipation theorem (no surprise, since the algorithm given so far is purely deterministic). In order to repair this inconsistency, the authors introduce a stochastic force $\vec{\tilde{F}}$, whose covariance is adjusted to the temperature according to the usual fluctuation–dissipation relation

$$\langle \tilde{F}_a(t)\, \tilde{F}_b(t') \rangle = 2K_B T \eta\, \delta(t - t')\, \delta_{ab}. \tag{12.79}$$

By adding this stochastic force to the friction force (12.74), the authors show that the corresponding Langevin dynamics achieves full compliance with the fluctuation dissipation theorem. For full details see the original reference.

12.7 Snow transport and deposition

Another interesting application of LBE with moving particles has recently been developed by Masselot *et al.* [204]. In this application, standard LBE schemes for turbulent flows are combined with *multiparticle* LGA schemes permitting description of the motion of solid particles (snowflakes) in a turbulent flow (wind) and their deposition on the ground. The solid particles move in the flow field and are assigned to the lattice sites with interpolation/distribution procedures similar to those described in the previous section. Each time a solid particle hits the ground, it gets a probability to 'freeze' and thus increase the solid fraction at that site. Once the solid fraction exceeds a prescribed threshold, the entire site becomes 'solid' and acts as a rigid wall boundary for subsequent fluid particles. Preliminary tests show close agreement with experimental data on snow deposition patterns [204].

Such type of hybrid LBE–particle scheme might prove very useful for the numerical simulation of complex phenomena such as corrosion and erosion in the presence of fluid flows.

12.8 A new paradigm for non-equilibrium statistical mechanics?

In a rather enthusiastic vein, it is sometimes argued that LBE-like equations with generalized sources may constitute a new paradigm in non-equilibrium statistical mechanics.

This is certainly a very bold, perhaps even somewhat provocative statement, but we hope that after reading through this chapter, the reader may be willing to concede that there is some basis for it.

First, as repeatedly observed, use of self-consistent forces/potentials offers the potential opportunity to describe flow complexities, say phase transitions, chemical reactions, particle–flow interactions and similar effects, at a more fundamental level than in classical mean field theory formulations.

To this end, it is worth recalling that LBE is in some respects even *more general than the Boltzmann equation itself.* For instance, unlike the true Boltzmann equation, LBE does *not* require any assumption of diluteness of the fluid. In fact, since particles live on the grid, they are forced to meet at lattice sites where they may undergo many-body collisions even if their interaction potential is zero-ranged.

Second, always in the spirit of going beyond one-body physics, the fluctuating LBE shows how to regain 'non-Boltzmann' correlation effects.[44]

Third, the (still unexplored) fact that LBE is basically a system of spins in $D+1$ (space–time) dimensions, makes it amenable to the powerful arsenal of analytical and numerical tools of modern statistical mechanics.

The picture emerging from the above arguments is undeniably rich.

This said, it is this author's opinion that most of this potential remains yet to be fully demonstrated because: First, a thermodynamically consistent treatment, possibly along with with an attendant H-theorem, remains to be proven.

Second, even if such a consistent one-body formalism can be found, we would still be left with the practical issue of assessing its numerical fidelity on a finite grid (the lattice Burnett issue discussed earlier in this chapter).

Further ideas, methods and techniques[45] are surely needed if generalized LBEs are to live up to the promise of becoming a new paradigm for non-equilibrium statistical mechanics. Nonetheless, the prospects opened up by the applications discussed in this section definitely look exciting.

[44] It is worth noticing that lattice fluctuations seem to share essential features with true physical fluctuations [234], another encouraging clue to a kinetic description of fluctuating hydrodynamics using a combination of LGCA and LBE.

[45] A recently emerging approach is based on the formulation of generalized LBEs guided by the BBGKY hierarchy [205,206].

12.9 New vistas

Due to its introductory character, as well as space limitations, this book leaves out several pieces of advanced LBE research, such as

- *LBE for classical wave propagation* [207];
- *LBE for magnetohydrodynamic flows* [208];
- *LBE for viscoelastic flows* [209];
- *LBE for granular flows* [210];
- *LBE for ocean circulation* [211],

and many others.

These schemes are based on ingenious extensions of the basic 'stream and collide' rule lying at the heart of LBE methods, and in most cases, they manage to do so without compromising the LBE assets of simplicity and amenability to parallel computing. Much remains to be done to prove that they really can compete with existing methods, but these ideas seem definitely worth being pursued further. The reader keen on learning more about these topics is kindly directed to the current literature.

Due to its peculiar character (quantum versus classical mechanics), we feel like making an exception for the lattice Boltzmann scheme for quantum mechanics, the subject of the next chapter.

12.10 Exercises

1. Estimate the fastest chemical reaction that can be simulated by a reactive LBE scheme in a cube 10 cm in size discretized with 100 grid points per side.
2. Write down the explicit expression of the force F_x, F_y on the liquid for the Shan–Chen model in a 6-speed hexagonal lattice.
3. Derive the Van der Waals expression of the free energy $W[\rho]$.
4. Derive the relation between the parameter G and the surface tension in the Shan–Chen model.
5. Prove that the stochastic source $\tilde{S}_i$ in fluctuating LBE does not affect mass and momentum.
6. Using eqn (12.60) compute the force exerted on a square particle of radius a by a fluid with speed $U = U_0\, x/L$. Use the equilibrium distribution f_i^{e} for the usual 9-speed square lattice.
7. Discuss the impact of interpolations such as (12.75) on the conservation properties of the LBE–polymer algorithm.

13

LBE FOR QUANTUM MECHANICS

The lattice Boltzmann concepts and applications described so far refer to classical (non-quantum) physics. Whether this limitation is strictly scientific or purely contingental in character, it is difficult to say. However, it appears plausible to speculate that at least part of the LBE assets should be exportable to the quantum world. In principle, this does not even require much of a conceptual jump, since it has been known for a long time that quantum mechanics can be formulated in a fluid dynamic language. This analogy is precisely the stepping stone for this final chapter: stay with us, LBE goes quantum mechanical!

13.1 Quantum mechanics and fluids

Intriguing analogies between quantum mechanics and fluid mechanics have been pointed out since the earliest days of quantum theory [212]. The orthodox tenet is that these analogies are purely formal in character and do not bear upon the basic physics of quantum phenomena. A less orthodox, albeit not minor, stream of thought insists instead that quantum mechanics and notably its philosophical hard-core, Heisenberg's uncertainty principle, are nothing but a mirror of our ignorance of the underlying (hidden) microscopic physical level. This leads to the puzzling theory of 'hidden variables' which can be traced back to Einstein and subsequently to D. Bohm and others [213]. It is not our intention here to enter this fascinating and still open subject [214]. We turn to a practical question instead: what can the analogy do for us in terms of numerical modeling of evolutionary quantum mechanical phenomena? The question is legitimate because, regardless of its philosophical implications, the fluid analogy certainly provides an intuitive and physical sound basis to develop numerical methods for time-dependent quantum mechanics. In particular, it appears reasonable to ask whether the advantages brought about by LBE in fluid dynamics can (by means of the fluid analogy) be exported to the context of quantum mechanics. Before we put forward our discrete kinetic theory version of the analogy, it is useful to provide a cursory survey of the main ideas behind the analogy itself. To this end, a short recap of basic notions of quantum mechanics is in order.

13.2 The fluid formulation of the Schrödinger equation

Let us begin with the Schrödinger equation for a non-relativistic quantum particle of mass m in an external potential $V(\vec{x})$:

$$\mathrm{i}\,\hbar\,\partial_t \Psi = \left[-\frac{\hbar^2}{2m}\Delta + V(\vec{x})\right]\Psi, \tag{13.1}$$

where $\Psi(\vec{x}, t)$ is the wavefunction of the material particle.

Upon multiplying (13.1) by the complex conjugate Ψ^*, and the complex conjugate of (13.1) by Ψ and then subtracting, we obtain

$$\mathrm{i}\,\hbar\,\partial_t|\psi|^2 = -\frac{\hbar^2}{2m}\left(\Psi^*\Delta\Psi - \Psi\Delta\Psi^*\right). \tag{13.2}$$

A simple integration by parts of the right-hand side delivers the following conservative form:

$$\partial_t|\Psi|^2 + \mathrm{i}\frac{\hbar}{m}\partial_a\left(\Psi\partial_a\Psi^* - \Psi^*\partial_a\Psi\right) = 0, \tag{13.3}$$

which is readily recognized as a familiar continuity equation for a quantum 'fluid' with density

$$\rho = |\Psi|^2 \tag{13.4}$$

and current density

$$J_a = \mathrm{i}\frac{\hbar}{m}\left[\Psi\partial_a\Psi^* - \Psi^*\partial_a\Psi\right]. \tag{13.5}$$

By a similar token, multiplying by $\partial_a\Psi^*$ eqn (13.1) and by $\partial_a\Psi$ its complex conjugate and summing up, we obtain the evolution equation for the current density:

$$\partial_t J_a + \partial_b P_{ab} = 0, \tag{13.6}$$

where the momentum flux tensor of the quantum fluid is given by

$$P_{ab} = \frac{\hbar^2}{2m}\left[\Psi^*\Delta\Psi + \Psi\Delta\Psi^* + V|\Psi|^2\right]\delta_{ab}. \tag{13.7}$$

The first two terms on the right-hand side correspond to the kinetic energy, plus a genuinely quantum potential (see below) while the third term is the classical fluid pressure resulting from the external potential. This configures quantum matter as an ideal (inviscid, dissipationless), irrotational (no vorticity) compressible fluid. The inviscid character of the quantum fluid stems from the reversible nature of the Schrödinger equation, a diffusion equation in *imaginary* time. The quantum analogy becomes even more revealing in the so-called eikonal representation [215]:

$$\Psi = R\mathrm{e}^{\mathrm{i}\theta}. \tag{13.8}$$

Upon inserting this expression in eqn (13.5), we obtain

$$J_a = \frac{\hbar}{m}\partial_a\theta \equiv \rho u_a \tag{13.9}$$

and

$$P = \rho\left(\frac{u^2}{2} + \frac{V}{m} + \frac{V^Q}{m}\right), \tag{13.10}$$

where

$$V^Q \equiv -\frac{\hbar^2}{2m}\frac{\Delta R}{R} \tag{13.11}$$

is the famous quantum potential advocated by Bohm and co-workers to support the picture of quantum mechanics as an intrinsically non-local description of the microscopic world [216].

This naturally invites the picture of a cloud (fluid) of particles moving about their center of mass with a spatial distribution $\rho(\vec{x})$, and a speed which is 'read off' as the space gradient of the phase $\theta(\vec{x})$ according to the expression (13.9). Hard to resist the appeal of the celebrated de Broglie's 'pilot wave' picture: the classical trajectory is just the average over the cloud of particles (hidden variables) in much the same way as fluid density and speed are the average over an ensemble of molecular trajectories.

So much for the analogy in the continuum world. What about the discrete lattice world? Interestingly enough, this analogy becomes even *more* compelling once transposed into the language of the lattice world. In fact, the lattice formulation naturally calls for an an 'upgrade' from the non-relativistic Schrödinger equation to its relativistic associate, the Dirac equation. Symbolically, the analogy goes as follows: DE: Dirac Equation, SE: Schrödinger Equation, LBE: Lattice Boltzmann Equation, NSE: Navier–Stokes Equation,

$$\text{DE} \rightarrow \text{SE}, \tag{13.12}$$

$$\text{LBE} \rightarrow \text{NSE}. \tag{13.13}$$

Before venturing into the task of substantiating this daring analogy, let us make a short recap of the Dirac equation for the sake of self-consistency.

13.2.1 *Relativistic quantum mechanics: the Dirac equation*

The Dirac equation is the 'Wunderkind' of the marriage of the two high points of modern physics: relativity and quantum theory. It builds upon a formal requirement of Lorentz invariance on quantum mechanics, which implies a symmetric balance between space and time derivatives. Such a balance is manifestly broken by the Schrödinger equation, in which time derivatives appear at first-order with the spatial ones at second-order. The foresighted reader might already have a clue to the analogy: like the Schrödinger equation, the Navier–Stokes equation shows the same imbalance. And similarly for the Dirac equation, in LBE the space and time derivatives are well balanced: both first-order.

But let us go on with the Dirac equation.

In standard form, and using atomic units ($\hbar = c = 1$), the Dirac equation reads as follows [215]:

$$\mathrm{i}\,\partial_t \psi_j = \left[\alpha^x_{jk}\partial_x + \alpha^y_{jk}\partial_y + \alpha^z_{jk}\partial_z + m\beta_{ij}\right]\psi_k. \tag{13.14}$$

Here ψ_j is a complex 4-*spinor* describing a particle–antiparticle pair with spins $\pm 1/2$. Relativistic invariance implies that each particle with momentum p_a is associated with a corresponding antiparticle of opposite momentum $-p_a$ (parity symmetry).

The streaming of this spinning pair is described by the three 4×4 complex matrices α^a_{jk} which are explicitly given in terms of the 2×2 Pauli matrices σ^a, $a = 1, 2, 3$:

$$\alpha^a = \begin{pmatrix} \sigma^a & 0 \\ 0 & -\sigma^a \end{pmatrix}, \tag{13.15}$$

$$\beta = \begin{pmatrix} 0 & I \\ I & 0 \end{pmatrix}, \tag{13.16}$$

with

$$\sigma^x = \begin{pmatrix} 0 & 1 \\ 1 & 0 \end{pmatrix}, \quad \sigma^y = \begin{pmatrix} 0 & -\mathrm{i} \\ \mathrm{i} & 0 \end{pmatrix}, \quad \sigma^z = \begin{pmatrix} 1 & 0 \\ 0 & -1 \end{pmatrix}, \tag{13.17}$$

where I is the 2×2 identity matrix and spinorial indices have been dropped for simplicity. Likewise the Schrödinger equation, the Dirac equation can also be written in terms of fluid motion. However, since we are dealing with a 4-component wavefunction, the relativistic picture points to a *four-component fluid mixture.* To unfold this picture, it proves expedient to recast the Dirac equation into a form where all streaming matrices, sometimes called Weil matrices, become real. This is the so-called *Maiorana form,* and it is easily obtained via the following unitary transformation: $\psi_j \to (\alpha^y_{jk} + \mathrm{i}\beta_{jk})\psi_k$.

In a compact four-dimensional notation, this yields

$$\left[W^\mu_{jk}\partial_\mu\right]\psi_k = \mathrm{i}M_{jk}\psi_k, \quad \mu = 0, \ldots, 3, \tag{13.18}$$

with

$$W^0_{jk} = \delta_{jk}, \quad W^1_{jk} = \alpha^x_{jk}, \quad W^2_{jk} = \beta_{jk}, \quad W^3_{jk} = -\alpha^z_{jk},$$
$$M_{jk} = -\mathrm{i}m\alpha^y_{jk} + qV\delta_{jk} + A^a_{jk}J_a.$$

Here $qV + J_a A^a$ is the interaction of the elementary charge q with an external electromagnetic field described by the 4-vector potential (V, A^a).

A scalar product of eqn (13.18) with ψ^*_j yields the desired set of continuity equations:

$$\partial_t \rho_j + \partial_a J^a_j = S_j, \quad j = 1, \ldots, 4, \tag{13.19}$$

where $\rho_j = \psi^*_j\psi_j$ is the partial density of the j-th fluid, $J^a_j = \psi^*_j\alpha^a_{jk}\psi_k$ the corresponding current density, and $S_j = \mathrm{i}\psi^*_j M_{jk}\psi_k$ is a 'chemical' source term transferring mass across the different components of the relativistic mixture. Note that in the above expressions only the index k is summed upon. Unitarity, read norm conservation, implies $\sum_j S_j = \mathrm{i}\sum_{jk}\psi^*_j M_{jk}\psi_k = 0$. This is automatically secured by the anti-Hermitian character of the mass matrix: $M_{kj} + M^*_{jk} = 0$.

As promised, the fluid analogy comes by *more* naturally than in the non-relativistic case, because the Dirac equation only involves first-order derivatives.

Another pleasing feature is that the external interaction is easily accommodated into a formal redefinition of the mass matrix, without compromising the local nature of the theory. The fluid interpretation of the Dirac equation is equally transparent: four types of *spinning particles* stream in space and, once on the same space–time location, they interact via the 'scattering matrix' M_{jk}. Again, a decided flavor of LBE.

A qualitative difference with classical particle motion is apparent, though. Classical particles have no internal structure, and consequently a type-1 particle at location x at time t with speed v propagates to $x + v\,\mathrm{d}t$ at time $t + \mathrm{d}t$ and it is still entirely of type 1.

A relativistic particle, however, undergoes mixing during free propagation because the rotation around the direction of motion mixes up the four spinorial components. This is why the streaming matrix is generally non-diagonal, echoing the fact that spin is not an ordinary vector. This suggests that the discrete space–time of a relativistic particle should be represented by a 'hypernetted network' in which each link is made up of four distinct but communicating channels, one per spinorial state [217].

This 'hypernetted lattice theory' is less of a joke than it seems. It has been realized recently that lattice formulations of field theory based upon spinning particle motion may offer potential advantages over more popular techniques such as path integration [218]. In a nutshell, this is because in quantum lattice models '*instead of seeking discretized versions of the Hamiltonian or the Lagrangian, a discretized version of the evolution operator is introduced*' [218]. In fact, what this author finds is that '*the rotation group, the Lorentz group and spin emerge automatically in the continuum limit from unitary dynamics on a cubic lattice.*' The reader fond of more details is directed to the original reference.

13.2.2 *Dirac to Schrödinger: the adiabatic approximation*

Having briefly reviewed the Dirac equation, we now proceed to substantiate the analogy DE : SE = LBE : NSE. As a first step in this direction, there is a useful lesson to learn by inspecting the way Schrödinger equation is obtained as a long wavelength (low energy) limit of the Dirac equation. We shall see in a moment that this involves a sort of adiabatic approximation which is formally very similar to the low Knudsen adiabatic expansion taking the Boltzmann equation into the Navier–Stokes equations. The formal parallel emerging from this analogy is

$$Kn = \frac{l_\mu}{l_M} \sim \beta = \frac{v}{c}, \tag{13.20}$$

where l_μ is the particle mean free path, l_M a typical coherence length of the macroscopic fluid and β is the relativistic particle to light speed ratio. To make the argument quantitative, let us consider the 1D version of the DE in Maiorana format:

$$\partial_t u_{1,2} - \partial_z u_{1,2} = m d_{2,1}, \tag{13.21}$$

$$\partial_t d_{1,2} + \partial_z d_{1,2} = -m u_{2,1}. \tag{13.22}$$

Here (u_j, d_j) represent a pair of up/down moving bi-spinors with spin up/down $(1, 2)$. They propagate independently (the streaming matrix α^z can always be diagonalized independently of the other two) and mix via the mass matrix alone. Now let us introduce slow/fast modes by mixing u/d as follows (spinorial indices are relaxed for the sake of simplicity):

$$\phi^{\pm} = 2^{-1/2}\,(u \pm \mathrm{i}d)\,\mathrm{e}^{\mathrm{i}mt}, \tag{13.23}$$

where m is the particle rest mass.[46]

The slow/fast modes evolve accordingly to

$$\partial_t \phi^+ + \partial_z \phi^- = 0, \tag{13.24}$$

$$\partial_t \phi^- - \partial_z \phi^+ = 2\mathrm{i}m\phi^-. \tag{13.25}$$

Now comes the 'adiabatic' assumption (the meaning of adiabaticity in the context of quantum mechanics will be clarified shortly):

$$\left|\partial_t \phi^-\right| \ll 2m \left|\phi^-\right|. \tag{13.26}$$

This delivers $\phi^- = \partial_z \phi^+/2\mathrm{i}m$.

By inserting this into eqn (13.24) we obtain

$$\partial_t \phi^+ = \mathrm{i}\frac{\hbar^2}{2m}\partial_z^2 \phi^+, \tag{13.27}$$

which is nothing but the sought-after (one-dimensional) Schrödinger equation for a free particle of mass m.

What is the lesson here?

The relativistic motion implies that any particle of momentum p_a is invariably associated with an antiparticle with opposed momentum $-p_a$. The symmetric combination of these two gives rise to a smooth, emergent field, whereas the antisymmetric combination defines a low amplitude, high frequency mode which decouples from the system dynamics in the limit $\beta \to 0$. More precisely, inspection of eqn (13.27) reveals that the amplitude of ϕ^-/ϕ^+ scales like β^2 whereas the frequency ratio goes like $1/\beta$. This means that, as β goes to zero, the antisymmetric mode becomes smaller and smaller in amplitude and faster and faster in frequency, so that it finally becomes unobservable on scales longer than $1/\omega^-$.

Isn't this just the picture of adiabatic elimination of kinetic modes in (discrete) kinetic theory?

The scenario is exactly the same, except for a key difference. Kinetic theory describes dissipative phenomena in which adiabatic elimination irons out the initial conditions, the transient modes die out, never to return. Quantum mechanics is reversible, and fast modes never die out: they just oscillate so fast that any observation on time-scales longer than their period of oscillation simply overlooks them. But they are still there and more resolved (higher energy) measurements could always bring them back again. Note that it is the fast mode, not

[46] In physical units it is the particle Compton frequency $\omega_c = mc^2/\hbar$.

the antiparticle mode that fades away; the particle–antiparticle twin-link does not dissolve even in the low energy limit.

Another interesting remark concerns the symmetry breaking induced by a nonzero mass m. If m is made zero the up and down walkers do not see each other and go across with no interaction, the result being the wave equation for photons. Manifestly this is a singular limit which cannot be described by the Schrödinger equation (diffusion coefficient goes to infinity). Any nonzero mass causes 'collisions' which slow down the wavepackets and confer on them a subluminal speed $v < c$ as befits material particles.

13.2.3 *The interacting case*

Interactions with an external or self-consistent fields are readily included by a minor extension of the 'collision operator'. They read as follows:

$$\partial_t u_{1,2} - \partial_z u_{1,2} = m d_{2,1} + igu_{2,1}, \tag{13.28}$$

$$\partial_t d_{1,2} + \partial_z d_{1,2} = -m u_{2,1} + igd_{1,2}, \tag{13.29}$$

where $g = eV/\hbar$ is the coupling frequency of the potential. Self-consistent potentials, such as those arising in connection with the nonlinear Schrödinger equation, are easily accommodated by making g a function of the local density $u^2 + d^2$.

13.3 The quantum LBE

After all this preparation, we are finally in the position to reformulate the basic analogy in quantitative terms. This is based on the following position [217, 219, 220]:

The discrete speed v_i is the analogue of the particle spin s_i.

From this assumption it follows that:

1. *The discrete population f_i is the analogue of the 4-spinor ψ_j.*
2. *The scattering matrix A_{ij} is the analogue of the mass matrix M_{jk}.*
3. *The local equilibrium f_j^e is the analogue of the spinorial local equilibrium defined by the relation $m\alpha_{jk}^y \psi_k^e = W_{jk}^a \psi_k A_a$.*

Position 1 follows from the observation that all LBE schemes work with a set of discrete speeds coming in pairs like spins. In quantum mechanics the spin is intrinsically discrete since it can only take two measurable values, all intermediate ones being unstable. That is real-world physics. In LBE the particle speed is made artificially discrete; even though it could take on any value, we know that the correct macroscopic physics does not ask for more than a few. In quantum mechanics spin quantization is a must, a true physical effect due to the impossibility of measuring the speed and position at a time (spin has dimensions of angular momentum $r \times p$). In LBE speed quantization is a convenient option with no physical meaning (at a macroscopic scale at least). The analogy is tempting, but a minute's thought reveals at least two severe flaws, related to a *dimensional* and a *structural* mismatch respectively between LBE and DE. The dimensional mismatch refers to the fact that while the 4-spinor (we consider $s = 1/2$ throughout) has always four components in any dimensions, the discrete population array is a set of b real functions with b a sensitive function of the dimensionality. The structural mismatch concerns the streaming operator: while for LBE this is always diagonal in momentum space, there is *no way* the three Weil matrices can be simultaneously diagonalized. Again, this is intimately related to the quantum nature of the spin variable. In the end, both flaws trace back to the simple fact that spins are not speeds, and they are not vectors altogether! This looks like a serious stumbling block.

13.3.1 *Extended operator splitting* $3 = 1 + 1 + 1$

These problems sound like a decided 'no go' to the quantum LBE in more than one dimension. How to get around this situation?

The claim is that in the limit of 'small' time-steps, actually much shorter than the inverse Compton frequency ω_c^{-1}, *both flaws can be circumvented by decomposing the three-dimensional particle motion into a sequence of three one-dimensional motions along the coordinate directions x, y, z.* The technical key to achieve this task is a well-known tool of the trade in computational fluid dynamics: 'operator splitting'. The main use of operator splitting in computational fluid dynamics is to handle 3D problems as a sequence of lower-dimensional ones. In quantum field theory, a very similar technique aimed at infinite-dimensional problems goes under a different name: 'Trotter formula': $\mathrm{e}^A = \left(\mathrm{e}^{A/n}\right)^n$ with n integer and A any 'reasonable' operator.

The principle of operator splitting is simple.

Consider the formal solution to the Dirac equation for a massless particle (the collisional operator plays no role at this stage):

$$\Psi_j(x^\mu + \mathrm{d}x^\mu) = \left[\mathrm{e}^{\mathrm{d}t \sum_{\mu=0}^{3} W_{jk}^\mu \partial_\mu}\right] \Psi_k\left(x^\mu\right). \tag{13.30}$$

Manifestly, the propagator taking the wavefunction from $x^\mu \equiv (t, x, y, z)$ to $x^\mu + \mathrm{d}x^\mu \equiv (t, x + \mathrm{d}x, y + \mathrm{d}y, z + \mathrm{d}z)$ is the direct product of three one-dimensional partial propagators $P^a \equiv \mathrm{e}^{\mathrm{d}t\left[\partial_t + W_{jk}^a \partial_a\right]}$, $a = x, y, z$ (no summation upon a implied).

This is the natural consequence of the additivity of the streaming operator. This expression is a good starting point for 'conventional' numerical treatment of the Dirac equation [221], but is definitely unsuitable to a quantum LBE formulation because spinorial states get mixed during the propagation step, something that would not occur to a classical particle.

Therefore, a naive application of operator splitting is not viable.

At this stage, we argue that we *do not* need to work with the same representation of the Dirac equation during the three separate streaming steps. As long as we are able to develop a recipe securing uniqueness of the representation in x^μ and $x^\mu + \mathrm{d}x^\mu$, we are free of choosing the representation that better fits our needs.

Once this notion of extended operator splitting is accepted, we get full leeway to cast eqn (13.30) in quantum LBE format.

The idea is to perform each 1D partial streaming in the representation where the corresponding Weil matrix is diagonal. This is certainly possible because we only require diagonalization of a single Weil matrix at a time, never the three of them together!

New errors are introduced in the numerical treatment, but we shall argue that they are $O(\mathrm{d}t^2)$, namely within the accuracy of the LBE method.

Let us describe the procedure in some more detail.

Suppose we start in the representation where the spin points along z: streaming along z proceeds naturally in LBE format since the Weil matrix along z is diagonal.

Next, we need to move along y (again we forget about collisions, for they are local hence irrelevant to the present discussion) and we ought to change the representation in order to prevent spinorial states to mix under the effect of W^y.

What we need is a representation in which W^y becomes diagonal. Call R^y the corresponding transformation matrix, the requirement is $W'^y = (R^y)^{-1}W^yR^y = \pm I$, where I denotes the identity. The transformed wavefunction $\Psi' = R^y\Psi$ evolves according to

$$\partial_t \Psi'_i + v_i \partial_y \Psi'_i = 0, \tag{13.31}$$

where $v_i = \pm 1$. This is a standard LBE for a pair of walkers along y.

All we need, at this stage, is to 'rotate back' the wavefunction Ψ', in order to retrieve the original wavefunction Ψ, only displaced at $(z + \mathrm{d}z, y + \mathrm{d}y, t + \mathrm{d}t)$.

Since the motion is reversible, no information is lost in the process and there is every reason to believe that this operation is mathematically well-posed.

13.3.2 *Quantum LBE: move, turn and collide*

The quantum LBE is just slightly more cumbersome than its classical counterpart, the only extra charge being the 'rotation' R and 'antirotation' R^{-1} steps needed to keep the spin aligned with the speed. This is the only trace left of the quantum nature of the moving particle. It should noted that quantum LBE bears many similarities with other quantum lattice schemes discussed in the recent [222, 223] and not so recent [224, 225] literature.

What sets it apart from all these schemes is the fact of insisting on a diagonal representation of the Weil matrices, so as *to retain the notion of classical trajectories as much as we can*. In fact, the 'turn' operator R can formally be interpreted as an 'internal scattering' between particle–antiparticle states, thus leaving the concept of quantum trajectory still well-defined. In a pictorial sense, we might say that while classical particles just 'stream and collide', quantum particles, somehow like swimmers, need a somersault to bounce back in space: they 'stream, turn and collide'! Again, the 'turn' step is a necessity induced by the internal structure of the relativistic particle.

13.3.3 *Time marching*

A final word on the procedure to march quantum LBE in time is in order. Again, an important difference with classical LBE is to be pointed out. Having agreed upon (extended) operator splitting, we shall stick to the one-dimensional case with no loss of generality. Simple Fourier analysis shows that an explicit light-cone marching such as the one used for fluid LBEs is unconditionally unstable. This relates to the imaginary nature of the diffusion coefficient. A simple and effective way out is to march eqns (13.21) and (13.22) with an implicit Crank–Nicolson method, whereby the right-hand side of eqns (13.21) and (13.22) are evaluated at mid-times as follows:

$$\hat{u} - u = m\left(\frac{\hat{d} + d}{2}\right) + ig\left(\frac{\hat{u} + u}{2}\right), \tag{13.32}$$

$$\hat{d} - d = -m\left(\frac{\hat{u} + u}{2}\right) + ig\left(\frac{\hat{d} + d}{2}\right), \tag{13.33}$$

where $\hat{u} \equiv u(z+1, t+1)$, $\hat{d} \equiv d(z-1, t+1)$ and $g = V_0 e/\hbar$ is the coupling to the external potential.

The scheme (13.32) and (13.33) is linearly implicit, but it does not pose any problem since we can easily invert it analytically site-by-site, to obtain the following *explicit* form:

$$\hat{u} = au + bd, \tag{13.34}$$

$$\hat{d} = ad - bu, \tag{13.35}$$

where

$$a = \frac{1 - \Omega/4}{1 + \Omega/4 - ig}, \tag{13.36}$$

$$b = \frac{m}{1 + \Omega/4 - ig}, \tag{13.37}$$

$$\Omega = m^2 - g^2. \tag{13.38}$$

This is the final form of our quantum lattice Boltzmann equation.

A few comments are in order.

First, with $g = 0$, implicit time marching translates into a mere redefinition of the particle mass $m \to m' = m/(1 - m^2/4)$. By reinstating the time-step Δt, it is easily recognized that $m' \to m$ in the limit $\Delta t \to 0$, which means that quantum LBE fulfils the requirement of numerical *consistency*.

Large time-steps $m\Delta t > 1$ lead to unphysical results, as is to be expected since the natural Compton frequency m (in atomic lattice units $\hbar = c = \Delta t = \Delta x = 1$) is no longer resolved. Simple algebra also shows that quantum LBE is *unconditionally stable and* norm-preserving (the all-important unitarity condition). This is fairly remarkable for an explicit numerical scheme [225], and ultimately traces back to the (implicit) light-cone discretization hidden behind quantum LBE (13.34). The light-cone discretization also provides freedom from phase errors and it is therefore likely to liberate the quantum LBE from the infamous 'fermion doubling' problem (see J. Kogut, The lattice gauge theory approach to quantum chromodynamics, *Rev. Mod. Phys.* **5**(3), 775, 1983). Finally, note that at no point in our treatment did we need to care about stringent symmetry requirements: apparently a simple cubic lattice is good enough for our purpose. This probably relates to the diagonal nature of the quantum mechanical pressure tensor and to the fact that, unlike fluid dynamics, the theory is *not* self-interacting. Finally, we observe that quantum LBE is as computationally lean and amenable to parallel processing as an explicit scheme can be.

All in all, a good set of credentials for a numerical scheme.

Let us now convince the reader that these credentials do work in practice as well.

13.4 Numerical tests

The quantum LBE scheme has been validated in a series of one-dimensional textbook calculations, including (i) *free particle propagation*, (ii) *harmonic oscillator* and (iii) *scattering from a rectangular barrier* [219]. In addition, the scheme has also been demonstrated for simple cases of nonlinear Schrödinger equations of direct relevance to Bose–Einstein condensation [220].

13.4.1 *Free particle motion*

The free motion of a quantum particle in the absence of any external potential is characterized by the well-known phenomenon of loss of coherence; that is, the spatial extent of the wavefunction grows indefinitely in time until any form of coherence is lost in dull uniformity. This phenomenon is best analyzed for the case of the so-called *minimum uncertainty wavepacket*, a Gaussian packet centered at $x = x_0$ propagating at speed $u_0 = p_0/m$ [215]:

$$\psi(x,0) = \left(2\pi\delta_0^2\right)^{-1/4} \mathrm{e}^{\mathrm{i}p_0 x/\hbar} \mathrm{e}^{-(x-x_0)^2/4\delta_0^2}. \tag{13.39}$$

Denoting by δ^2 the variance of the wavefunction $\delta^2 = \int \psi^* x^2 \psi \, \mathrm{d}x$, the analytical expression for this loss of coherence reads as:

$$\delta^2(t) = \delta_0^2 + \frac{D^2 t^2}{\delta_0^2}, \tag{13.40}$$

where $D = \hbar/2m$ is the quantum diffusion coefficient.

Note that, at variance with classical diffusion ($\delta \sim t^{1/2}$), the expression (13.40) implies a ballistic loss of coherence $\delta \sim t$ in the asymptotic regime $t \gg \delta_0/D$. The analytical expression (13.40) has been checked for a series of simulations with different values of δ_0, u_0 and number of grid points N.

A typical set of results is shown in Figs 13.1 and 13.2, from [219].

These tests prove that quantum LBE reproduces accurately the analytical results as long as the spatial uncertainty δ is well resolved, typically above 16 lattice units.[47] Narrower packets perforce contain high wavelengths which fail to obey the adiabaticity assumption underlying the non-relativistic limit. Whether the non-adiabatic behavior of the shortest wavelengths obeys faithful non-relativistic dynamics is an interesting question that remains to be explored.

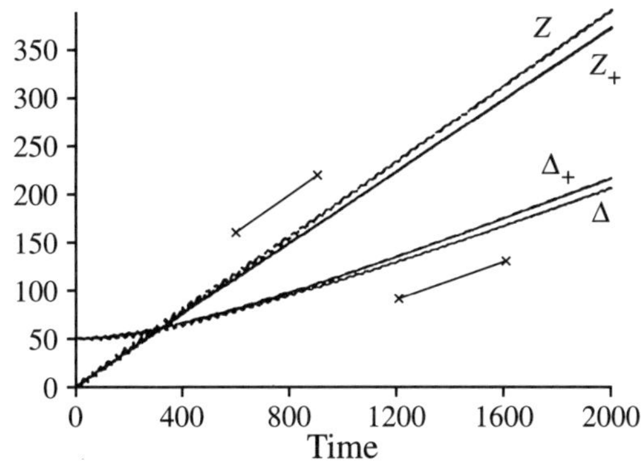

FIG. 13.1. Mean displacement and variance of the wavefunction as a function of time for $\beta = 0.2$ and $m = 0.1$. The suffix + indicates the average over the density of the 'hydrodynamic' mode. From [219].

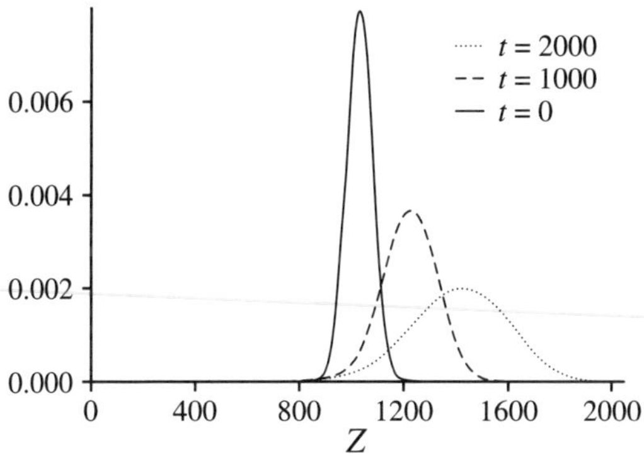

FIG. 13.2. Probability distribution of the 'hydrodynamic' mode at three successive instants in time. Parameters are the same as in Fig. 13.1. From [219].

[47]This figure may be contrasted to the 1–2 lattice units required by fluid applications. This seems to indicate that non-relativistic quantum wave behavior on the lattice requires an order of magnitude more lattice sites than fluid behavior. The author does not have any cogent explanation for this.

13.4.2 *Harmonic oscillator*

The next test is the familiar harmonic oscillator, a particle bound in a quadratic harmonic potential:

$$V_H(x) = 4V_0 \frac{(x - x_0)^2}{L^2}. \tag{13.41}$$

As is well known, a wavepacket with variance $\delta_0 = \sqrt{\hbar/2m\Omega}$, $\Omega^2 = 8V_0/mL^2$, behaves like a classical particle, bouncing back and forth in the potential with a speed $u_H = \delta_0\Omega/2\pi$ with no loss of coherence.

The coherent motion of such a wavepacket has been reproduced by quantum LBE for a series of values of the parameters V_0, L and N. A typical result is shown in Fig. 13.3 (from [219]).

Again, quantitative match with analytical result is obtained only for sufficiently broad packets. Packets narrower than 16 lattice units are slightly distorted by relativistic, non-adiabatic, motion of the shortest wavelengths.

13.4.3 *Scattering over a rectangular barrier*

As a final exercise, the quantum LBE has been tested on a simple scattering problem over a rectangular potential of the form:

$$V_s(x) = \begin{cases} V_0 & -w/2 < x < w/2, \\ 0 & \text{otherwise.} \end{cases}$$

A typical manifestation of quantal effects is a finite probability for the quantum particle to go through the potential barrier even if is energy E is lower than the potential height V_0 (*quantum tunneling*).

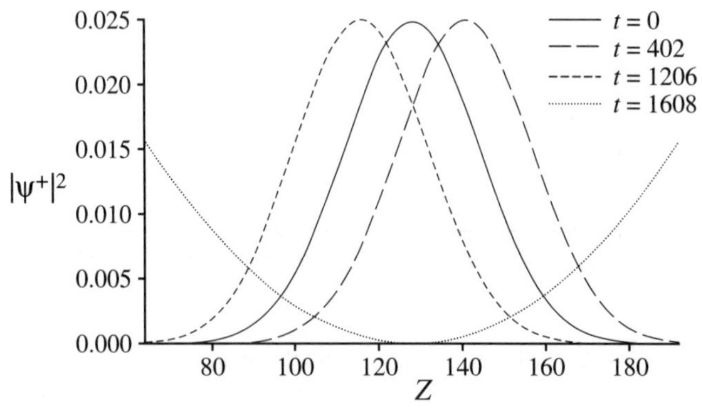

FIG. 13.3. Wavefunction in a harmonic potential at $t = 0$, $T/4$, $3T/4$ and T, where $T = 1608$ is the oscillation period in lattice units. The vertical axis displays the probability density $|\psi^+|^2$ of the 'hydrodynamic' mode. From [219].

Under tunneling conditions, an incident plane wave $\psi_{\text{in}} = A\mathrm{e}^{\mathrm{i}k_{\text{in}}x}$ at $x = -w/2$ develops transmitted and reflected components of the form

$$\psi(x) = T\mathrm{e}^{\mathrm{i}k_t x} + R\mathrm{e}^{-\mathrm{i}k_r x}, \tag{13.42}$$

where the reflected and transmitted wavenumbers are given by

$$k_t = \frac{\sqrt{2m(E - V_0)}}{\hbar}, \quad k_r = \frac{\sqrt{2mE}}{\hbar}, \tag{13.43}$$

respectively.

Inside the classically forbidden region $-w/2 < x < w/2$, the wavenumber of the transmitted wave becomes purely imaginary, corresponding to the well-known attenuation effect due to the barrier itself: $T/A \sim \mathrm{e}^{-2|k_t|w}$.

The existence of a reflected and transmitted waves has been clearly detected in a series of simulations with different values of the parameters V_0, w and N (see Fig. 13.4, from [219]).

To sum up, all of these tests provide evidence of the viability of quantum LBE, at least in one dimension. The scheme performs efficiently and, what is more, provides *stability* and *unitarity* at the same time, a very valuable property for an explicit scheme. As we said, this is related to the peculiar light-cone space–time marching technique inherent in quantum LBE. To date, higher-dimensional versions of the present quantum LBE scheme are still waiting for numerical validation. However, multi-dimensional quantum lattice schemes (not based on the spin-speed analogy, though) have been developed systematically by a number of authors [222, 223].

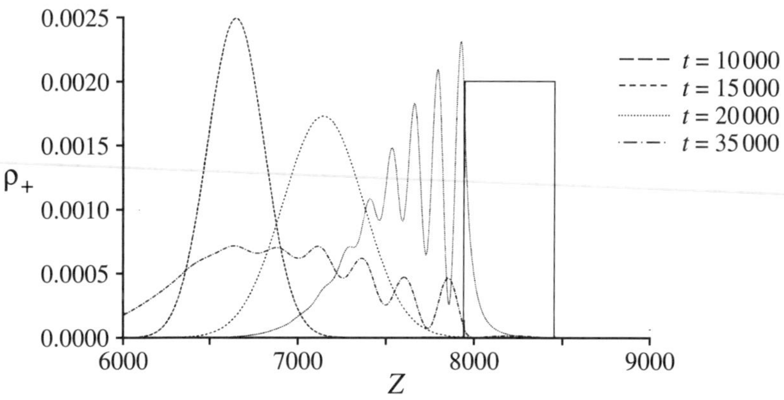

FIG. 13.4. Probability density of the 'hydrodynamic' mode at different time instants for the case of the scattering against a rectangular barrier (solid line). The exponential decay within the barrier is clearly visible, and so are reflected waves in front of it. From [219].

13.5 The quantum N-body problem

Solving the single particle Schrödinger, or Dirac, equation in three dimensions is certainly no mean feat from the computational point of view. This task, however, pales in the face of the ordeal of solving the Schrödinger equation for a collection of, say, N particles (typically ions and electrons). This is the goal of computational quantum chemistry, a cutting edge of modern scientific computing. The target is the N-body Schrödinger equation:

$$\mathrm{i}\,\hbar\,\partial_t \Phi = \sum_{n=1}^{N} \left[-\frac{\hbar^2}{2m} \Delta_n + V\left(X_n\right) \right] \Phi, \tag{13.44}$$

where $X_n = (x_n, y_n, z_n)$ denotes the spatial coordinate of the n-th particle, $\Phi\left(X_1, \ldots, X_N\right)$ is the N-body wavefunction and V is the interparticle potential, typically in a two-body format $V\left(X_n\right) = \sum_{m \neq n} V\left(|X_n - X_m|\right)$. Given its importance in quantum chemistry, biology and material science, (13.44) makes the object of a whole industry of powerful numerical techniques to compute the ground and excited states of quantum many-body systems [226].

One of the most popular techniques for solving eqn (13.44) considers the time-independent version:

$$H\Phi = E\Phi \tag{13.45}$$

and numerically solves the corresponding eigenvalue problem. This is more easily said than done, since it implies at best[48] the diagonalization of a matrix of rank $G \times N$ where G is the number of grid points used for the spatial discretization. Typical desired values are in the order of $G \sim 100^3$ and $N \sim 100$, which means a 'monster' matrix of rank 10^8!

An alternative class of methods is based on the idea of minimizing the energy functional $E = (\Phi, H\Phi)$ by means of current iterative methods such as steepest descent or more recently, conjugate gradient techniques. Particularly successful is a special form of *dynamic minimization* known as the Car–Parrinello method. In a nutshell, the Car–Parrinello method replaces the true Lagrangian associated with the N-body energy functional, with a fictitious one in which the ions move classically, according to Newton's equations of motion, whereas the electrons obey a fictitious dynamics with the key property that the trajectory generated by this dynamics, although fictitious, does capture the true electron dynamics in the limit where electrons can be assumed adiabatically enslaved to the ions (Born–Oppenheimer approximation) [227]. The 'magic' Car–Parrinello algorithm permits tracking of ions and electrons at the same time, whence its alias of *ab-initio* molecular dynamics.

What could (quantum) lattice techniques contribute to this all-important research field?

[48] Whenever the N-body problem can be factorized into a series of N one-body problems, coupled via an effective potential. The corresponding $G \times N$ nonlinear eigenvalue problem can be solved by a repeated, iterative, sequence of N eigenvalue problems, each of size G.

In a broad sense, the putative N-body quantum LBE would also fall within the same class of 'fictitious dynamical methods' as the Car–Parrinello method. The mathematical formulation is obviously different, but the general strategy is somehow similar: replace the true N-body equation with a stylized, and yet faithful, surrogate. In fact, the analogy is even slightly more poignant, since the Car–Parrinello Lagrangian makes use of a fictitious electronic mass much larger than the physical one, but still much smaller than the ionic mass, to keep ionic and electronic motion well scale-separated. This is very similar to the lattice gas/Boltzmann trick of using a fictitiously large Knudsen number to give kinetic theory access to larger scales. The author is not aware of any LBE quantum many-body algorithm destined to electronic computers, although a hypothetical strategy will shortly be outlined at the end of this chapter.

Instead, some interesting thoughts have been spent in the direction of quantum computing, a really important topic these days [228].

13.5.1 *Quantum lattices for quantum computers*

It has recently been pointed out [222] that quantum lattice algorithms constitute excellent candidates as numerical schemes for hypothetical, yet to come, quantum computers. In the N-body quantum LBE, each quantum particle is represented by bG walkers, b being the coordination number of the lattice, namely the number of discrete speeds attached to each lattice site. These walkers move around according to a fictitious microdynamics whose macroscopic limit is precisely the N-body Schrödinger equation.

What would this hypothetical N-body quantum LBE algorithm look like?

'Simply' evolve N replicas of the single particle quantum LBE scheme and tie them up together via a two-body potential collecting the sum of all contributions $V_g^n = \sum_{g'} \sum_m V\left(X_g^n - X_{g'}^m\right)$ at each given site X_g^n.

If one does not insist on the idea of a particle generalized trajectory, and turns instead to an 'information network' picture, a generic quantum lattice algorithm would take the form of a first-order, explicit, non-local map for the complex array Φ_i:

$$\phi_j(X,t) = \sum_k T_{jk}\phi_k\left(X - V_k\Delta t, t - \Delta t\right), \tag{13.46}$$

where V_k scans the $3N$-dimensional neighborhood of $X_n = (x_n, y_n, z_n)$, $n = 1, \ldots, N$, and T_{jk} is the complex transfer matrix fulfilling the unitarity condition $\sum_l T_{jl}^* T_{lk} = \delta_{jk}$.

The kinetic energy operator is sweet since any walker in a given single particle state can be moved independently of the others, resulting in a linear $O(bGN)$ complexity. Unfortunately, the long-range potential generates an unforgiving quadratic complexity, $(bGN)^2$, because any walker interacts with all others. To say nothing of course of the $(bG)^N$ requirement in computer storage ... The scheme meets with an 'exponential complexity wall' which rules out conventional electronic computers [229].[49]

Then back to quantum computers. Since the concept of particle trajectory can safely be abandoned in quantum mechanics (here 'can' means 'can', not 'must'!), it makes no sense to talk about a left-moving or right-moving particle. Every particle moves left *and* right with given probabilities which vary continuously in space and time. Hence, the quantum particle takes all possible intermediate states between the extreme move left, move right options (pure states). This links up with Feynman path integral formulation of quantum mechanics, where a particle traveling from, say, X to Y tries out all possible paths between these two points, if with a different probability e^{-S}, S being the action of the system. Cellular automata are ill-suited to describe such a situation, accounting as they do for only two extreme, pure states: move left or move right, *tertium non datur*. Even more so lattice gas CA, with their Boolean occupation numbers n_i specifying unambiguously both space and momentum coordinates of the moving particle. To describe quantum motion we need to replace these 'sharp' objects with smoother ones, capable of providing graded responses between all black, all white extremes. This change in perspective should not come that new to the reader, remnant as it is of the paradigm shift from LGCA to LBE in classical mechanics. Here, like there, the idea is to smooth out single particle trajectories in 'ensemble-averaged' clouds of trajectories. The physical observable is no longer a particle here or there, but only its probability to be here and there. Of course, real numbers serve us well in achieving this task (as we learned with LBE), but other options, more intimately linked to the principles of quantum mechanics, are also available.

13.5.2 *Bits, trits and q-bits*

It appears that quantum bits (q-bits) have precisely what it takes to implement quantum lattice algorithms (and more ...). Let us make a quick acquaintance with the q-bit concept. As is well known, bits (binary digits) encode just two possible outcomes: yes or no. Similarly, 'trits' would allow three possible outcomes: yes, no or perhaps. The q-bits encode a graded response encompassing all possible states between yes and no, actually any real number between 0 and 1. More precisely, any couple of real numbers (p, q), $0 < p, q < 1$, $p + q = 1$, expressing the probability of 'yes' (p) and 'no' (q). Obviously this corresponds to an infinite amount of information, whereas in practice one would expect to assign a finite number say m, of ordinary bits to represent the finite sequence of 2^m values $2^{-m}(0, 1, \ldots, 2^m - 1)$ which the q-bit can take on.

Consider for instance a 1D lattice with $m = 2$ q-bits attached to each site. The interpretation is as follows:

[49] For more than a few hundred particles.

```
00:   move right (pure state)
01:   1/2 move right, 1/2 move left
10:   1/2 move right, 1/2 move left
11:   move left  (pure state)
```

Each of the four 'replicas' is equally entitled to evolve according to the local map (13.46), which means that the system unfolds along four distinct trajectories at the same time. This evokes quite closely the Feynman path integral view of quantum mechanics [224]. On a classical computer, the task of tracking the evolution of an m-bit long q-bit generates a 2^m complex problem, but on a hypothetical quantum computer all of these replicas are tracked simultaneously so that the cost of moving the whole bunch of trajectories is just what it takes a conventional computer to track a single one! Thus, a quantum computer would most naturally solve N-body quantum problems in basically the same time it takes today an electronic computer to solve the single particle Schrödinger equation! This wonderland is surely not an exclusive privilege of lattice applications. Still, it is good to know that the day quantum computers will be with us, quantum lattice schemes will be ready to use them right from day one!

13.5.3 *Quantum LBE and density functional theory*

As previously discussed, numerical algorithms for the full quantum many-body wavefunction are probably hopeless (on electronic computers) because of the exponential complexity wall. However, effective one-body theories have been developed in the last forty years which permit us to learn a great deal about the properties of quantum many-body systems without ever invoking the explicit knowledge of many-body wavefunctions.

Particularly successful in this respect is the density functional theory developed in the mid-60s by Hohenberg–Kohn and Kohn–Sham [230]. The core idea of density functional theory is that the ground-state of a many-electron wavefunction (nuclei are regarded as classical particles on account of their higher mass) is uniquely determined by the electronic density $n(\vec{x}) = \sum_j |\phi_j|^2 (\vec{x})$, where ϕ_j are one-particle orbitals. The ground-state energy can then be obtained by summing up the single particle orbital energy obtained by solving the Kohn–Sham equations:

$$H_{\mathrm{KS}}\phi_j = E_j\phi_j, \tag{13.47}$$

where the Kohn–Sham Hamiltonian consists of four contributions:

$$H_{\mathrm{KS}} = -\frac{\hbar^2}{2m}\Delta_j + V_{\mathrm{ext}}(\vec{x}) + e^2 \int \frac{n(\vec{y})}{|\vec{y}-\vec{x}|}\,\mathrm{d}\vec{y} + V_{\mathrm{ex}}[n]. \tag{13.48}$$

The first two contributions are the usual kinetic energy and external potential operators, the third one relates to the self-consistent Hartree–Fock potential. Finally, the fourth one is an effective 'exchange' energy functional which collects the effects of N-body interactions. The idea is that an effective functional of the electron density exists (whose exact form remains to be educatively guessed) such that the ground-state energy of a fictitious system of *independent* electrons moving in such a potential is *exactly the same* ground-state energy of the interacting system! Describing how such a magic comes about is certainly beyond the scope of this book. Here, we shall simply remark that density functional theory leans heavily on the intuitive picture of a quantum many-body system as a backbone of ions tied up together by a very mobile *electronic fluid.* In this respect, it certainly puts a premium on efficient real space solvers for the one particle (nonlinear) Schrödinger equation, both in the time-independent (ground-state) and time-dependent (excited states) form. Whence, perhaps, scope for quantum LBE techniques.

13.6 Exercises

1. Derive the Maiorana form of the Dirac equation.
2. Prove the commutativity of the streaming operator.
3. Prove the stability of the quantum LBE scheme (13.34).
4. Plot the modified mass m' as a function of m for various Δt. For what values of Δt does the scheme operate safely?
5. Do you see any analogy between the antisymmetric component ϕ^- and the ghost modes of hydrodynamics? Please comment.

14

THERMOHYDRODYNAMIC LBE SCHEMES

The LBE discussed thus far did not address the issue of a self-consistent coupling between temperature dynamics and heat transfer within the fluid flow. Fully thermohydrodynamic LBE schemes represent a standing challenge to LBE research and despite several brilliant attempts, to date, a consistent thermodynamic LBE scheme working over a wide range of temperatures is still missing. The difficulty is that heat and temperature dynamics require more kinetic momenta and consequently they probe the discrete space–time 'fabric' of the lattice more keenly than isothermal flows. At the time of writing, this is one of the most challenging issue left with (basic) LBE research. In this chapter we shall give an account of this important ongoing topic.

14.1 Isothermal and athermal lattices

It is known from kinetic theory that the internal energy density of an ideal fluid is defined as the average kinetic energy around the mean flow at a local equilibrium:

$$\rho e = \frac{1}{2} \int f^{\mathrm{e}} \left(\vec{v} - \vec{u}\right)^2 \mathrm{d}\vec{v}. \tag{14.1}$$

Similarly, kinetic temperature T is defined as (twice) the average kinetic energy per spatial degree of freedom. In D spatial dimensions:

$$\rho T = \frac{1}{D} \int f^{\mathrm{e}} \left(\vec{v} - \vec{u}\right)^2 \mathrm{d}\vec{v}, \tag{14.2}$$

yielding the following equality:

$$T = \frac{2e}{D}. \tag{14.3}$$

This, in turn, delivers the equation of state of an ideal fluid

$$P = \rho T \tag{14.4}$$

and the attendant expression for the sound speed[50]

$$c_s^2 = \frac{\partial P}{\partial \rho} = T. \tag{14.5}$$

Thermodynamic consistency requires that temperature, pressure and density changes be related through the above equation throughout the entire evolution

[50] As usual we work in units of $K_B = m = 1$.

of the flow. This is asking a lot of a system living in a lattice with a handful of discrete speeds.

A necessary, yet not sufficient, prerequisite for thermodynamic consistency is that these discrete speeds do *not* share the same magnitude (*multi-energy lattices*), for otherwise temperature is trivialized to a frozen constant. Although trivial, this issue is worth being spelt out.

For a discrete distribution function consisting of a set of monochromatic beams:

$$f^B(\vec{v}) = \frac{\rho}{b} \sum_{i=1}^{b} f_i \delta(\vec{v} - \vec{c}_i), \tag{14.6}$$

with $|\vec{c}_i| = c$, it is readily seen that even moments $\langle v^{2n} \rangle \equiv \int f^B v^{2n} \, d\vec{v}$ obey the relation $\langle v^{2n} \rangle / \langle v^2 \rangle^n = 1$, showing that temperature, as well as any higher-order moment, trivializes to a constant value (we assumed $\int f^B \, d\vec{v} = 1$).

We also wish to emphasize that while in the continuum kinetic and thermodynamic temperature (as defined by the Boltzmann exponential weight) come to the same at local equilibrium, in a discrete lattice this is *not* the case. A glance at eqn (14.6) shows that, according to the thermodynamic definition, the lattice temperature is, strictly speaking, zero because of the Dirac delta lattice representation. This is why some authors find it more appropriate to qualify mono-energetic LBE schemes as *athermal* (zero temperature) instead of *isothermal.*

This is more than mere semantics.

Indeed, as pointed out by M. Ernst [231], '*many authors have introduced quantities called "temperatures" ... which do not coincide with the true temperature in the sense of thermodynamics and statistical mechanics.*'

Although Ernst's criticism was directed to cellular automata fluids, it applies to LBE as well since the discrete speeds are basically the same. The point is that the kinetic temperature, as defined by eqn (14.2) and the thermodynamic temperature defined via the Boltzmann weight $e^{-E/T}$ do *not* generally coincide in discrete momentum space.

Mathematically, this stems from the inequality

$$\frac{1}{\beta} \neq \sum_{i=1}^{b} c_i^2 e^{-\beta c_i^2} \Big/ \sum_{i=1}^{b} e^{-\beta c_i^2}, \tag{14.7}$$

which turns into an equality only in the limit where the number of discrete speeds is sent to infinity (this is an exponentially fast convergent limit, though).

A subtle description of the ensuing paradoxes and related solutions can be found in Cercignani's original paper [232]. A thorough study of momentum discreteness effects on lattice thermohydrodynamics is given in [233]. Other remarkable properties of multi-energy lattice gas automata, such as the properties of spontaneous thermodynamic fluctuations, are discussed by Grosfils *et al.* [234]. For a more recent review of thermal LBEs as discrete velocity models, the reader is directed to the instructive paper by L. S. Luo [235].

Having at least brought up the issue, we shall feel free to keep sticking to the simpler, if less rigorous, definition of 'isothermal' models.

14.2 Thermodynamic equilibria and multi-energy lattices

In principle, our desiderata on a thermodynamically consistent LBE are as follows:

1. Cover a wide range of temperatures.
2. Stable numerical operation within this range.

Thermal LBE schemes are best organized in terms of *energy shells* according to a two index notation:

$$\epsilon_j = \frac{1}{2} c_{ji}^2, \tag{14.8}$$

where j labels the energy level and i the discrete speed 'momentum eigenstate' within energy level j.[51]

Multi-energy lattices have been met before in this book. For instance, the D2Q9 model discussed in Part I, splits into the direct sum of three sub-lattices E0, E1, E2 with energy 0, 1/2, 1, respectively, where the notation Ej means $2\epsilon = j$.

With this definition, the internal energy density ρe becomes a weighted sum over three energy levels:

$$\rho e = \rho_1 \epsilon_1 + \rho_2 \epsilon_2 + \rho_3 \epsilon_3, \tag{14.9}$$

where ρ_j are the partial densities of three shells.

A given lattice with minimum/maximum speeds c_{min}, c_{max} can only support temperatures in the range

$$\frac{c_{\text{min}}^2}{D} < T < \frac{c_{\text{max}}^2}{D}. \tag{14.10}$$

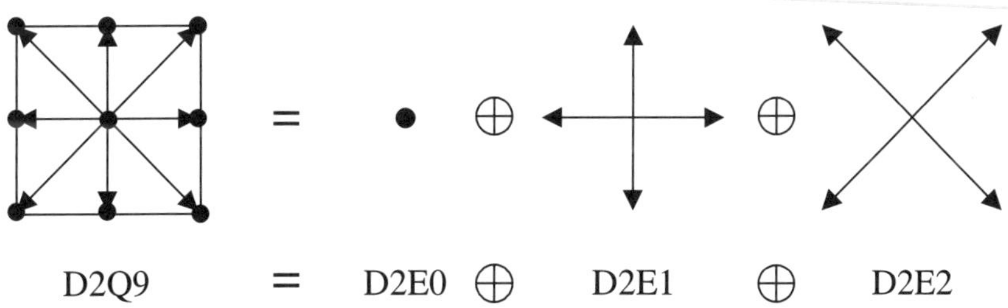

FIG. 14.1. The decomposition of D2Q9 in three energy levels, 0, 1 and 2.

[51] A useful mnemonic is to think of a momentum–energy pair of quantum numbers.

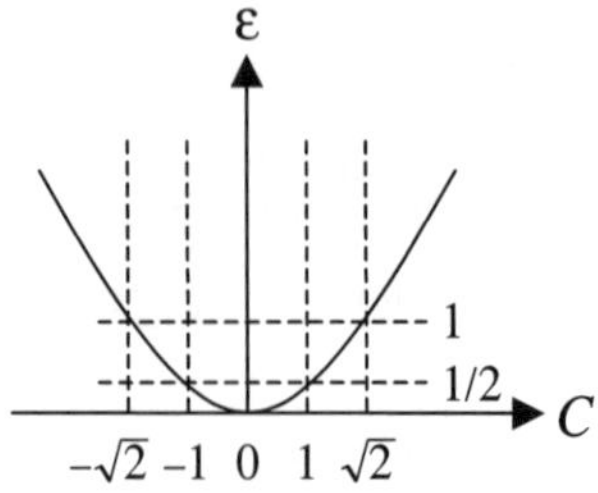

FIG. 14.2. Sketch of the parabolic energy–momentum relation with three discrete energy levels.

For instance, with reference to D2Q9, $T = 0$ means that all discrete populations have 'condensed' into the stopped particle state $f_0 = 1$, $f_i = 0$, $i > 1$, whereas in the 'hot' phase $T = 1$ only long-jumper states $c = \sqrt{2}$ are populated.

It should be clearly borne in mind that these are only realizability constraints, which say little about the actual stability of the corresponding thermal LBE.

Intuitively, we expect safer navigation away from the realizability edges, but there is no way to decide this *a priori* without first developing a more precise notion of discrete thermodynamic equilibria. This requires the inspection of the set of thermohydrodynamic constraints to be fulfilled by thermal LBE equilibria. That was precisely the starting point of earliest efforts in the field.

14.3 Extended parametric equilibria

One of the main lessons learned from LBE theory is that local LBE equilibria need *not* correspond to any of the three distributions familiar from statistical physics for fermions, bosons or classical particles. They can be chosen 'freely', so long as they reproduce the desired form of the macroscopic moments, *possibly* along with a H-theorem ensuring compliance with the second principle of thermodynamics (no mean feat). We have also learned how to exploit this freedom in order to replace true physical distributions with simpler (quadratic) polynomials in the Mach number. It is therefore natural to seek for the inclusion of compressibility effects via higher-order expansions in the Mach number. Thermal effects can subsequently be accounted for by letting the coefficients of the expansion depend smoothly on the temperature T of the flow.[52]

To widen the range of accessible Mach numbers, as well as to enforce the correct thermodynamic constraints, Alexander, Chen and Sterling propose the following $6J$ parameter (each parameter accounts for J degrees of freedom, one per energy level) expansion of the local equilibrium [236]:

[52]In the continuum, temperature enters the game in a beautifully self-similar form, via the dimensionless relative speed $(v-u)/v_T$. This expression encodes a further symmetry of continuum Maxwellian equilibria, namely scale-invariance under space and time dilatations/contractions: $x \to \lambda x$, $t \to \mu t$. Out of local equilibrium this invariance is broken by dissipative effects.

$$f_i^{\mathrm{e}} = A_i + B_i u_i + C_i u_i^2 + D_i u^2 + E_i u_i^3 + F_i u_i u^2, \tag{14.11}$$

where $u_i = c_{ia} u_a$ are the covariant components of the flow speed along the discrete directions of the lattice. The idea is to adjust the Lagrange multipliers $A_i, \ldots, F_i$ in such a way as to recover the desired set of macroscopic momenta. Owing to the wide degree of latitude provided by this set parameters, this looks like a fairly comfortable situation, so long as one is able to pick up physically meaningful solutions out of a redundant set of possible ones.

Let us focus on the task of putting this huge freedom to work to impose the set of usual hydrodynamic constraints:

$$\sum_i f_i^{\mathrm{e}} = \rho, \tag{14.12}$$

$$\sum_i f_i^{\mathrm{e}} c_{ia} = \rho u_a, \tag{14.13}$$

$$\sum_i f_i^{\mathrm{e}} c_{ia} c_{ib} = \rho \left(u_a u_b + T \delta_{ab} \right), \tag{14.14}$$

$$\sum_i f_i^{\mathrm{ne}} c_{ia} c_{ib} = \rho \nu \left[\tfrac{1}{2} \left(D_{ab} + D_{ba} \right) - \left(\partial_c u_c \right) \delta_{ab} \right], \tag{14.15}$$

where $D_{ab} = \partial_a u_b$ is the strain tensor.

This set must be augmented with the additional constraints ensuring that the internal energy (temperature) e obeys the correct energy transport equation:

$$\partial_t \rho e + \partial_a \rho e u_a = -\partial_a q_a + \Phi, \tag{14.16}$$

where $\Phi = P_{ab} D_{ab}$ is the dissipation function resulting from work done by the pressure tensor over the strained flow configuration.

The first new entry is the *heat flux vector*

$$q_a = \frac{1}{2} \sum_i f_i \, c'^2_i \, c'_{ia}, \tag{14.17}$$

where

$$c'_{ia} = c_{ia} - u_a \tag{14.18}$$

is the speed of the molecules relative to the fluid ('peculiar' speed).

The requirements on the heat flux vector are:

$$q_a^{\mathrm{e}} = \frac{1}{2} \sum_i f_i^{\mathrm{e}} c'^2 c'_{ia} = 0, \tag{14.19}$$

$$q_a^{\mathrm{ne}} = \frac{1}{2} \sum_i f_i^{\mathrm{ne}} c'^2 c'_{ia} = -\chi \partial_a T, \tag{14.20}$$

where χ is the thermal diffusivity of the fluid.

Manifestly, the set of constraints (14.12)–(14.15), (14.19) is outnumbered by the free parameters at our disposal, so that the problem of fixing the discrete thermodynamic equilibria is an underdetermined one: more solutions than unknowns.

Out of various admissible solutions, Alexander *et al.* propose the following one ($\rho = 1$):

$$A_0 = 1 - \tfrac{5}{2}e + 2e^2, \quad A_1 = \tfrac{4}{9}\left(e - e^2\right), \quad A_2 = \tfrac{4}{9}\left(e - e^2\right), \tag{14.21}$$

$$B_1 = \tfrac{4}{9}\left(1 - e\right), \quad B_2 = \tfrac{1}{36}\left(-1 + 4e\right), \tag{14.22}$$

$$C_1 = \tfrac{4}{9}\left(2 - 3e\right), \quad C_2 = \tfrac{1}{72}\left(-1 + 6e\right), \tag{14.23}$$

$$D_0 = \tfrac{1}{4}\left(-5 + 8e\right), \quad D_1 = \tfrac{2}{9}\left(-1 + e\right), \quad D_2 = \tfrac{1}{72}\left(1 - 4e\right), \tag{14.24}$$

$$E_1 = -\tfrac{4}{27}, \quad E_2 = \tfrac{1}{108}. \tag{14.25}$$

This refers to a two-dimensional triangular lattice with three energy levels, $\epsilon = 0, 1, 2$, and thirteen speeds:

$$c_0 = 0, \quad c_{jix} = j\cos\theta_i, \quad c_{jiy} = j\sin\theta_i, \tag{14.26}$$

$$\theta_i = \frac{2\pi i}{6}, \quad j = 1, 2, \quad i = 1, \ldots, 6. \tag{14.27}$$

The transport parameters take the following expressions:

$$\mu = \rho e\left(\tau - \frac{1}{2}\right), \quad \chi = 2\mu. \tag{14.28}$$

Note that within this model, the Prandtl number (momentum diffusivity versus heat diffusivity) is frozen to the value $Pr = 1/2$. This is about half the value of ideal gases, and far from other important fluid heat carriers (for instance liquid metals, which feature $Pr \sim 0.01$).

This limitation can be partially removed, see [238], by using a *two-time* relaxation operator of the form:

$$C_{\mathrm{BGK},2} = -\frac{f_i - f_i^{\mathrm{e}}}{\tau_1} - \frac{f_{i^*} - f_{i^*}^{\mathrm{e}}}{\tau_2}, \tag{14.29}$$

where the star denotes mirror conjugation.

With this extension, the Prandtl number takes the following expression:

$$Pr = \frac{2\tau_v - 1}{2\tau_k - 1}, \tag{14.30}$$

where $\tau_v = \tau_1\tau_2/(\tau_1 + \tau_2)$ and $\tau_k = \tau_1\tau_2/(\tau_1 - \tau_2)$.

Capabilities and limitations of this two time relaxation model for heat transfer problems are discussed in [237, 238].

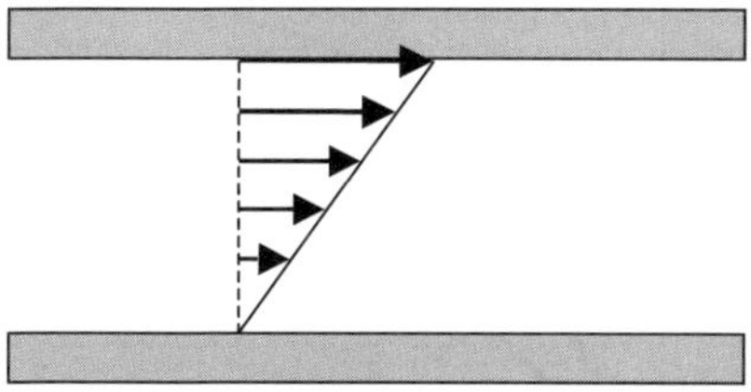

FIG. 14.3. Thermal Couette flow.

The derivation of the thermohydrodynamic equation of motion from the multi-energy LBGK proceeds along the same general lines described for isothermal flows: multiscale expansion as combined with systematic neglect of terms $O(Kn\, Ma^2)$.

This derivation is rather lengthy and will not be reported here. The reader interested in full details is referred to the relevant literature [233, 239]. Suffice it to say that, as pointed out by many authors, anomalies are obviously more troublesome than in the isothermal case simply because sizable Mach numbers and temperature dynamics are deeper probes of the discrete fabric of the lattice.

This is after all a common trait of statistical mechanics: temperature is precisely the property allowing the system to explore its phase-space. It is therefore not surprising that many anomalies which simply lay dormant in the ordered state $T \to 0$ do raise their face as soon as $T \neq 0$.

Alexander *et al.* do not discuss the theoretical properties of their model (positivity, stability and related issues) any further, but move on to demonstrate its practical viability on a number of test cases, including *(i) isothermal Poiseuille flow, to test the fluid viscosity* μ *as a function of the relaxation parameter* τ, *(ii) heat transfer across a transversal longitudinal temperature jump*, to compute the thermal conductivity χ.

In both cases, the authors report excellent agreement with theoretical expressions. In addition they successfully check that density waves propagate according to the desired expression $c_s^2 = 2e/D$, in compliance with the ideal gas equation of state. They also simulate a Couette shear flow with a temperature gradient between the boundaries and check against the analytical temperature profile:

$$e^* \equiv \frac{e - e_0}{e - e_1} = \frac{1}{2}(1 + y) + Br\frac{(1 - y^2)}{8}, \tag{14.31}$$

where y is the normalized distance from the center, e_0 and e_1 are the upper and lower wall internal energies, respectively, and Br is the Brinkman number, a measure of dissipative versus conductive heat effects.

The energy profile for the case $U_1 = 0.1$ when $Br = 5$ and $Br = 10$ are reported in Fig. 14.4 (from [236]).

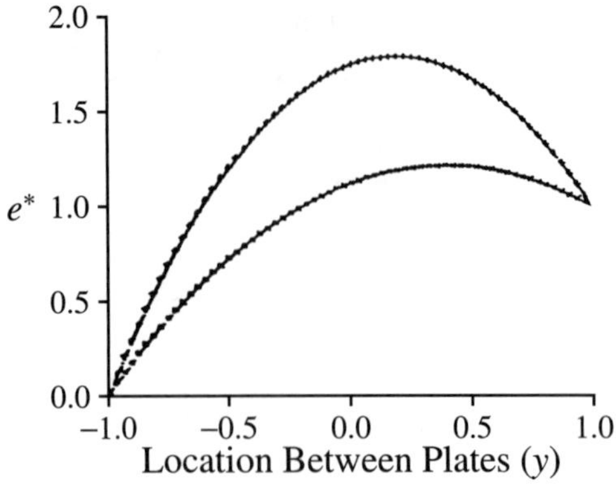

FIG. 14.4. Internal energy (in lattice units) for the Couette flow with heat transfer, for Brinkman numbers $Br = 5$ and 10. The upper wall is moving with speed $u = 0.1$ and the lower wall is at rest. The solid lines represent the analytical results (from [236]).

These tests bring convincing evidence that the Alexander *et al.* thermohydrodynamic LBGK scheme reproduces the basic mechanisms of momentum and heat transfer. This is fairly interesting, but does not cover the thorny issue of *nonlinear* momentum and heat transfer, which is left unprobed by the tests discussed above. Nor do these tests touch upon the issue of density dynamics, e.g., compressibility effects, in the nonlinear regime. Alexander, Chen and Sterling's pioneering work is a very valuable stepping stone for further exploration of nonlinear thermohydrodynamic behavior, but unfortunately, not the final curtain on the subject.

14.4 Thermal LBE models without nonlinear deviations

In the previous section we introduced the heat flux vector as a further local constraint for the discrete equilibrium distribution. We have then made acquaintance with the Alexander *et al.* equilibria, one of the possible solutions to the enlarged system of thermohydrodynamic constraints.

As anticipated, the Alexander *et al.* scheme is exposed to anomalies coming from higher-order terms in the double Knudsen–Mach expansion [238]. To remove these anomalies, Yu Chen [239] proposes higher-order parametric equilibria fulfilling the *full set* of thermohydrodynamic constraints:

$$\rho = \sum_i f_i^{\mathrm{e}}, \tag{14.32}$$

$$J_a = \sum_i f_i^{\mathrm{e}} c_{ia} \qquad = \rho u_a, \tag{14.33}$$

$$P_{ab} = \sum_i f_i^{\mathrm{e}} c_{ia} c_{ib} \quad = \rho \left(u_a u_b + T \delta_{ab} \right), \tag{14.34}$$

$$Q_{abc} = \sum_i f_i^{\mathrm{e}} c_{ia} c_{ib} c_{ic} = \rho \left(u_a u_b u_c + T t_{abc} \right), \tag{14.35}$$

$$K_{ab} = \sum_i f_i^{\mathrm{e}} c_{ia} c_{ib} c^2 = \rho \left[u_a u_b \left(u^2 + (D+4) T \right) + u^2 T \delta_{ab} + (D+2) T^2 \right], \tag{14.36}$$

where $t_{abc} = u_a \delta_{bc} + u_b \delta_{ac} + u_c \delta_{ab}$.

We shall refer to Q_{abc} and K_{ab} as to momentum flux triple tensor (third-order) and energy flux tensor (second-order).

Yu Chen proves that in order for these constraints to be fulfilled, the parametric equilibria should contain terms up to the *sixth*-order in the Mach number. To ensure isotropy of six order tensors, he introduces an $8J$ parameter family, fourth-order polynomial expressions of the form

$$f_i^{\mathrm{e}} = A_i + B_i u_i + C_i u^2 + E_i u_i^2 + F_i u_i u^2 + G_i u_i^3 + H_i u_i^2 u^2 + L_i u^4. \tag{14.37}$$

As for Alexander *et al.* the coefficients $[A_i, \ldots, L_i]$ depend linearly on the fluid density and at most quadratically on the internal energy e, so that the generic coefficient $X_i \equiv [A_i, \ldots, L_i]$ can be cast in the form

$$X_i = \rho \sum_{j=0}^{2} X_{ji} e^j. \tag{14.38}$$

Matching the coefficients X_{ji} to the above set of constraints generates a linear system with $N_K = 24b$ kinetic degrees of freedom (8 parameters, 3 energy levels each, b discrete speeds each) to be matched by a total of

$$N_c = 1 + D + D(D+1) + \frac{1}{6} D(D+1)(D+2) \tag{14.39}$$

constraint equations. This means 5, 13, 26 equations in one, two and three dimensions, respectively.

It is perhaps instructive to compare the number of isothermal and thermal constraints in various dimensions, see Table 14.1.

Free parameters and constraints cannot be expected to be in exact balance for a generic lattice, an unsatisfactory aspect of the parametric approach, betraying its empirical nature, to which we shall return shortly.[53]

Nonetheless, Yu Chen provides a number of solutions at all dimensionalities $D = 1, 2, 3$. In one dimension, he finds the 5-speed model (D1V5) depicted in Fig. 14.5.

[53] Beware: a strict N_c versus N_K count would be deceptive because the problem has many symmetries which strongly reduce the effective number of kinetic degrees of freedom.

TABLE 14.1. Number of isothermal and thermal constraints in dimension D.

D	N_c^{iso}	N_c^{the}
1	3	6
2	6	15
3	10	26

FIG. 14.5. The D1V5 lattice (5 speeds, 3 energies).

The parameters (normalized to $\rho = 1$) are:

$$A_0 = 1 - \tfrac{5}{2}e + 3e^2, \quad A_1 = \tfrac{4}{3}e - 2e^2, \quad A_2 = -\tfrac{1}{12}e + \tfrac{1}{2}e^2, \tag{14.40}$$
$$B_1 = \tfrac{2}{3}e - e^2, \quad B_2 = -\tfrac{1}{24}e + \tfrac{1}{4}e^2, \tag{14.41}$$
$$C_1 = -\tfrac{5}{4} + 3e, \quad C_2 = -\tfrac{1}{3} + \tfrac{1}{2}e, \tag{14.42}$$
$$D_1 = 1 - \tfrac{5}{2}e, \quad D_2 = -\tfrac{1}{64} - \tfrac{5}{32}e, \tag{14.43}$$
$$E_1 = -\tfrac{17}{12}, \quad E_2 = \tfrac{1}{24}, \tag{14.44}$$
$$F_1 = \tfrac{5}{4}, \tag{14.45}$$
$$G_1 = -\tfrac{1}{4}, \quad G_2 = \tfrac{1}{64}, \tag{14.46}$$
$$H_0 = \tfrac{1}{4}, \quad H_1 = \tfrac{1}{12}, \quad H_2 = -\tfrac{1}{48}. \tag{14.47}$$

In two dimensions the solution is the 16-speed lattice depicted in Fig. 14.6 (note the absence of rest particles).

The explicit expression of the corresponding parameters is:

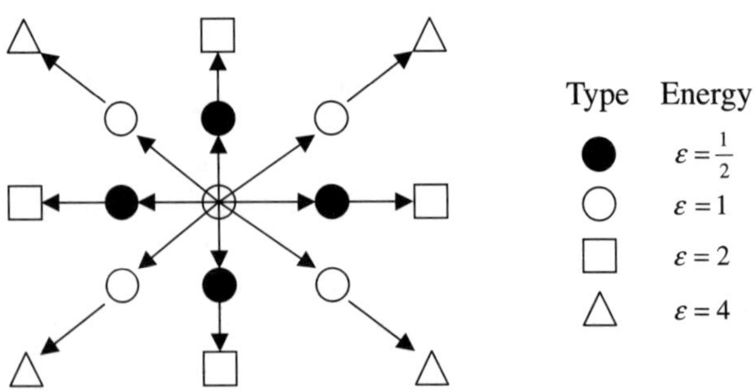

FIG. 14.6. The D2V16 (16 speeds, 4 energies) thermal lattice.

$$A_1 = \tfrac{8}{15} - \tfrac{2}{3}e + \tfrac{1}{3}e^2, \tag{14.48}$$
$$A_2 = -\tfrac{1}{30} + \tfrac{1}{24}e + \tfrac{1}{24}e^2, \tag{14.49}$$
$$A_3 = -\tfrac{4}{15} + \tfrac{2}{3}e - \tfrac{5}{12}e^2, \tag{14.50}$$
$$A_4 = \tfrac{1}{60} - \tfrac{1}{24}e + \tfrac{1}{24}e^2, \tag{14.51}$$

$$B_1 = \tfrac{2}{3}e - e^2, \tag{14.52}$$
$$B_2 = -\tfrac{1}{24}e + \tfrac{1}{8}e^2, \tag{14.53}$$
$$B_3 = \tfrac{1}{4}e, \tag{14.54}$$

$$C_1 = -\tfrac{2}{3} + \tfrac{5}{6}e, \tag{14.55}$$
$$C_2 = \tfrac{1}{24} + \tfrac{5}{12}e, \tag{14.56}$$
$$C_3 = \tfrac{1}{6} - \tfrac{7}{4}e, \tag{14.57}$$
$$C_4 = -\tfrac{1}{96} + \tfrac{1}{96}e, \tag{14.58}$$

$$D_1 = \tfrac{2}{3} - e, \tag{14.59}$$
$$D_2 = -\tfrac{1}{48} + \tfrac{1}{16}e, \tag{14.60}$$
$$D_3 = \tfrac{1}{6} - \tfrac{1}{8}e, \tag{14.61}$$
$$D_4 = -\tfrac{1}{384} + \tfrac{1}{128}e, \tag{14.62}$$

$$E_1 = -\tfrac{1}{2}, \tag{14.63}$$
$$E_2 = 0, \tag{14.64}$$
$$E_3 = -\tfrac{1}{8}, \tag{14.65}$$
$$E_4 = 0, \tag{14.66}$$

$$F_1 = \tfrac{1}{3}, \tag{14.67}$$
$$F_2 = \tfrac{1}{96}, \tag{14.68}$$
$$F_3 = \tfrac{1}{8}, \tag{14.69}$$
$$F_4 = 0, \tag{14.70}$$

$$G_1 = -\tfrac{1}{6}, \tag{14.71}$$
$$G_2 = \tfrac{1}{96}, \tag{14.72}$$
$$G_3 = -\tfrac{1}{48}, \tag{14.73}$$
$$G_4 = \tfrac{1}{768}, \tag{14.74}$$

$$H_1 = \tfrac{1}{18}, \tag{14.75}$$
$$H_2 = -\tfrac{1}{64}, \tag{14.76}$$
$$H_3 = -\tfrac{1}{32}, \tag{14.77}$$
$$H_4 = 0. \tag{14.78}$$

Finally in $D = 3$ the minimal thermodynamic lattice requires 40 speeds (D3V40), organized as follows:

- 6 face centers: $(\pm 1, 0, 0)$ and permutations thereof;
- 6 far face centers: $(\pm 2, 0, 0)$ and permutations thereof;
- 12 edge centers: $(\pm 1, \pm 1, 0)$ and permutations thereof;
- 8 vertices: $(\pm 1, \pm 1, \pm 1)$;
- 8 far vertices: $(\pm 2, \pm 2, \pm 2)$.

The corresponding set of parameters X_{ji} can be found in Yu Chen's thesis work where he demonstrates his thermal BGK schemes for mildly supersonic one- and two-dimensional shocks at $Ma \sim 1.1$. These tests consist of preparing a gas

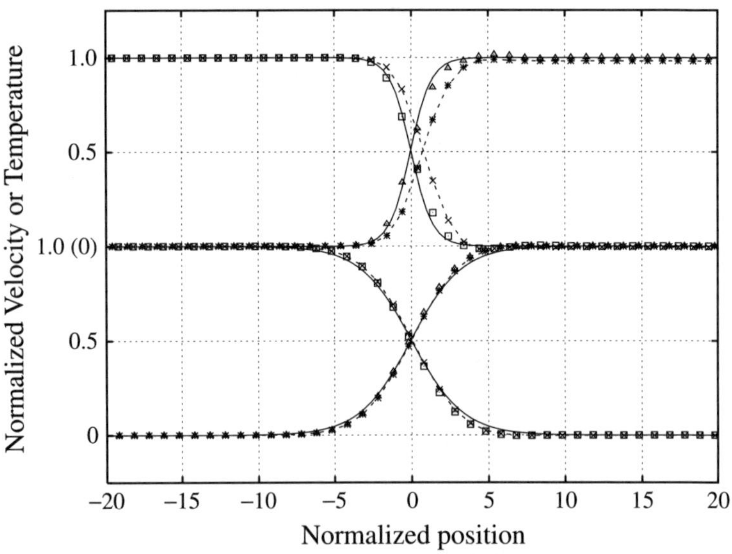

FIG. 14.7. Comparison of the structure of the shock wave fronts. Squares and triangles correspond to the D2V16 model; lines with symbols (dashed line with crosses and stars) are the results of the D2V13 model; solid lines represent the analytical solution. The left/right steps represent the sharp temperature increase, while the right/left step is the velocity increase. The upper half of the graph is for $Ma = 1.14$ while the lower half corresponds to $Ma = 1.02$. (After [239], courtesy of Y. Chen.)

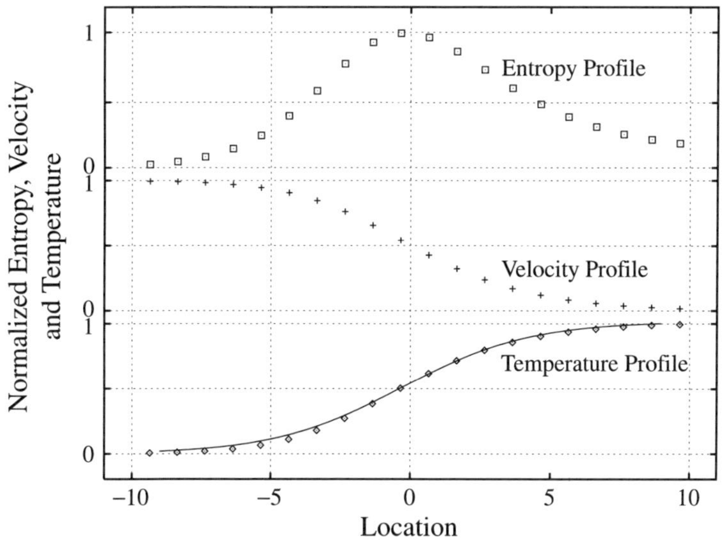

FIG. 14.8. Profiles of the shock wave front. The solid line is obtained by numerical integration of the Navier–Stokes equations, while symbols correspond to the results of the D1V5 model. The Mach number is 1.01. (After [239], courtesy of Y. Chen.)

with two distinct sets of thermodynamic conditions (ρ_l, P_l, T_l) and (ρ_r, P_r, T_r) on the left/right side of a separating sect. Upon removing the sect, a shock wave starts propagating rightwards out of the high-pressure (left) region. The density and pressure jumps across the shock front are related by the Rankine–Hugoniot equations (subscripts r, l mean right and left, respectively):

$$\frac{\rho_l}{\rho_r} = \frac{(\gamma+1)\,M^2}{(\gamma-1)\,M^2+2}, \tag{14.79}$$

$$\frac{T_l}{T_r} = \frac{2\gamma M^2 - (\gamma-1)\left[(\gamma-1)\,M^2+2\right]}{(\gamma+1)\,M^2}, \tag{14.80}$$

where $\gamma = C_p/C_v = (D+2)/D$, as it befits a mono-atomic ideal gas in dimension D.

Excellent agreement with analytical results is reported for various values of the Mach number Ma and other thermodynamic parameters.

These results prove that this thermodynamic LBGK does possess a non-empty 'working window' of thermodynamic parameters also in the nonlinear compressible regime. This is a remarkable result, which furthers the frontier opened up by Alexander, Chen and Sterling. However, it still tells us very little about the actual size of the working window in thermodynamic parameter space. In other words, given a triplet of values $[\rho, P, u]$ for which the method is known to work, we still have not much feeling for how far we can sail away from this triplet without incurring numerical troubles.

Again, not a very comfortable situation.

This mixed state of affairs was pinpointed quite sharply in early work by B. Alder and collaborators, which we now discuss in some detail.

14.5 Reduced thermohydrodynamic schemes

McNamara, Garcia and Alder start from Grad's celebrated 13-moment approach to fluid dynamics, based on hyperbolic superset of the Navier–Stokes–Fourier equation for the following 13 quantities: density (1), momentum (3), momentum flux tensor (6) and heat flux (3) [240]. They observe that in order to recover the 13-moment Grad equations, the lowest 26 moments of the discrete distribution should match those of the Maxwell–Boltzmann distribution (see previous section).

As a result, at least 26 discrete speeds are required to fit the bill. It so happens that a discrete set consisting of 12 speeds $\sqrt{2}$, 8 speeds $\sqrt{3}$ and 4 speed 2 particles provides exactly the 26 sought degrees of freedom.

Unfortunately, such a minimal scheme proves very unstable for all but the largest transport coefficients, typically of order $O(1)$ in lattice units, far too high for most applications.

Surprise? Not really.

We already have clues from isothermal fluids that kinetic space, the one spanned by the discrete speeds, ought to be 'larger' than the hydrodynamic one if the correct macroscopic symmetries are to be fulfilled with some room still left for entropy minimization! In this light, it is pretty plausible that a minimal kinetic space matching exactly the size of thermohydrodynamic space leaves no hope for stability.

A natural move is therefore to extend the set of discrete speeds so as to allow more freedom (as done by Yu Chen). Unfortunately, the authors report no success in this direction [241]. Best guess is that adding speeds brings more freedom to match symmetries, but also higher exposure to numerical instabilities driven by high energy populations going negative. To the best of this author's knowledge, the exact nature of these instabilities has not been understood in depth, but it is again quite plausible to expect the type of numerical instabilities associated with high-order finite difference schemes: long jumpers are keener probes of the lattice discreteness and their dynamics is more easily distorted by the interaction with the lattice.[54]

[54] A trendy expression for this type of behavior is *molecular individualism*: situations where the system does not settle for just a handful of kinetic moments, but requires a whole pile of them. This situation is commonplace in the kinetic theory of ultra-fast phenomena, say, wave–matter interactions [242].

A possible way out, suggested by McNamara *et al.*, is to go just the other way around, namely *reduce*, instead of raising, the number of discrete speeds, in an attempt to *trade thermohydrodynamic consistency* on the higher moments for better stability. This is what we might call the *reduced thermohydrodynamic* approach: accept errors on higher moments, hoping they will not corrupt the physical observables to any significant extent. (Always the same story, don't throw out the baby with the bath water ...).

For this purpose, McNamara *et al.* use a 21-speed lattice consisting of

- 1 stopped particles (energy 0);
- 6 particles at energy 1/2 (face centers);
- 8 particles at energy 3/2 (vertices);
- 6 particles at energy 2 (far face centers).

With this lattice, the following 21 moments are reproduced

$$\langle 1 \rangle = \rho, \tag{14.81}$$

$$\langle c_{ia} \rangle = \rho u_a, \tag{14.82}$$

$$\langle c_a^2 \rangle = \rho \left(u^2 + 2e \right), \tag{14.83}$$

$$\langle c_a c_b - \tfrac{1}{3} c_a^2 \delta_{ab} \rangle = \rho u_a u_b - \tfrac{1}{3} u^2 \delta_{ab}, \tag{14.84}$$

$$\langle c_i^2 c_{ia} \rangle = \rho u_a \left(u^2 + \tfrac{10}{3} e \right), \tag{14.85}$$

$$\langle c_{ia} c_{ib} c_{ic} - \tfrac{1}{5} c_i^2 t_{iabc} \rangle = 0, \tag{14.86}$$

$$\langle c_i^2 Q_{iab} \rangle = 3 \rho u_a u_b - \rho u^2 \delta_{ab}, \tag{14.87}$$

$$\langle c_i^4 \rangle = \tfrac{20}{3} \rho e^2, \tag{14.88}$$

$$\langle c_i^4 - 5 c_{ix}^2 c_{iy}^2 + c_{ix}^2 c_{iz}^2 + c_{iy}^2 c_{iz}^2 \rangle = 0, \tag{14.89}$$

where each $\langle \cdot \rangle$ denotes $\sum_i f_i^{\mathrm{e}} (\cdot)$.

The first five moments are the usual conserved densities and they are automatically conserved by the collision operator since they appear explicitly in the local equilibrium. Entries four and five correspond to the viscous and thermal transport properties. By making Q_{iab} eigenvectors of the scattering matrix (McNamara *et al.* use a matrix based LBE instead of LBGK), one generates separate values of the viscosity and thermal conductivity thereby allowing non-unit Prandtl numbers. This completes the list of Grad's 13 moments. The remaining moments are purely kinetic modes, which should not couple to the Navier–Stokes–Fourier dynamics. The LBE scattering matrix is tuned so as to wipe them out as fast as possible (remember ghost modes in isothermal LBE). The equilibrium component of these kinetic modes does couple, however, to Navier–Stokes–Fourier dynamics and this is the cause of nonlinear deviations scaling like Ma^2. Since McNamara *et al.* focus on incompressible thermal applications, typically Rayleigh–Bénard convection, these errors are no source of major concern. Still, we are again left with a rather unsatisfactory situation. And indeed McNamara *et al.* conclude on a rather discouraging note that they '*have not discovered any advantage to employing LBE over conventional Navier–Stokes solvers for thermal systems.*'

They report on beneficial effects of Lax–Wendroff fractional streaming for three-dimensional Rayleigh–Bénard convection (see original reference), but this is another story.

End of the line for LBGK thermal models?

Maybe, maybe not, stay with us a little longer.

14.6 Attempts to rescue thermal LBE

In this final part, we provide a cursory view of the numerous attempts to improve on the the gloomy picture emerging from McNamara *et al.* analysis.

14.6.1 *Tolerance to realizability violations*

A set of mild pain-easing results were provided by de Cicco *et al.* [243], who showed that positivity of the discrete equilibria, definitely not a sufficient condition for stability, might not be necessary either. We refer to this property as 'tolerance to lack of realizability'. At each point of the fluid domain, let us introduce the realizable hydrodynamic space $\mathcal{H}_\mathcal{R}$ defined as the set of triples $(\rho, \vec{u}, T)$ such that the discrete equilibria are realizable; that is,

$$0 < f_i^{\mathrm{e}} < 1. \tag{14.90}$$

In principle, this is a superset of the stable hydrodynamic subspace $\mathcal{H}_\mathcal{S}$ spanned by the triples $(\rho, \vec{u}, T)$ such that the scheme is numerically stable. The finding by de Cicco *et al.* is that the hydrodynamic behavior can be reasonable even in the presence of *negative* populations (mildly) breaking the realizability constraint (14.90). Whether this is a lucky break or a general property of the scheme remains to be seen.

14.6.2 *The kinetic closure approach*

Along the spirit of reduced thermohydrodynamics, Renda *et al.* put forward an alternative approach based upon the idea of representing high energy (secondary) equilibria in terms of low energy (primary) ones via a *kinetic closure* [244]. Primary equilibria are then computed by solving exactly a lower-dimensional system of constraints. The guiding idea is that *secondary equilibria cannot go negative if the primary ones do not.* This cannot secure automatic stability, but certainly makes life harder for numerical errors driving high energy populations to negative values. Even though these thermal LBEs never become unstable, they deliver *quantitatively* correct results only for a limited range of values of the thermodynamic parameters.

More research work is needed to assess on quantitative grounds which regions of thermodynamic space are correctly represented by these schemes in two and three dimensions. And, more generally, whether an optimal closure can be found which would minimize thermodynamic inconsistency.

14.6.3 *Non space-filling lattices*

Another interesting proposal is to use non space-filling lattices, typically octagons, offering a higher degree of isotropy. For instance, a 17-speed lattice defined by

$$\vec{c}_{ji} = j\left(\cos\left(\frac{\pi i}{8}\right), \sin\left(\frac{\pi i}{8}\right)\right), \quad j = 1, 2, \quad i = 0, 1, \ldots, 7$$

plus one rest-particle, can be shown to provide isotropy up to sixth-order tensors [245].

The price to pay is that the corresponding tiling is no longer space-filling, so that particle microdynamics needs to be supplemented with an interpolation step in order to latch onto the 'crystal' grid.

The authors demonstrate the method for isothermal plane Poiseuille flow and two-dimensional Rayleigh–Bénard convection, and report good quantitative agreement with previous literature, as well as enhanced numerical stability.

14.6.4 *Models with internal energy*

LBE models with pure kinetic energy necessarily lead to the ideal gas relation $\gamma = c_p/c_v = (D+2)/D$. This upsets gas dynamic simulations in less than three dimensions since the desired value 5/3 is replaced by 3 and 2 in one and two dimensions, respectively. In addition, specific numerical values such as $\gamma = 1.4$ (air) are not realizable either. It is therefore desirable to extended thermal models so as to include the capability of a tunable γ parameter. An intuitive way of achieving this goal is to make allowance for a different rest mass energy ϵ_{0j} for each energetic level j. This corresponds to the following energy–momentum relation:

$$\epsilon_j = \epsilon_{0j} + \frac{1}{2}c_{ji}^2. \tag{14.91}$$

The internal gas energy then becomes

$$e = \sum_j \left(\epsilon_{0j} + \frac{1}{2}c_j^2\right)\rho_j, \tag{14.92}$$

where $\rho_j = \sum_j f_{ji}$ is the partial density of the j-th energy shell.

By allowing distinct rest energies for different energy levels, one can manipulate the equation of state of the LBE gas. It can be shown that by appropriate tuning of the rest energies, the range of density and temperature gradients supported by the model, i.e., the range of nonlinear stability, is significantly extended. This idea was successfully demonstrated by Shouxin *et al.* [246]. The theoretical arguments are supported by convincing simulations of one- and two-dimensional shock wave propagations at density ratios of order 10.

In a similar vein, other authors [247] have recently proposed a new thermal LBE in which the internal energy is modeled by a scalar field using a separate distribution function. The energy is coupled to the fluid density and momentum by means of an appropriate repartition between rest and moving particles in

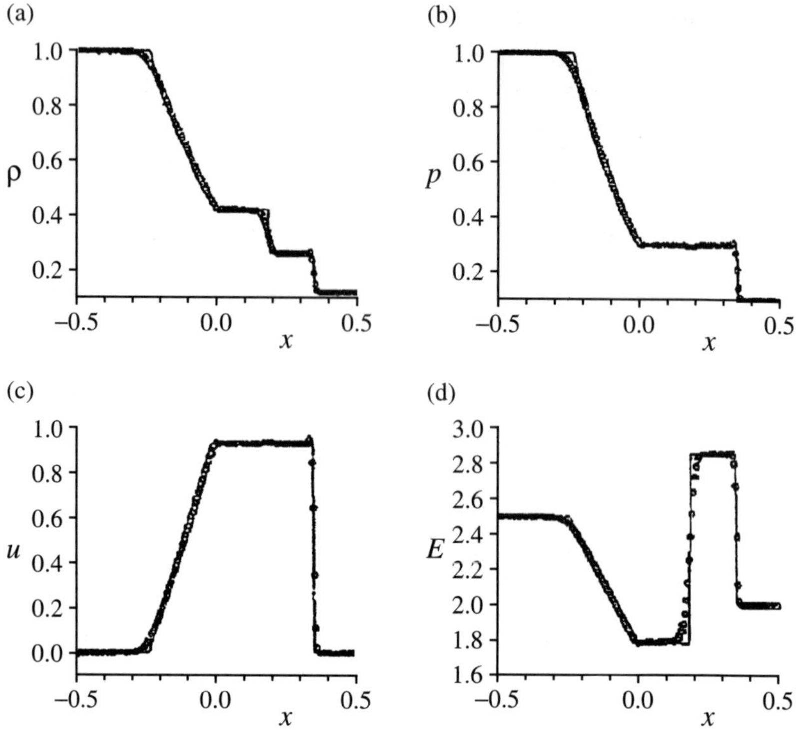

FIG. 14.9. Density, pressure, velocity and energy profiles as obtained by the compressible LBE model by Shouxin *et al.* The solid lines represent the analytical solution. (From [246]).

conventional two speed models. This repartition amounts to making the density of rest particles a suitable function of density and energy. The model does not account for heat dissipation effects but, in principle, it can deal with non-ideal thermodynamics.

A model with heat dissipation effects has been recently proposed by other authors [248]. Internal energy is again described by a separate distribution, say g, but this distribution is linked up to the particle distribution f by the relation

$$g = \frac{1}{2}\left(\vec{v} - \vec{u}\right)^2 f, \tag{14.93}$$

so that the energy condition is automatically fulfilled: $\int \frac{1}{2}(\vec{v} - \vec{u})^2 f \, d\vec{v} = \rho e$.

The g distribution is then postulated to obey its own LBGK equation:

$$\partial_t g + v_a \partial_a g = -\frac{g - g^{\mathrm{e}}}{\tau_g} - Hf, \tag{14.94}$$

where $H \equiv (v_a - u_a)[\partial_t + v_b\partial_b]u_a$ is the heat dissipation term and g^e is related to the particle equilibrium f^e by the condition (14.93).

This pair of equations is subsequently cast into discrete LBGK form using the Hermite–Gaussian projection-quadrature approach described in [59]. The time discretization is obviously more involved than in standard LBEs due to the time derivative entering the expression of the heat dissipation H. In fact, a *second-order* time marching is required, the details of which can be found in the original work.

Since heat dissipation is explicitly accounted for, this system of LBGKs is in principle capable of tracking energy dynamics in a self-consistent way. The scheme is numerically tested for the two-dimensional Couette flows with temperature gradients and Rayleigh–Bénard convection, providing good quantitative agreement with the previous literature in both cases. The authors do not mention performance issues, but they do advocate better numerical stability as compared to previous thermal LBEs.

14.7 The Digital Physics approach

We wish to recall that the problem of thermodynamic consistency is significantly softened by going back to a nonlinear collision operator with underlying microdynamics. This is the approach taken by Molvig and co-workers [249] within the so-called Digital Physics (DP) paradigm.[55]

To achieve thermodynamic consistency over a sizable range of temperatures, $1/3 \leqslant T \leqslant 2/3$, these authors show that as many as 54 speeds, distributed over three energy shells, are needed.

The minimal model consists of three energy levels, $\epsilon_j = 0,\, 1,\, 2$ organized as follows:

- 6 *stopped particles, energy zero*;
- 24 *energy 1 particles distributed across the FCHC lattice, the* 24 *permutations of* $(\pm1, \pm1, 0, 0)$;
- 24 *energy* 2 *particles, distributed across two lattices,*
  8 *in the hyperoctahedron defined by permutations of* $(\pm2, 0, 0, 0)$ *and*
  16 *in* $(\pm1, \pm1, \pm1, \pm1)$.

This makes a total of 54 (yes, fifty-four!) discrete speeds.

The DP technical hallmark is the use of multi-bit (integers) populations N_{ji} going through multibody collisions.

The corresponding discrete equilibria read as follows:

$$N^e_{ji} = \rho g_j(T)\left[1 + \frac{1}{T}u_{ji} + \frac{1}{2T^2}\left(u_{ji}^2 - u^2T\right) + \frac{1}{6T^3}\left(u_{ji}^3 - u^2u_{ji}T\right)\right], \quad (14.95)$$

where $u_{ji} = u_a c_{jia}$ and the notation N_{ji}, in lieu of f_{ji}, highlights the integer (multi-bit) nature of the discrete populations.

[55] Trademark EXA Corporation.

This is an appealing expression, as it corresponds to a set of Maxwellians expanded to third-order in the Mach number, each with a partial density $\rho_j = \rho g_j(T)$. The partial densities are fixed by the following conditions:

$$\sum g_j d_j = 1, \tag{14.96}$$

$$\sum g_j \epsilon_j^n d_j = \frac{d}{2}\left(\frac{d}{2}+1\right)\cdots\left(\frac{d}{2}+n-1\right)T^n, \quad n = 1, 2, 3, \tag{14.97}$$

where d_j is the number of momentum states at energy level ϵ_j and $d = \sum_j d_j$.

The first two equations are just mass and energy conservation and the other two conditions result from the requirement of anomaly cancellation.

Full details, admittedly requiring a great deal of lengthy but rewarding algebra, are found in Teixeira's thesis [233]. The method is clean and elegant; the coefficients of the expansion have a definite physical meaning and so do the partial densities ρ_j. Another pleasing feature of the DP equilibria is the uniqueness of the solution, in the sense that four energy levels deliver a one-to-one solution to the set of thermodynamics constraints.

As proven by H. Chen [250], these equilibria are even equipped with a local H-theorem, although this local H-theorem cannot be turned into a global one in the case of a space-dependent temperature profile. A detailed account of the subject can be found in [250].

14.8 Fake temperature schemes

We would not leave this chapter without reminding the reader that LBE techniques can be successfully employed for flows with *mild* compressibility and dissipative heating effects where temperature behaves like a passive/active scalar.

A typical example is Rayleigh–Bénard convection within the Boussinesq approximation (temperature-driven only, minor density changes $\delta\rho/\rho \ll 1$),

$$\partial_t T + u_a \partial_a T = \chi \partial_a^2 T + \alpha g \left(T - T_0\right). \tag{14.98}$$

Here α is the thermal dilatation coefficient, g the gravitational acceleration, and T_0 a reference temperature.

This equation can be handled very efficiently using isothermal methods with passive scalars, and buoyancy effects mimicked by external sources. This approach has been pioneered by Massaioli *et al.* [251] and further developed by several other authors [252,253] for the case of three-dimensional Rayleigh–Bénard turbulence.

14.9 Summary

Despite many promising attempts, none of the thermal LBEs presented so far provides a watertight solution to the problem of propagating heat and temperature in a discrete lattice over an *extended* range of values the way they do in the continuum. This is an open frontier of LBE research. At this stage, it is very

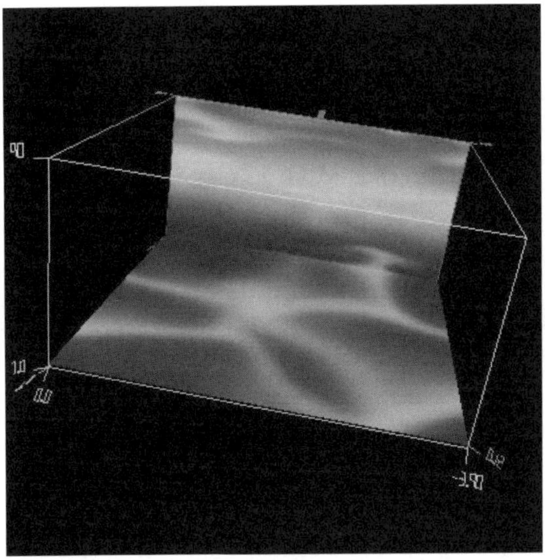

FIG. 14.10. LBE calculation of a Rayleigh–Bénard turbulent cell. Cuts of temperature isosurfaces, the clear spots represent high temperature regions. (Courtesy of F. Toschi.)

hard to foresee whether we are facing an unforgiving 'no go' barrier or if instead, as often in the past of LGCA and LBEs adventures, some smart intuition will finally save the day.

14.10 Exercises

1. Draw the realizability domain of the D1V5 lattice.
2. Compute the temperature for the D2V9 lattice as a function of the partial densities ρ_j, $j = 0, 1, 2$. At a given ρ_0 for which ratio ρ_2/ρ_1 do we get population inversion (negative temperature)?
3. Discuss the realizability domain of the D2V9 model in the limit $u = 0$.

15

FINALE: WHO NEEDS LBE?

I am conscious of being only an individual strugging weakly against the stream of time. But it still remains in my power to contribute in such a way that, when the theory of gases is again revived, not too much will have to be rediscovered.

L. Boltzmann

After so much ado about lattice Boltzmann it is worth taking a final step back and asking a very simple question: **who needs LBE**?

The views can be quite disparate. Some authors, including one of the originators, seem to regard it as just 'yet another finite difference scheme'.[56] In a much more enthusiastic vein, some others claim that LBE and LGA might even represent a new paradigm in non-equilibrium statistical physics.

In this short epilog, the author would like to offer his personal—hence very limited—view of this state of affairs. To a wide degree of arbitrariness, I pigeonhole LBE fluid dynamic (and beyond) applications in the following four general classes:

- *MU* (*Must Use*);
- *SU* (*Should Use*);
- *CU* (*Can Use*);
- *DU* (*Don't Use it!*).

Let us proceed from the bottom up so as to pave the way to glorious conclusions.

15.1 DontUse class

As it stands, the LBE it is probably not yet ready to compete with state of the art solvers for the simulation of flows with *strong* compressibility and *substantial* heat transfer effects. Further efforts and new ideas are required to move this frontier ahead.

15.2 CanUse class

This is of course the most widely populated class of applications, simply because any fluid problem can somehow be moulded into LBE format. We are typically in the 'Navier–Stokes in disguise' situation, where using LBE instead

[56] 'You can do what you want with LBE, the point is whether this is most efficient way of doing', J. Jimenez, priv. communication.

of other methods is basically a matter of taste. Sometimes, as with this author, parallel computing can make the difference. Consider for instance the 3D turbulent flow computations described in Part II. These computations could have certainly been performed with comparable efficiency using finite difference or finite volume schemes. However, LBE was the preferred choice mainly on account of its simplicity, hence rapidity of implementation on the parallel computer at hand. The epitome of CU class are turbulent flows in simple geometries (cubes, channels, steps), whereas for more complicated geometries, finite difference or finite volume extensions of LBE probably still lag slightly behind the state of the art (a statement that only applies to the academic environment). Reaching full maturity is just a matter of time and labour, no conceptual hurdles in sight.

15.3 ShouldUse class

The SU class includes those problems in which use of LBE is recommended on account of either efficiency, friendliness or both. Typical examples are single and (even more so) multiphase flows in grossly irregular geometries (porous media). Features to be leveraged: easy handling of irregular boundary conditions, easy incorporation of mesoscopic forces driving phase transitions and other related complexities hard to describe within a continuum approach. Broadly speaking, LBE is at a potential vantage point whenever hydrodynamic equations are not well settled, the typical territory of soft matter research.

Outside the academia, real world geometries (cars, airplanes, and the like) seem to be setting a more and more convincing SU, possibly even MU class case for LBE and related techniques. The main highlight here is ease of use and quick set-up of very complex geometries, a major issue in commercial fluid dynamics.

15.4 MustUse class

Top of the line: problems for which LBE stands head and shoulders above any other technique: the cradle of computational breakthroughs.

In this author's opinion, this class is presently still rather 'dilute'.

Possible exceptions to the above statement might come from fluctuating microhydrodynamics, such as the flows with suspended particles described in Part III.[57]

Natural candidates to MU class promotion are all prominent SU class problems, on condition that faithful and still economic ways to feed additional physics into the scheme are found.

Other prospective candidates are all CU class problems with *substantial* potential for massively parallel computing, including the complex geometry situations previously mentioned.

Is MU class going to be well populated some day soon? Only time will tell: the subject is up for grabs!

[57]We should also mention numerical experiments on basic issues of statistical mechanics, like long time tails, although this pertains more to LGCA than LBE.

No matter what, we feel like closing this book with a tribute of gratitude to Ludwig Boltzmann, without whose work not a single bit of this subject would have ever come into existence.

16

APPENDICES

A Integer LBE

The LBE pays a twofold price for the ability to access macroscopic scales:

1. Loss of non-Boltzmann effects.
2. Loss of round-off freedom.

Item 1 is a fundamental physical limitation; item 2 is technical, but surely not a minor one either. The latter problem can be partially recouped by representing the distribution function $\langle n_i \rangle$ by means of integer, rather than floating-point, variables:

$$N_i = \langle n_i \rangle. \tag{A.1}$$

The corresponding ILBE (Integer LBE) represents an appealing compromise between LBE and Boolean LGCA.

The appeal is that, as for lattice gas, conservation laws and round-off freedom can be restored exactly. At variance with LG however, ILBE may still be exposed to numerical problems in the form of *underflows* ($N < 0$) and *overflows* ($N > N_{\max}$) if the collision operator is taken in relaxation rather than truly nonlinear many body form.

Let N_i be the integer coarse-grained average of the Boolean occupation numbers; assuming an l-bit integer representation, N_i takes on all integer values between 0 and $N_{\max} = 2^l - 1$. Normalizing variables to $[0, 1]$ yields a corresponding accuracy $\epsilon_l \sim 2^{-l}$, showing that at a given value of bits per variable, typically $l = 8, 16$, ILBE achieves a lowly resolution (less than 3 and 5 decades, respectively, $1/256$ and $\sim 1/65\,536$), as compared to the floating-point representation. Thus in principle it would appear that ILBE is doomed to failure for actual computations of practical interest. Fortunately, this is too hasty a conclusion.

It is well-known that most computations related to the dynamics of complex systems involve a huge number of operations, typically above 1 Teraflop $= 10^{12}$ floating-point operations per simulation. These simulations are in principle exposed to potentially dangerous error accumulation. Call ϵ_0 the initial inaccuracy on initial data, the error accumulated after T time-steps is likely to grow like T^{α} with α some power close to 0.5 if error propagation is Gaussian (diffusive) or larger in the presence of persistent correlations induced by, say, phase-space instabilities. This is why floating-point numbers strive to keep round-off at a minimum and prevent error build-up in the course of time. Typically $\epsilon = 10^{-6}$, 10^{-14} in simple (32 bits) and double precision (64 bits), respectively. The situation with ILBE is radically different. With ILBE, the tolerance bar to inaccuracy

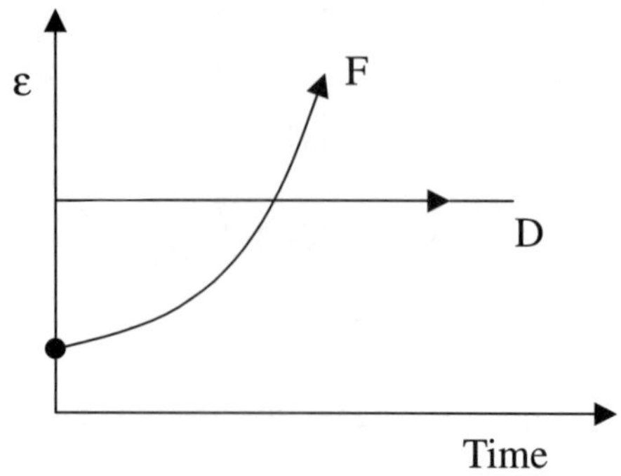

FIG. A.1. Error build-up in the floating-point (F) and integer-digital (D) representation.

on the single variable can be raised significantly thanks to the ability of securing *exact* conservation. This means that even if the single site accuracy is low, since it does not deteriorate in time, statistical averaging can finally offset the poor single site accuracy. This observation lies at the heart of a number of ingenious integer/multi-bit approaches [254] including the Digital Physics (DP) paradigm discussed earlier in this book.[58]

B The pseudospectral method

The pseudospectral method is based on the idea of solving the Navier–Stokes equations by means of Fourier transforms in reciprocal (wavevector) space. For the sake of simplicity we shall refer to the two-dimensional case, where fluid equations are most conveniently recast in the *vorticity–streamfunction* representation. Vorticity has already been introduced as the curl of the velocity field, and in two dimensions it reduces to the z component ω, a pseudo-scalar. The streamfunction ψ is defined by the condition that surfaces ψ = constant define lines tangents to the velocity field u, v:

$$\partial_y \psi = u, \tag{B.1}$$

$$\partial_x \psi = -v. \tag{B.2}$$

From the definition, the streamfunction follows from vorticity via a static Poisson equation $\Delta\psi = \omega$.

The pseudospectral method is based on a discrete Fourier representation of both vorticity and streamfunction:

[58] On a practical note, it is worth mentioning that the constraint of securing immunity from under/overflows makes integer programming somewhat more cumbersome than its floating-point counterpart.

$$\omega(x,y,t) = \sum_{m,n} \omega_{mn}(t)\, \mathrm{e}^{-\mathrm{i}(k_{x,m}x + k_{y,n}y)}, \tag{B.3}$$

$$\psi(x,y,t) = \sum_{m,n} \psi_{mn}(t)\, \mathrm{e}^{-\mathrm{i}(k_{x,m}x + k_{y,n}y)}, \tag{B.4}$$

$$k_{x,m} = \frac{2\pi m}{L}, \quad k_{y,n} = \frac{2\pi n}{L}, \quad m,n = 0,1,\ldots,N-1. \tag{B.5}$$

The vorticity equation reads:

$$\partial_t \omega + J(\omega,\psi) = \nu \Delta \omega, \tag{B.6}$$

where $J(f,g)$ is the Jacobian operator $J(f,g) \equiv \partial_x f \partial_y g - \partial_x g \partial_y f$.

Upon Fourier transforming (B.6), one obtains

$$\partial_t \omega_{mn} + \mathrm{FT}\left[J(\omega,\psi)\right] = -\nu k^2 \omega_{mn}, \tag{B.7}$$

where the symbol FT stands for Fourier Transform. Equation (B.7) is very appealing because all space derivatives are turned into algebraic operators according to $\partial_a = -ik_a$, $a = x, y$. However, the FT of the Jacobian involves a lengthy convolution between the vorticity and streamfunction transforms. Such a convolution would require $O(N^2)$ operations, thus considerably taxing the computational efficiency.

The pseudospectral method of Orszag–Patterson [255] turns around this problem by first computing the Jacobian products in real space and only once these are available, taking the FT of these products. This cuts down complexity to $O(N \log_2 N)$.

In the process spurious modes, known as aliases, are introduced. They correspond to wavevectors $\vec{k'}$, $\vec{k''}$ outside the resolved range which couple back to the resolved spectrum via nonlinear beatings $\vec{k'} \pm \vec{k}''$. These aliases are usually removed by cutting away all modes with magnitude $k > k_N/\sqrt{2}$.

The pseudospectral method is very accurate and very fast for periodic flows, the latter property being related to the availability of the celebrated Fast Fourier Transform (FFT) algorithm, which computes the Fourier transform, in principle a N^2 operation, in $N \log_2 N$ steps.

C A primer on parallel computing

Parallel computing is a fast-expanding area of high performance computing [256, 257]. The idea is only about 2000 years old, based as it is on the time-honored Roman principle *divide et impera* (divide and conquer). To solve a huge problem it is often convenient to break it down in small parts, solve them concurrently, and glue partial solutions together to finally obtain the global solution. Ideally, using P computers each solving a $1/P$-th of the original problem, we cut the solve time by a juicy factor P. Make P as high as you can and even the largest problems fall down on their knees ... To make this blue sky scenario really work, a series of conceptual and practical hurdles need to be overcome. The primary issues relate to

- *Fraction of parallel content.*
- *Communication to computation ratio.*
- *Load balance.*

C.1 Fraction of parallel content

Fraction of parallel content refers to the practical observation that not all portions of a given algorithm might be available to parallelization. Take the simple problem of summing N sumbers using $P = N/K$ processors where K is an integral number. Each processor can compute its share of K numbers in happy unawareness of all others, but in the end, all partial sums computed by each processor need to be accumulated to produce the final result. Summing the partial sums is a task for just one processor, two being already one too many. This is a so-called 'serial bottleneck'.

An algorithm with many such bottlenecks will not work well on parallel computers. In fact, it is simple matter to realize that if f is the fraction of work that can be performed in parallel, the time spent using P processors is

$$T(P) = T(1)\left(\frac{f}{P} + (1-f)\right). \tag{C.1}$$

Consequently, the benefit of using P computers (parallel speed-up) is given by

$$S(P) \equiv \frac{T(1)}{T(P)} = \frac{1}{f/P + (1-f)}. \tag{C.2}$$

This simple equation shows that even in the limit of an infinite number of processors $P \to \infty$, the asymptotic speed-up flattens at $S(\infty) = 1/(1-f)$. Unfortunately, this a fairly sharp function of f only in the close vicinity of $f = 1$.

For instance, suppose we have only a lowly ten per cent of serial code. What the above equation shows is that no matter how many computers we can pay for, the maximum pay-off will never exceed a factor of 10. Therefore 10 processors is approximately the threshold above which parallel computing becomes a waste. These simple and yet far-reaching considerations are known as Amdahl's law, after Gene Amdahl a pioneer in computer architecture design. LBE is extremely well placed *vis-à-vis* Amdahl's law. Large-scale computations easily place more than 99 per cent of the computer demand on the collision phase which is perfectly parallel. This means that there is often scope for $O(100)$ processors and more.

C.2 Communicativity

The next constraint is communicativity. As the simple example of summing N numbers shows already, at some point the independent actions of the P processors need be coordinated to glue together a sensible answer. In a real world, this takes time—time to communicate results either among other processors or to a supervisor, a glorious word for a slower machine doing nothing but collecting other processors' work (justly named 'slaves') and perform a lightweight duty (summing the partial sums). In more democratic environments the supervisor may well be sharing the slaves' task as well.

Whatever the arrangement, the fact remains that communication should be kept to a minimum because it constitutes a sheer improductive overhead (too much talking ...). Such an overhead is sharpened by technological issues: interprocessor communication is usually significantly slower, typically a factor of 10, than intraprocessor calculations.

The figure of merit here is the so-called 'Communication to Computation Ratio', CCR for short. CCR can be measured in bytes transmitted off-processor over in-processor executed flops.

Numerical schemes with low CCRs are dear to parallel computers. And again, LBE does quite well. The only communication comes from the streaming step, which is only nearest neighbor. LBE features CCRs in excess of 10, to be contrasted to CCR $O(1)$ for most explicit finite difference schemes [258].

C.3 Load balancing

Finally, we come to the issue of load balancing.

Assuming the parallel computer is homogeneous, namely all processors work at the same speed, it is clear that to avoid wasteful delays, they should all get the same amount of work, on pain of being pace-made by the slowest one.

Load balancing is a trivial matter with regular geometries, where a simple geometric domain decomposition assures good performance. It can become fairly tricky (actually NP-complex) when the geometry is complex and possibly even changing in time (think of piston-valve motion in engine combustion) [259]. Again, LBE with its simple data structure is well positioned to steer clear of load balancing issues for many practical applications. However, advanced LBEs with moving grids (yet to come) would perforce be faced with load balancing issues as well.

D From lattice units to physical units

In this book we repeatedly emphasized the fact that LBE is most often directed to 'synthetic' matter studies, meaning by this that the focus is more on dimensionless numbers than on the actual values of physical properties. This is practical for theoretical studies, but it may easily lead to embarassing mumbles as an answer to the request for communicating results in good old centimeters, seconds and grams!

In the following we remind the basic elementary steps which permit conversion of the result of LBE simulations in actual physical units. For the sake of convenience, we shall refer to the cgs system.

D.1 Length

Most LBE simulations assume the lattice spacing Δx as the space unit. Therefore, upon using N grid points to represent, say, a linear size of L centimeters, the space unit is

$$S_l \equiv \Delta x = \frac{L}{N} \quad \text{(cm)}. \tag{D.1}$$

D.2 Time

In a similar way, the time unit of LBE simulations is the elementary lattice time-step. Its physical value can be defined via the sound speed as follows $C_s = c_s \Delta x / \Delta t$, where the capital denotes the physical value of the sound speed. The result is

$$S_t \equiv \Delta t = \frac{c_s}{C_s} \Delta x \quad \text{(seconds)}. \tag{D.2}$$

In view of (D.1), the relation (D.2) can also be written as:

$$S_t = c_s \frac{T}{N} \quad \text{(seconds)}, \tag{D.3}$$

where T is the time (in seconds) it takes for sound waves to travel a distance L (sound transit time).

D.3 Mass

Since lattice particles are typically given a unit mass, the mass unit is

$$S_m = m \quad \text{(grams)}, \tag{D.4}$$

where m is the physical mass of the fluid molecules.

All other physical quantities can then be derived by the knowledge of the three scale factors S_l, S_t, S_m.

It is of some interest to discuss the scale factor for the number of particles, so as to clarify the often asked question: how many physical molecules does a LBE particle represent?

The count goes as follows.

At a given *reduced density* d (between 0 and 1) with b discrete speeds per site, the lattice number density is bd. Calling n the corresponding physical number density (molecules per cubic centimeter), and recalling that a lattice cell corresponds to a volume of S_l^3 cubic centimeters, we obtain:

$$S_N = \frac{n S_l^3}{bd}. \tag{D.5}$$

To conclude with some actual numbers, let us consider air in standard conditions in a cube one meter in size. With 100 lattice nodes per linear dimension, we obtain $S_l = 1\,\text{cm}$. Taking approximately $300\,\text{m/s}$ for the sound speed in standard conditions and $c_s = 1$ for simplicity, we obtain $S_t = 1/(3 \times 10^4) \times 1 \sim 3.3 \times 10^{-5}$ seconds. At a reduced density of, say, $d = 0.1$, on a, say, 15-speed scheme, each lattice particle corresponds to about $S_N = n_L/1.5 \sim 1.8 \times 10^{19}$ molecules, $n_L = 2.687 \times 10^{19}$ being the Loschmidt number (see Chapter 1).

The same figures, as applied to a smaller cube of, say, 1 cm in linear size, would yield $S_N \sim 2 \times 10^{13}$, again a very large number as befits a mesoscopic method.

REFERENCES

[1] L. Boltzmann, *Vorlesungen über Gastheorie*, Lectures on Gas Theory, unab. Dover, 1995.

[2] C. Cercignani, *Theory and Application of the Boltzmann Equation*, Elsevier, New York, 1975; and
Mathematical Methods in Kinetic Theory, Plenum Press, New York, 1969.

[3] R. Liboff, *Kinetic Theory, Classical, Quantum and Relativistic Description*, Wiley, New York, 1998.

[4] T. Bohr, M. Jensen, G. Paladin and A. Vulpiani, *Dynamical Systems Approach to Turbulence*, Cambridge Nonlinear Science Series, 1998.

[5] D. Ruelle, *Hasard et Chaos*, Editions Odile Jacob, Paris, 1991.

[6] P. Gaspard, Chaos and hydrodynamics, *Physica A* **240**(1/2), 54, 1997.

[7] K. Huang, *Statistical Mechanics*, Wiley, New York, 1987.

[8] B. Alder and J. Wainwright, Decay of the velocity autocorrelation function, *Phys. Rev. A* **1**, 18, 1970.

[9] S. Goldstein, *Classical Mechanics*, Addison–Wesley, London, 1959.

[10] C. Cercignani, *Ludwig Boltzmann: The Man Who Trusted Atoms*, Oxford University Press, 1998.

[11] E. Cohen, Boltzmann and statistical mechanics, in *Boltzmann's Legacy 150 Years After His Birth, Atti dei Convegni Lincei*, Vol. 131, p. 9, Accademia Nazionale dei Lincei, Rome, 1997.

[12] V. Bogoliubov, Problems of a dynamical theory in statistical physics, in *Studies in Statistical Mechanics*, J. de Boer and G. Uhlenbeck eds, Vol. I, p. 11, North–Holland, Amsterdam, 1962.

[13] U. Frisch, D. d'Humières, B. Hasslacher, P. Lallemand, Y. Pomeau and J. P. Rivet, Lattice gas hydrodynamics in two and three dimensions, *Complex Systems* **1**, 649, 1987, reprinted in *Lattice Gas Methods for Partial Differential Equations*, p. 11, Addison–Wesley, Reading, Massachusetts, 1989; and
U. Frisch, Lectures in turbulence and lattice gas hydrodynamics, in *Lecture Notes on Turbulence, NCAR–GTP Summer School, June, 1987*, J. R. Herring and J. C. McWilliams eds, p. 219, World Scientific, 1989.

[14] L. Landau and E. Lifshitz, *Fluid Mechanics*, Pergamon Press, Oxford, 1987.

[15] A. de Masi and E. Presutti, *Mathematical Methods for Hydrodynamical Limits*, Lecture Notes in Mathematics 1501, Springer–Verlag, Heidelberg, 1991.

[16] U. Frisch and S. Orszag, Turbulence: challenges for theory and experiment, *Phys. Today* **46**(3), 24, 1993.

[17] P. Bhatnagar, E. Gross and M. Krook, A model for collisional processes in gases I: small amplitude processes in charged and neutral one-component system, *Phys. Rev.* **94**, 511, 1954.

[18] U. Frisch, B. Hasslacher and Y. Pomeau, Lattice gas automata for the Navier–Stokes equations, *Phys. Rev. Lett.* **56**, 1505, 1986.

[19] G. McNamara and G. Zanetti, Use of the Boltzmann equation to simulate lattice gas automata, *Phys. Rev. Lett.* **61**, 2332, 1988.

[20] F. Higuera and J. Jimenez, Boltzmann approach to lattice gas simulations, *Europhys. Lett.* **9**, 663, 1989.

[21] R. Benzi, S. Succi and M. Vergassola, The lattice Boltzmann equation: theory and applications, *Phys. Rep.* **222**(3), 145, 1992.

[22] D. Rothman and S. Zaleski, *Lattice Gas Cellular Automata: Simple Models of Complex Hydrodynamics*, Cambridge University Press, 1997; and
B. Chopard and M. Droz, *Cellular Automata Modeling of Physical Systems*, Cambridge University Press, 1998; and
J. P. Rivet and J. P. Boon, *Lattice Gas Hydrodynamics*, Cambridge University Press, 2000.

[23] J. Hardy, O. de Pazzis and Y. Pomeau, Molecular dynamics of a lattice gas: transport properties and time correlation functions, *Phys. Rev. A* **13**, 1949, 1973; and
J. Hardy, O. de Pazzis and Y. Pomeau, Time evolution of a two-dimensional model system I: invariant states and time correlation functions, *J. Math. Phys.* **14**, 1746, 1976.

[24] I. Foster, *Designing and Building Parallel Programs*, Addison–Wesley, Menlo Park, 1995.

[25] M. Creutz, *Quarks, Gluons and Lattices*, Cambridge University Press, 1983.

[26] S. Orszag and V. Yakhot, Reynolds number scaling of cellular automaton hydrodynamics, *Phys. Rev. Lett.* **56**, 1691, 1986.

[27] S. Succi, P. Santangelo and R. Benzi, High resolution lattice gas simulation of two-dimensional lattice gas turbulence, *Phys. Rev. Lett.* **60**, 2738, 1988.

[28] D. d'Humières, A. Cloqueur and P. Lallemand, Lattice gas and parallel processors, *Il Calcolo* **25**, 129, 1987.

[29] N. Margolus, G. Vichniac and T. Toffoli, Cellular automata supercomputers for fluid dynamics modelling, *Phys. Rev. Lett.* **56**, 1694, 1986.

[30] P. Grosfils, J. P. Boon and P. Lallemand, *Phys. Rev. Lett.* **68**, 1077, 1992.

[31] D. d'Humières, P. Lallemand and U. Frisch, Lattice gas models for 3D hydrodynamics, *Europhys. Lett.* **2**, 291, 1986.

[32] J. P. Rivet, M. Henon, U. Frisch and D. d'Humières, Simulating fully three-dimensional external flow by lattice gas methods, *Europhys. Lett.* **7**, 231, 1988.

[33] P. Rem and J. Somers, Cellular automata on a transputer network, in *Discrete Kinetic Theory, Lattice Gas Dynamics and Foundations of Hydrodynamics*, R. Monaco ed., p. 268, World Scientific, Singapore, 1989.

[34] G. Zanetti, Counting hydrodynamic modes in lattice gas automata models, *Physica D* **47**, 30, 1991; and
M. Ernst, Linear response theory for cellular automata fluids, in *Fundamental Problems in Statistical Mechanics*, Vol. 7, p. 321, H. van Beijeren ed., North–Holland, Amsterdam, 1990; and
R. Cornubert, D. d'Humières and D. Levermore, A Knudsen layer theory for lattice gases, *Physica D* **47**, 241, 1991.

[35] M. Ernst, Mode coupling theory and tails in CA fluids, *Physica D* **47**, 198, 1991.

[36] B. Boghosian and W. Taylor, *Renormalization of Lattice Gas Transport Coefficients*, Thinking Machines Tech. Rept TMC-208, 1991; and Correlations and renormalization in lattice gases, *Phys. Rev. E* **52**(1), 510, 1995.

[37] F. Higuera, Lattice gas simulation based on the Boltzmann equation, in *Discrete Kinetic Theory, Lattice Gas Dynamics and Foundations of Hydrodynamics*, R. Monaco ed., p.162, World Scientific, Singapore, 1989.

[38] F. Higuera and S. Succi, Simulating the flow around a circular cylinder with a lattice Boltzmann equation, *Europhys. Lett.* **8**, 517, 1989.

[39] E. Foti and S. Succi, Three-dimensional flows in complex geometries with the lattice Boltzmann method, *Europhys. Lett.* **10**(5), 433, 1989.

[40] F. Higuera, S. Succi and R. Benzi, Lattice gas dynamics with enhanced collisions, *Europhys. Lett.* **9**, 345, 1989.

[41] S. Succi, R. Benzi and F. Higuera, The lattice Boltzmann equation: a new tool for computational fluid dynamics, *Physica D* **47**, 219, 1991.

[42] W. Press, B. Flannery, S. Teukolsky and W. Vetterling, *Numerical Recipes*, Cambridge University Press, 1986.

[43] S. Wolfram, Cellular automaton fluids I: basic theory, *J. Stat. Phys.* **45**, 471, 1986.

[44] U. Frisch, Relation between the lattice Boltzmann equation and the Navier–Stokes equations, in *Lattice Gas Methods for PDEs, September 6–9, 1989*, G. Doolen ed., special issue of *Physica D* **47**, 231, 1991.

[45] I. Aitchinson and A. Hey, *Gauge Theories in Particle Physics*, Adam Hilger, Bristol, 1982.

[46] S. Succi, H. Chen and I. Karlin, Internal symmetries of lattice kinetic equations, *J. Phys. France, IV* **8**, 271, 1998.

[47] J. Green, R. Schwartz and E. Witten, *Supersting Theory*, Cambridge University Press, 1989.

[48] R. Benzi, S. Succi and M. Vergassola, Turbulence modeling with non-hydrodynamic variables, *Europhys. Lett.* **13**(8), 727, 1990.

[49] Y. Qian, S. Succi and S. Orszag, Recent developments in lattice Boltzmann computing, *Ann. Rev. Comp. Phys.* **3**, 195, 1995.

[50] S. Chen and G. Doolen, Lattice Boltzmann method for fluid flows, *Ann. Rev. Fluid Mech.* **30**, 329, 1998.

[51] R. Benzi and S. Succi, Two-dimensional turbulence with the lattice Boltzmann equation, *J. Phys. A* **23**, L1, 1990.

[52] I. Karlin, A. Ferrante and H. C. Oettinger, Perfect entropies in the lattice Boltzmann method, *Europhys. Lett.* **62**(3), 1999.

[53] A. Cancelliere, E. Chan, E. Foti, D. Rothman and S. Succi, The permeability of a random media: comparison of simulation with theory, *Phys. Fluids Lett. A* **2**, 2085, 1990.

[54] S. Chen, H. Chen, D. Martinez and W. Matthaeus, Lattice Boltzmann model for simulation of magnetohydrodynamics, *Phys. Rev. Lett.* **67**(27), 3776, 1991; and
J. M. Koelman, A simple lattice Boltzmann scheme for Navier–Stokes fluid flow, *Europhys. Lett.* **15**(6), 603, 1991; and
Y. Qian, D. d'Humières and P. Lallemand, Lattice BGK models for the Navier–Stokes equation, *Europhys. Lett.* **17**(6), 479, 1992.

[55] P. Bhatnagar, E. Gross and M. Krook, A model for collision processes in gases I: small amplitude processes in charged and neutral one-component system, *Phys. Rev.* **94**, 511, 1954.

[56] Y. Chen, H. Ohashi and M. Akiyama, Thermal lattice Bhatnagar–Gross–Krook model without nonlinear deviations in macrodynamic equations, *Phys. Rev. E* **50**, 2776, 1994.

[57] D. d'Humières, Generalized lattice Boltzmann equations, *Rarefied Gas Dynamics: Theory and Simulation, Prog. Astronaut. Aeronaut.*, Vol. 159, p. 450, AIAA Washington, 1992; and
D. d'Humières, Beyond BGK, Lecture notes delivered at the DFG–KONWIHR workshop on '*Lattice Boltzmann Methods, Theory and Applications in Fluid Mechanics*, Erlangen–Nürnberg University, March 26–28, Erlangen, Germany, 2001.

[58] T. Abe, Derivation of the lattice Boltzmann method by means of the discrete ordinate method for the Boltzmann equation, *J. Comp. Phys.* **131**, 241, 1997.

[59] X. He and L. S. Luo, *A priori* derivation of the lattice Boltzmann equation, *Phys. Rev. E* **55**, 6333, 1997.

[60] X. Shan and X. He, Discretization of the velocity space in the solution of the Boltzmann equation, *Phys. Rev. Lett.* **80**(1), 65, 1998.

[61] J. Broadwell, Study of the rarefied shear flow by the discrete velocity method, *J. Fluid Mech.* **19**, 401, 1964; and
R. Gatignol, *Theorie Cinetique des Gaz a Repartition Discrete de Vitesses*, Lecture Notes in Physics 36, Springer–Verlag, Berlin, 1975; and
H. Cabannes, The discrete models of the Boltzmann equation, *Transp.*

Theory and Stat. Phys. **16**, 809, 1987; and
C. Cercignani, *Mathematical Methods in Kinetic Theory*, Plenum Press, New York, 1969; and
R. Monaco and L. Preziosi, *Fluid Dynamic Applications of the Discrete Boltzmann Equation*, Series on Advances in Mathematics for Applied Sciences 3, World Scientific, Singapore, 1991.

[62] K. Molvig, P. Donis, R. Miller, J. Myczkowski and G. Vichniac, Removing the discreteness artifacts in 3D lattice gas fluids, in *Discrete Kinetic Theory, Lattice Gas Dynamics and Foundations of Hydrodynamics*, R. Monaco ed., p. 409, World Scientific, Singapore, 1989.

[63] C. Cercignani and M. Lampis, Kinetic models for gas–surface interactions, *Transp. Theory and Stat. Phys.* **2**, 101, 1971.

[64] O. Filippova and D. Haenel, Lattice Boltzmann simulation of gas–particle flows in filters, *Comput. Fluids* **26**(7), 697, 1997; and
O. Filippova and D. Haenel, Grid refinement for Lattice BGK models, J. Comp. Phys. **147**, 219, 1998.

[65] R. Mei, L. S. Luo and W. Shyy, An accurate curved boundary treatment in the lattice Boltzmann method, *J. Comp. Phys.* **155**, 307, 1999.

[66] G. Strumolo and V. Babu, New directions in computational aerodynamics, *Physics World*, 45, August, 1997.

[67] A. Anagnost, A. Alajbegovic, H. Chen, D. Hill, C. Teixeira and K. Molvig, DIGITAL PHYSICS: Analysis of the Morel body in ground proximity, SAE Tech. Paper 970139, p. 31, Int. Congress and Exposition, Detroit, 1997.

[68] P. Lavallée, J. P. Boon and A. Noullez, Boundaries in lattice gas flows, *Physica D* **47**, 233, 1991.

[69] J. Koeplik and J. Banavar, Continuum deductions from molecular hydrodynamics, *Ann. Rev. Fluid Mech.* **27**, 257, 1995.

[70] O. Inamuro, M. Yoshino and F. Ogino, A non-slip boundary condition for lattice Boltzmann simulations, *Phys. Fluids* **7**(12), 2928, 1995.

[71] R. Maier, R. Bernard and D. Grunau, Boundary conditions for the lattice Boltzmann method, *Phys. Fluids* **8**, 1788, 1996.

[72] Q. Zou and X. He, On pressure and velocity boundary conditions for the lattice Boltzmann BGK model, *Phys. Fluids* **9**, 1591, 1997.

[73] T. Ladd, Numerical simulations of particulate suspensions via a discretized Boltzmann equation. Part I: theoretical foundations, *J. Fluid Mech.* **271**, 285, 1994.

[74] R. Cornubert, D. d'Humières and D. Levermore, A Knudsen layer theory for lattice gases, *Physica D* **47**, 241, 1991.

[75] I. Ginzbourg and P. M. Adler, Boundary flow condition analysis for the three-dimensional lattice Boltzmann model, *J. Phys. II France* **4**, 191, 1994.

[76] I. Ginzbourg and D. d'Humières, Local second-order boundary methods for lattice Boltzmann models, *J. Stat. Phys.* **84**(5/6), 927, 1996.

[77] H. Chen, Volumetric formulation of the lattice Boltzmann method for fluid dynamics: basic concept, *Phys. Rev. E* **58**(3), 3955, 1998.

[78] S. Chen, D. Martinez and R. Mei, On boundary conditions in lattice Boltzmann methods, *Phys. Fluids* **8**, 2527, 1996.

[79] X. He and L. S. Luo, Lattice Boltzmann model for the incompressible Navier–Stokes equations, *J. Stat. Phys.* **88**(3/4), 927, 1997; and D. Noble, S. Chen, J. Georgiadis and R. Buckius, A consistent hydrodynamic boundary condition for the lattice Boltzmann method, *Phys. Fluids* **7**, 203, 1996.

[80] S. Patankar, *Numerical Heat Transfer and Fluid Flow*, McGraw–Hill, New York, 1980.

[81] L. Kadanoff, G. McNamara and G. Zanetti, A Poiseuille viscosimeter for lattice gas automata, *Complex Systems* **1**, 791, 1987, reprinted in *Lattice Gas Models for PDEs*, G. Doolen ed., p. 383, Addison–Wesley, Menlo Park, 1989.

[82] A. Chorin, A numerical method for solving incompressible viscous flows problems, *J. Comp. Phys.* **2**, 12, 1967.

[83] H. Tennekes and J. Lumley, *A First Course in Turbulence*, MIT Press, 1972.

[84] S. Succi, R. Benzi, E. Foti, F. Higuera and F. Szelenyi, Lattice Boltzmann computing on the IBM 3090/VF, *Cellular Automata and Modelling of Complex Physical Systems*, P. Manneville *et al.* eds, p. 178, Springer–Verlag, Berlin, 1989.

[85] L. S. Luo, *Lattice Gas Automata and Lattice Boltzmann Equations for Two-Dimensional Hydrodynamics*, Ph.D. Thesis, Georgia Institute of Technology, 1993.

[86] S. Orszag, *Lectures on the Statistical Theory of Fluid Turbulence*, MIT University Press, 1976.

[87] Q. Zou, S. Hou, S. Chen and G. Doolen, An improved incompressible lattice Boltzmann model for time-independent flows, *J. Stat. Phys.* **81**(1/2), 35, 1995.

[88] D. Stauffer and A. Aharony, *Introduction to Percolation Theory*, Taylor & Francis, London, 1992.

[89] M. Sahimi, Flow phenomena in rocks: from continuum models to fractals, percolation, cellular automata and simulated annealing, *Rev. Mod. Phys.* **65**(4), 1393, 1993.

[90] J. Bear, *Dynamics of Fluids in Porous Media*, Elsevier, New York, 1972.

[91] S. Succi, E. Foti and M. Gramignani, Flow through geometrically irregular media with lattice gas automata, *Meccanica* **25**, 253, 1990. See also [84,94].

[92] S. Chen, K. Diemer, G. Doolen, K. Eggert, C. Fu, S. Gutman and B. Travis, Lattice gas models for non-ideal fluids, *Physica D* **47**, 97, 1991.

[93] M. Pilotti, Generation of realistic porous media by grains sedimentation, *Transp. Porous Media* **33**, 257, 1998.

[94] E. Foti, S Succi and F. Higuera, Three-dimensional flows in complex geometries with the lattice Boltzmann method, *Europhys. Lett.* **10**(5), 433, 1989.

[95] A. Cancelliere, C. Chang, E. Foti, D. Rothman and S. Succi, Permeability of a three-dimensional random media: comparison of simulation with theory, *Phys. Fluids A* **2**, 2085, 1990.

[96] A. Gunstensen, D. Rothman, S. Zaleski and G. Zanetti, Lattice Boltzmann model of immiscible fluids, *Phys. Rev. A* **43**(8), 4320, 1991.

[97] P. Adler, *Porous Media*, Butterworth–Heinemann, London, 1992.

[98] A. Koponen, D. Kandhai, E. Hellen, M. Alava, A. Hoekstra, M. Kataja, K. Niskasen and P. Sloot, Permeability of three-dimensional random fiber webs, *Phys. Rev. Lett.* **80**(4), 716, 1998.

[99] M. Mareschal, Microscopic simulations of complex flows, *Ann. Rev. Chem. Phys.* **100**, 317, 1997.

[100] G. Bird, A contemporary implementation of the direct simulation Monte Carlo method, in *Microscopic Simulations of Complex Flows*, NATO ASI Series B: Vol. 292, 239, 1991.

[101] A. N. Kolmogorov, The local structure of turbulence in incompressible viscous fluid for very large Reynolds numbers, *Dokl. Akad. Nauk SSSR* **30**, 9, 1941.

[102] U. Frisch, *Turbulence, the Legacy of A. Kolmogorov*, Cambridge University Press, 1996.

[103] S. Orszag and D. Gottlieb, *Numerical Analysis of Spectral Methods*, SIAM, Philadelphia, 1977.

[104] J. Ferziger, Large eddy simulation, in *Simulation and Modelling of Turbulent Flows*, ICASE/LarC Series in Computational Science and Engineering, p. 109, T. Gatski, M. Hussaini and J. Lumley eds, Oxford University Press, 1996.

[105] Y. Qian and S. Orszag, Lattice BGK models for the Navier–Stokes equations: nonlinear deviations in compressible regimes, *Europhys. Lett.* **21**, 255, 1993.

[106] D. Martinez, W. Matthaeus, S. Chen and D. Montgomery, Comparison of spectral methods and lattice Boltzmann simulations of two-dimensional hydrodynamics, *Phys. Fluids* **6**, 1285, 1994; and
S. Chen, Z. Wang, X. Shan and G. Doolen, Lattice Boltzmann computational fluid dynamics in three dimensions, *J. Stat. Phys.* **68**, 379, 1992.

[107] J. W. Cooley and J. W. Tukey, An algorithm for the machine calculation of complex Fourier transforms, *Math. Comput.* **19**, 297, 1964.

[108] G. Punzo, F. Massaioli and S. Succi, High resolution lattice Boltzmann computing on the IBM SP-1 scalable parallel computer, *Comp. Phys.* **8**, 705, 1994.

[109] S. Succi, E. Osoy and B. Ayat, *A Six Lectures Primer on Parallel Computing*, University of Chicago, Tech. Rept CS 96-11, 1996.

[110] A. Bartoloni *et al.*, LBE simulations of Rayleigh–Bénard convection on the APE-100 parallel processor, *Int. J. Mod. Phys. C* **4**, 993, 1993.

[111] R. Benzi, S. Ciliberto, R. Tripiccione, F. Massaioli, C. Baudet and S. Succi, *Phys. Rev. E* **48**, R29, 1993.

[112] J. Smagorinski, General circulation experiments with the primitive equations I: the basic experiments, *Monthly Weather Rev.* **91**, 99, 1963.

[113] J. Eggels and J. Somers, Direct and large eddy simulations of turbulent fluid using the lattice Boltzmann scheme, *Int. J. Heat Fluid Flow* **17**, 307, 1996.

[114] C. Speziale, Modeling of turbulent transport equations, in *Simulation and Modeling of Turbulent Flows*, ICASE/LarC Series in Computational Science and Engineering, T. Gatski, M. Hussaini and J. Lumley eds, p. 185, Oxford University Press, 1996.

[115] B. Launder, An introduction to single-point closure methodology, in *Simulation and Modeling of Turbulent Flows*, ICASE/LarC Series in Computational Science and Engineering, T. Gatski, M. Hussaini and J. Lumley eds, p. 243, Oxford University Press, 1996.

[116] S. Succi, G. Amati and R. Benzi, Challenges in lattice Boltzmann computing, *J. Stat. Phys.* **81**(1/2), 5, 1995.

[117] S. Succi, G. Bella, H. Chen, K. Molvig, C. Teixeira and A. de Maio, An integer Lax–Wendroff algorithm for multicomponent transport, *J. Comp. Phys.* **152**, 493, 1999.

[118] C. Teixeira, Incorporating turbulence models into the lattice Boltzmann method, *Int. J. Mod. Phys. C* **9**(8), 1159, 1999.

[119] F. Hayot and L. Wagner, A non-local modification of a lattice Boltzmann model, *Europhys. Lett.* **33**, 435, 1996.

[120] G. Alexander, H. Chen, S. Kandasamy, S. Mallick, Y. Qian, R. Shock, R. Zhang and V. Yakhot, Turbulent flow simulations via extended lattice Boltzmann algorithm, EXA preprint, 2001.

[121] H. Chen, S. Succi and S. Orszag, Analysis of sub-grid fluid turbulence using the Boltzmann Bhatnagar–Gross–Krook kinetic equation, *Phys. Rev. E* **59**(3), R1, 1999.

[122] J. Koelman, A simple lattice Boltzmann scheme for Navier–Stokes fluid flow, *Europhys. Lett.* **15**, 603, 1991.

[123] F. Nannelli and S. Succi, The lattice Boltzmann equation in irregular lattices, *J. Stat. Phys.* **68**, 401, 1992.

[124] N. Cao, S. Chen, S. Jin and D. Martinez, Physical symmetry and lattice symmetry in lattice Boltzmann method, *Phys. Rev. E* **55**, R21, 1997.

[125] X. He, L. S. Luo and M. Dembo, Some progress in lattice Boltzmann method: part I, non-uniform grids, *J. Comp. Phys.* **129**, 357, 1996.

[126] R. van der Sman, *Lattice Boltzmann Schemes for Convection–Diffusion Phenomena; Application to Packages of Agricultural Products*, Ph.D. Thesis, University of Wageningen, Holland, 1999.

[127] F. Nannelli and S. Succi, The finite volume formulation of the lattice Boltzmann equation, *Transp. Theory and Stat. Phys.* **23**, 163, 1994.

[128] G. Amati, S. Succi and R. Benzi, Turbulent channel flow simulation using a coarse-grained extension of the lattice Boltzmann method, *Fluid Dyn. Res.* **19**, 289, 1997; and
G. Amati, S. Succi and R. Piva, Massively parallel lattice Boltzmann simulation of turbulent channel flow, *Int. J. Mod. Phys. C* **4**(8), 869, 1997.

[129] P. Woodword and P. Colella, The numerical simulation of two-dimensional fluid flow with strong shocks, *J. Comp. Phys.* **54**, 125, 1984.

[130] G. Peng, H. Xi and C. Duncan, Lattice Boltzmann method on irregular meshes, *Phys. Rev. E* **58**(4), R4124, 1998; and
H. Xi, G. Peng and S.-H. Chou, Finite volume lattice Boltzmann schemes in two and three dimensions, *Phys. Rev. E* **60**(3), 3380, 1999.

[131] R. Hockney and J. Eastwood, *Computer Simulation Using Particles*, McGraw–Hill, London, 1979.

[132] T. Pang, *An Introduction to Computational Physics*, Cambridge University Press, 1997.

[133] X. He, X. Shan and G. Doolen, Lattice Boltzmann method on curvilinear coordinates system: vortex shedding behind a circular cylinder, *Phys. Rev. E* **57**, R13, 1998.

[134] W. Strang and G. Fix, *An Analysis of the Finite Element Method*, Prentice–Hall, Englewood Cliffs, New Jersey, 1973.

[135] N. H. Christ, R. Friedberg and T. D. Lee, Random lattice field theory: general formulation, *Nucl. Phys. B* **202**, 89, 1982.

[136] I. Karlin, S. Succi and S. Orszag, The lattice Boltzmann method on irregular lattices, *Phys. Rev. Lett.* **82**(26), 5245, 1999.

[137] J. Toelke, M. Krafczyk, M. Schulz, E. Rank and R. Berrios, Implicit discretization and non-uniform mesh refinement approaches for FD discretizations of LBGK models, *Int. J. Mod. Phys. C* **9**(8), 1143, 1999.

[138] R. Verberg and A. Ladd, Simulation of low Reynolds number flow via a time-independent lattice Boltzmann method, *Phys. Rev. E* **60**(3), 3366, 1999.

[139] F. Mazzocco, C. Arrighetti, L. Spagnoli, G. Bella and S. Succi, Multiscale lattice Boltzmann schemes, a preliminary application to axial turbomachine flows, *Int. J. Mod. Phys. C* **11**(2), 233, 2000.

[140] B. Elton, D. Levermore and G. Rodrigue, Convergence of convective–diffusive lattice Boltzmann methods, *SIAM J. Num. Anal.* **32**(5), 1327, 1995.

[141] D. Potter, *Computational Physics*, Wiley, London, 1977.

[142] J. Boris, New directions in computational fluid dynamics, *Ann. Rev. Fluid Mech.* **21**, 695, 1987.

[143] J. Sterling and S. Chen, Stability analysis of lattice Boltzmann methods, *J. Comp. Phys.* **123**, 196, 1996; and

P. Pavlo, G. Vahala, L. Vahala and M. Soe, Linear stability analysis of thermo-lattice Boltzmann models, *J. Comp. Phys.* **139**, 79, 1998; and [164].

[144] I. Karlin and S. Succi, Equilibria for discrete kinetic equations, *Phys. Rev. E* **58**, R4053, 1998.

[145] B. Boghosian, P. Coveney, J. Yepez and A. Wagner, Entropic lattice Boltzmann methods, BU-CCS-990802 preprint, 2000.

[146] H. Chen and C. Teixeira, H-theorems and origins of instability in thermal lattice Boltzmann models, *Proceedings of the 7th International Conference on Discrete Simulation of Fluids*, Tokyo, July, 1999, North–Holland, Amsterdam; and
Comp. Phys. Comm. **129**, 21, 2000.

[147] A. Wagner, An H-theorem for the lattice Boltzmann approach to hydrodynamics, *Europhys. Lett.* **44**(2), 144, 1998.

[148] L. S. Luo, Some recent results on discrete velocity models and ramifications for lattice Boltzmann equation, *Proceedings of the 7th International Conference on Discrete Simulation of Fluids*, Tokyo, July, 1999, North–Holland, Amsterdam; and
Comp. Phys. Comm. **129**, 63, 2000.

[149] S. Ansumali and I. Karlin, Single relaxation time model for entropic lattice Boltzmann methods, Swiss Polytech., Zürich, preprint, 2000.

[150] H. Chen, I. Karlin and S. Succi, Why lattice Boltzmann works and sometimes still doesn't: the role of the H-theorem, submitted to *Rev. Mod. Phys.*

[151] G. Amati, S. Succi and R. Piva, Massively parallel lattice Boltzmann simulation of turbulent channel flow, *Int. J. Mod. Phys. C* **8**(4), 869, 1997.

[152] S. Chen and X. Shan, High resolution turbulent simulations using the Connection Machine 2, *Comp. Phys.* **6**, 643, 1992.

[153] D. Kandhai, A. Koponen, A. Hoekstra, M. Kataja, J. Timonen and P. Sloot, Lattice Boltzmann hydrodynamics on parallel systems, *Comp. Phys. Comm.* **111**, 14, 1998.

[154] J. Bernsdorf and M. Schäfer, Comparison of cellular automata and finite volume techniques for the simulation of incompressible flows in porous media, Dept. of Fluid Mechanics (LSTM), University of Erlangen, preprint, 1993; and
J. Bernsdorf, F. Durst and M. Schäfer, Comparison of cellular automata and finite volume techniques for simulation of incompressible flows in complex geometries, *Int. J. Num. Meth. Fluids* **29**(3), 251, 1999; and
J. Bernsdorf, Th. Zeiser, G. Brenner and F. Durst, Simulation of a 2D channel flow around a square obstacle with lattice Boltzmann (BGK) automata, *Int. J. Mod. Phys. C* **9**(8), 1129, 1998.

[155] D. Kandhai, D. Vidal, A. Hoekstra, H. Hoefsloot, P. Iedema and P. Sloot, A comparison between lattice Boltzmann and finite element simulations of fluid flow in static mixer reactors, *Int. J. Mod. Phys. C* **9**(8), 1123, 1998.

[156] J. Ferziger and M. Peric, *Computational Methods for Fluid Dynamics*, Springer–Verlag, Berlin, 1995.

[157] X. Shan and X. He, Discretization of the velocity space in the solution of the Boltzmann equation, *Phys. Rev. Lett.* **80**, 65, 1998; and
X. He and L. S. Luo, *A priori* derivation of the lattice Boltzmann equation, *Phys. Rev. E* **55**, 6333, 1997.

[158] M. Ancona, Fully Lagrangian and lattice Boltzmann methods for solving systems of conservation equations, *J. Comp. Phys.* **115**, 107, 1994.

[159] F. Coquel and B. Perthame, Relaxation of energy and approximate Riemann solvers for general pressure laws in fluid dynamic equations, *SIAM J. Num. Anal.* **35**(6), 2223, 1998; and
E. Godlewski and P. Raviart, *Numerical Approximation of Hyperbolic Systems of Conservation Laws*, Springer–Verlag, Berlin, 1996; and
D. Aregba-Driollet and R. Natalini, Discrete kinetic schemes for multidimensional systems of conservation laws, *SIAM J. Num. Anal.* **37**(6), 1973, 2000.

[160] K. Xu, A new class of gas–kinetic relaxation schemes for the compressible Euler equations, *J. Stat. Phys.* **81**(1/2), 147, 1995; and
K. Xu, L. Martinelli and A. Jameson, Gas–kinetic finite volume methods, flux vector splitting and artificial diffusion, *J. Comp. Phys.* **120**, 48, 1995.

[161] M. Junk, *A finite difference interpretation of the lattice Boltzmann method*, Kaiserslautern University preprint, 1999; and
M. Junk and A. Klar, Discretization for the incompressible Navier–Stokes equations based on the lattice Boltzmann method, *SIAM J. Sci. Comp.* **22**(1), 1, 2000.

[162] B. Boghosian and P. Coveney, Inverse Chapman–Enskog derivation of the thermodynamic lattice BGK model for the ideal gas, *Int. J. Mod. Phys. C* **9**(8), 1231, 1998.

[163] Y. H. Qian and Y. Zhou, Higher-order dynamics in lattice based models using the Chapman–Enskog method, *Phys. Rev. E* **61**(2), 2103, 2000.

[164] L. S. Luo and P. Lallemand, Theory of the lattice Boltzmann method: dispersion, dissipation, isotropy, Galilean invariance and stability, *Phys. Rev. E* **61**(6), 6546, 2000.

[165] *Proceedings of the Vth International Symposium on Discrete Simulations of Fluid Dynamics*, B. M. Boghosian, F. Alexander and P. Coveney eds, Boston, July, 1997, special issue of *Int. J. Mod. Phys. C* **8**, 1997.

[166] *Proceedings of the VIth International Symposium on Discrete Simulations of Fluid Dynamics*, B. M. Boghosian ed., Oxford, July, 1998, special issue of *Int. J. Mod. Phys. C* **9**, 1998.

[167] *Proceedings of the VIIth International Symposium on Discrete Simulations of Fluid Dynamics*, H. Ohashi and Y. Chen eds, Tokyo, August, 1999, Special issue of *Comp. Phys. Comm.* **129**(1–3), 2000.

[168] F. Williams, *Combustion Theory*, 2nd Edn, Benjamin/Cummings, Menlo Park, 1985; and
R. Strehlow, *Combustion Fundamentals*, McGraw–Hill, New York, 1984.

[169] J. P. Boon, D. Dab, R. Kapral and A. Lawniczak, Lattice gas automata for reactive systems, *Phys. Rep.* **271**(2), 58, 1996.

[170] M. Ponce Dawson, S. Chen and G. Doolen, Lattice Boltzmann computations for reaction–diffusion equations, *J. Chem. Phys.* **98**(2), 1514, 1993.

[171] S. Succi, G. Bella and F. Papetti, Lattice kinetic theory for numerical combustion, *J. Sci. Comp.* **12**(4), 395, 1997.

[172] O. Filippova, Lattice BGK model for low Mach number combustion, *Int. J. Mod. Phys. C* **9**(8), 1439, 1998.

[173] G. Bella, M. Presti and S. Succi, Mass transfer improvements in catalytic converter channels: a hybrid BGK–finite volume numerical simulation method, Society Automotive Engineering, SAE paper 972907, 1997.

[174] J. Glimm, E. Isaacson, D. Marchesin and O. McBryan, Front tracking for hyperbolic systems, *Adv. Appl. Math.* **2**, 91, 1981.

[175] R. Scardovelli and S. Zaleski, Direct numerical simulation of free surface and interface flow, *Ann. Rev. of Fluid Mech.* **31**, 567, 1999.

[176] D. Rothman and S. Zaleski, Lattice gas models of phase separation, *Rev. Mod. Phys.* **66**, 1417, 1994.

[177] A. Gunstensen and D. Rothman, Lattice Boltzmann studies of immiscible two phase flows through porous media, *Phys. Rev. A* **43**, 4320, 1991; and E. Flekkoy, T. Rage, U. Oxaal and J. Feder, Hydrodynamic irreversibility in creeping flows, *Phys. Rev. Lett.* **77**, 4170, 1996.

[178] D. Rothman and J. Keller, Immiscible cellular automaton fluids, *J. Stat. Phys.* **52**, 1119, 1988.

[179] H. Chen, Discrete Boltzmann systems and fluid flows, *Comp. Phys.* **7**, 632, 1993; and
X. Shan and H. Chen, Lattice Boltzmann model for simulating flows with multiple phases and components, *Phys. Rev. E* **47**, 1815, 1993.

[180] See Section 7.3 of [49]. The first lattice gas cellular automaton for liquid–gas transitions is due to: C. Appert and S. Zaleski, *Phys. Rev. Lett.* **64**, 1, 1990.

[181] H. Shan and H. Chen, Simulation of non-ideal gas and liquid–gas phase transitions by the lattice Boltzmann equations, *Phys. Rev. E* **49**, 2941, 1995.

[182] M. Swift, W. Osborne and J. Yeomans, Lattice Boltzmann simulations of non-ideal fluids, *Phys. Rev. Lett.* **75**, 830, 1995.

[183] X. Shan, Simulation of Rayleigh–Bénard convection using a lattice Boltzmann method, *Phys. Rev. E* **55**, 2780, 1997.

[184] S. Chapman and T. Cowling, *The Mathematical Theory of Non-Uniform Gases*, 3rd Edn, Cambridge University Press, 1990.

[185] L. S. Luo, Unified theory of lattice Boltzmann models for non-ideal gases, *Phys. Rev. Lett.* **81**(8), 1618, 1998.

[186] E. Cohen, Fifty years of kinetic theory, *Physica A* **194**, 229, 1993.

[187] G. Uhlenbeck, *Lectures in Statistical Mechanics*, Chapter V, American Mathematical Society, Proceedings of the Summer Seminar, Boulder, Colorado, 1960.

[188] W. Osborne, E. Orlandini, J. Yeomans and J. Banavar, Lattice Boltzmann study of hydrodynamic spinodal decomposition, *Phys. Rev. Lett.* **75**, 4031, 1995; and
A. Lamura, G. Gonnella and J. Yeomans, A lattice Boltzmann model of ternary fluid mixtures, *Europhys. Lett.* **45**(3), 314, 1999.

[189] F. Alexander, S. Chen and D. Grunau, Hydrodynamic spinodal decomposition: growth kinetics and scaling, *Phys. Rev. B* **48**, 634, 1993.

[190] D. Grunau, S. Chen and K. Eggert, A lattice Boltzmann model for multiphase fluid flows, *Phys. Fluids A* **5**, 2557, 1993; and
D. Grunau, T. Lookman, S. Chen and A. Lapedes, Domain growth, wetting and scaling in porous media, *Phys. Rev. Lett.* **71**(25), 4198, 1993.

[191] A. Cieplak, Rupture and coalescence in 2D cellular automaton fluids, *Phys. Rev. E* **51**, 4353, 1995.

[192] B. Boghosian, P. Coveney and A. Emerton, Three-dimensional lattice gas model for amphiphilic fluid dynamics, *Proc. R. Soc. London A* **452**, 1221, 1996; and
M. Nekovee, P. Coveney, H. Chen and B. Boghosian, A lattice Boltzmann model for interacting amphiphilic fluids, preprint, submitted to *Phys. Rev. E*, 2000.

[193] G. Bossis and J. Brady, Brownian and Stokesian dynamics, in *Microscopic Simulations of Complex Hydrodynamic Phenomena*, M. Mareschal and B. Holian eds, p. 255, NATO ASI Series B 292, 1991.

[194] A. Gast and W. Russell, Simple ordering in complex fluids, *Phys. Today* **51**(12), 24, 1998.

[195] L. Landau and E. Lifshitz, *Statistical Physics*, Addison–Wesley, Reading, Massachusetts, 1975.

[196] S. de Groot and P. Mazur, *Non-Equilibrium Thermodynamics*, unab. Dover, 1983.

[197] A. Ladd, Numerical simulations of particulate suspensions via a discretized Boltzmann equation. Part 1: theoretical foundations, *J. Fluid Mech.* **271**, 285, 1994.

[198] B. Alder and T. Wainwright, Molecular dynamics by electronic computers, in *Proceedings of the International Symposium on Statistical Mechanical Theory of Transport Processes* (1956, Brussels), I. Prigogine ed., p. 97, Wiley Interscience, New York, 1958.

[199] G. Ciccotti, private communication.

[200] A. Ladd, Numerical simulations of particulate suspensions via a discretized Boltzmann equation. Part 2: numerical results, *J. Fluid Mech.* **271**, 311, 1994.

[201] P. Ahlrichs and B. Dunweg, Lattice Boltzmann simulation of polymer-solvent system, *Int. J. Mod. Phys. C* **9**(8), 1429, 1998.

[202] D. Frenkel and B. Smit, *Understanding Molecular Dynamics Simulation*, North–Holland, London, 1997.

[203] D. Rapaport, *The Art of Molecular Dynamics*, Cambridge University Press, 1995.

[204] A. Masselot, *A New Numerical Approach to Snow Deposition by Wind: a Parallel Lattice Gas Model*, Ph.D. Thesis, Geneva University, 2000; and references therein.

[205] N. Martys, Energy conserving discrete Boltzmann equation for non-ideal systems, *Int. J. Mod. Phys. C* **10**(7), 1367, 1999.

[206] T. Ihle and D. Kroll, Thermal lattice Boltzmann method for non-ideal gases with potential energy, *Comp. Phys. Comm.* **129**(1/2), 1, 2000.

[207] D. Rothman, Modeling seismic P waves with cellular automata, *Geophys. Res. Lett.* **14**, 17, 1987; and
P. Mora, The lattice Boltzmann phononic lattice solid, *J. Stat. Phys.* **68**(3/4), 591, 1992; and
B. Chopard and M. Droz, *Cellular Automata Modeling of Physical Systems*, Chapter VII, Cambridge University Press, 1998; and
P. Luthi, B. Chopard and J. F. Wagen, Wave propagation in urban microcells: a massively parallel approach using the TLM method, in *Proceedings of Applied Parallel Computing: Computations in Physics, Chemistry and Engineering Science*, J. Dongarra, K. Madsen and J. Wasnieski eds, Notes in Computer Science, Vol. 1, p. 429, Springer–Verlag, New York, 1996.

[208] S. Chen, H. Chen, D. Martinez and W. Matthaeus, Lattice Boltzmann model for simulation of magnetohydrodynamics, *Phys. Rev. Lett.* **67**(27), 3777, 1991; and
D. Martinez, S. Chen and W. Matthaeus, Lattice Boltzmann magnetohydrodynamics, *Phys. Plasmas* **1**(6), 1850, 1994; and
H. Chen and W. Matthaeus, New cellular automaton model for magnetohydrodynamics, *Phys. Rev. Lett.* **58**, 1845, 1987; and
V. Sofonea, Lattice Boltzmann approach to collective-particle interactions in magnetic fluids, *Europhys. Lett.* **25**(5), 385, 1994; and
S. Succi, M. Vergassola and R. Benzi, Lattice Boltzmann scheme for two-dimensional magnetohydrodynamics, *Phys. Rev. A* **43**, 4521, 1991; and
G Fogaccia, R. Benzi and F. Romanelli, Lattice Boltzmann algorithm for three-dimensional simulations of plasma turbulence, *Phys. Rev. E* **54**, 4384, 1996.

[209] L. Giraud, D. d'Humières and P. Lallemand, A lattice Boltzmann model for viscoelasticity *Int. J. Mod. Phys. C* **8**(4), 805, 1997; and

Y. Qian and F. Deng, A lattice BGK model for viscoelastic media, *Phys. Rev. Lett.* **79**(14), 2742, 1997; and
A. Aharonov and D. Rothman, Non-Newtonian flow (through porous media): a lattice Boltzmann method, *Geophys. Res. Lett.* **20**, 679, 1993.

[210] Y. Qian, S. Succi and J. Wang, Cluster instability in granular gases, *Int. J. Mod. Phys. C* **8**(4), 999, 1997.

[211] R. Salmon, The lattice Boltzmann method as a basis for ocean circulation modeling, *J. Mar. Res.* **57**, 503, 1999.

[212] D. Bohm, *Quantum Theory*, Englewood Cliffs, New Jersey, 1951.

[213] A. Messiah, *Quantum Mechanics*, North–Holland, Amsterdam, 1962.

[214] G.T'Hooft, Equivalence relations between deterministic and quantum mechanical systems, *J. Stat. Phys.* **53**(1/2), 323, 1988.

[215] L. Landau and E. Lifshitz, *Non-Relativistic Quantum Theory*, Pergamon Press, Oxford, 1960.

[216] J. Bell, *Speakable and Unspeakable in Quantum Mechanics*, Cambridge University Press, 1987.

[217] S. Succi and R. Benzi, Lattice Boltzmann equation for quantum mechanics, *Physica D* **69**, 327, 1993.

[218] I. Bialynicki–Birula, Weil, Dirac and Maxwell equations on a lattice as a unitary cellular automata, *Phys. Rev. D* **49**(12), 6920, 1994.

[219] S. Succi, Numerical solution of the Schrödinger equation with discrete kinetic theory, *Phys. Rev. E* **53**(2), 1969, 1996.

[220] S. Succi, Lattice quantum mechanics: an application to Bose–Einstein condensation, *Int. J. Mod. Phys. C* **9**(8), 1577, 1998.

[221] J. Wells, V. Oberacker, M. Strayer and A. Umar, Parallel implementation of the Dirac equation in three Cartesian dimensions, *Proceedings 6th International Conference on Physics Computing*, R. Gruber and M. Tomassini eds, p. 655, European Phys. Soc., Geneva, 1994.

[222] B. Boghosian and W. Taylor, Quantum lattice gas models for the many body Schrödinger equation, *Int. J. Mod. Phys. C* **8**(4), 705, 1997.

[223] D. Meyer, Quantum lattice gases and their invariants, *Int. J. Mod. Phys. C* **8**(4), 717, 1997.

[224] R. Feynman and A. Hibbs, *Quantum Mechanics and Path Integrals*, McGraw–Hill, New York, 1965.

[225] K. Kulander ed., *Comp. Phys. Comm.* **63**, 1991, special issue on time-dependent methods for quantum dynamics.

[226] E. Kaxiras, Review of atomistic simulations of surface diffusion and growth on semiconductors, *Comp. Mat. Sci.* **6**, 158, 1986.

[227] R. Car and M. Parrinello, Unified approach for molecular dynamics and density functional theory, *Phys. Rev. Lett.* **55**(22), 2471, 1985.

[228] D. P. di Vincenzo, Quantum computing, *Science* **270**, 255, 1995.

[229] W. Kohn, Nobel lecture: electronic structure of matter–wave functions and density functionals, *Rev. Mod. Phys.* **71**(5), 1253, 1999.

[230] P. Hohenberg and W. Kohn, Inhomogeneous electron gas, *Phys. Rev.* **136**, B864, 1964; and
W. Kohn and L. Sham, Self-consistent equations including exchange and correlation effects, *Phys. Rev.* **140**, A1133, 1965.

[231] M. Ernst, Temperature and heat conductivity in cellular automata fluids, in *Discrete Models of Fluid Dynamics*, A. Alves ed., p. 186, World Scientific, Singapore, 1991.

[232] C. Cercignani, On the thermodynamics of a discrete velocity gas, *Transp. Theory and Stat. Phys.* **23**(1–3), 1, 1994.

[233] C. Teixeira, *Continuum Limit of Lattice Gas Fluid Dynamics*, Ph.D. Thesis, Nuclear Engineering Dept, MIT, 1992.

[234] P. Grosfils, J. P. Boon, R. Brito and M. Ernst, Statistical hydrodynamics of lattice gas automata, *Phys. Rev. E* **48**(4), 2655, 1993.

[235] K. Xu and L. S. Luo, Connection between lattice Boltzmann equation and beam scheme, *Int. J. Mod. Phys. C* **9**(8), 1177, 1998.

[236] F. Alexander, S. Chen and J. Sterling, Lattice Boltzmann thermohydrodynamics, *Phys. Rev. E* **53**, 2298, 1993.

[237] Y. Chen, H. Ohashi and M. Akiyama, Prandtl number of lattice Bhatnagar–Gross–Krook fluid, *Phys. Fluids* **7**, 2280, 1995.

[238] Y. Chen, H. Ohashi and M. Akiyama, Thermal lattice Bhatnagar–Gross–Krook model without nonlinear deviations in macrodynamic equations, *Phys. Rev. E* **50**, 2776, 1994.

[239] Y. Chen, *Lattice Bhatnagar–Gross–Krook Method for Fluid Dynamics: Compressible, Thermal and Multiphase Models*, Ph.D. Thesis, Dept. of Quantum Engineering and System Science, University of Tokyo, 1994.

[240] H. Grad, On the kinetic theory of rarefied gas, *Comm. Math. Phys.* **2**, 331, 1949.

[241] G. McNamara, A. Garcia and B. Alder, Stabilization of thermal lattice Boltzmann models, *J. Stat. Phys.* **81**, 395, 1995.

[242] I. Muller and T. Ruggeri, *Rational extended thermodynamics*, Springer Tracts in Natural Philosophy, Vol. 37, Springer–Verlag, New York, 1998.

[243] M. de Cicco, S. Succi and G. Bella, Nonlinear stability of thermal lattice BGK models, IAC Rept 5/97, published in *SIAM J. Sci. Comp.* **21**(1), 366, 1999.

[244] A. Renda, S. Succi, I. Karlin and G. Bella, Thermohydrodynamic lattice BGK equilibria with non-perturbative equilibria, *Europhys. Lett.* **41**(3), 279, 1998.

[245] P. Pavlo, G. Vahala and L. Vahala, Higher-order isotropic velocity grids in lattice methods, *Phys. Rev. Lett.* **80**(18), 3960, 1998.

[246] S. Hu, G. Yan and W. Shi, A lattice Boltzmann model for compressible perfect gas, *Acta Mech. Sinica* **13**, 3, 1997.

[247] G. Yan, Y. Chen and S. Hu, Simple lattice Boltzmann model for simulating flows with shock waves, *Phys. Rev. E* **59**(1), 454, 1999.

[248] X. He, S. Chen and G. Doolen, A novel thermal model for the lattice Boltzmann method in the incompressible limit, *J. Comp. Phys.* **146**, 282, 1998.

[249] K. Molvig, Advising fluid dynamicists to be discrete, an interview with Kim Molvig, *Comp. in Phys.* **11**(2), 126, 1997.

[250] H. Chen, C. Teixeira and K. Molvig, Digital physics approach to computational fluid dynamics: some basic theoretical features, *Int. J. Mod. Phys. C* **8**(4), 675, 1997.

[251] F. Massaioli, R. Benzi and S. Succi, Exponential tails in two-dimensional Rayleigh–Bénard convection, *Europhys. Lett.* **21**(3), 305, 1993.

[252] X. Shan, Simulation of Rayleigh–Bénard convection using a lattice Boltzmann method, *Phys. Rev. E* **55**, 2780, 1997.

[253] G. Vahala, P. Pavlo, L. Vahala and N. Martys, Thermal lattice Boltzmann models for compressible flows, *Int. J. Mod. Phys. C* **9**(8), 1247, 1998.

[254] B. Chopard and A Masselot, Multiparticle lattice gas model for a fluid: application to ballistic annihilation, *Phys. Rev. Lett.* **81**(9), 1845, 1998; and

B. Boghosian, J. Yepez and F. Alexander, Integer lattice gases, *Phys. Rev. E* **55**, 4137, 1997; and

A. Malavanets and R. Kapral, Continuous velocity lattice gas model for fluid flow, *Europhys. Lett.* **44**(5), 552, 1998.

[255] S. Orszag and G. Patterson, Numerical simulation of three-dimensional homogeneous isotropic turbulence, *Phys. Rev. Lett.* **28**, 76, 1972.

[256] T. G. Lewis and H. El-Rewini, *Introduction to Parallel Computing*, Prentice–Hall, Englewood Cliffs, New Jersey, 1992.

[257] V. Kumar, A. Grama, A. Gupta and G. Karypis, *Introduction to Parallel Computing*, Benjamin/Cummings, Redwood City, 1994.

[258] S. Succi and F. Papetti, *An Introduction to Parallel Computational Fluid Dynamics*, Nova Science, New York, 1996.

[259] G. Fox and W. Furmanski, The physical structure of concurrent problems and concurrent computers, in *Scientific Applications of Multiprocessors*, Int. Series in Comput. Sci., R. J. Elliott and C. A. R. Hoare eds, p. 55, Prentice–Hall, New York, 1989.

INDEX